# SOLAR HOUSES IN EUROPE

## *How They Have Worked*

*Edited by*
W. Palz and T. C. Steemers
*Commission of the European Communities*

*Compiled and Written by*
W. Houghton-Evans
D. Turrent
C. Whittaker

Published for the
COMMISSION OF THE EUROPEAN COMMUNITIES
by
**PERGAMON PRESS**

OXFORD · NEW YORK · TORONTO · SYDNEY · PARIS · FRANKFURT

| | |
|---|---|
| U.K. | Pergamon Press Ltd., Headington Hill Hall, Oxford OX3 0BW, England |
| U.S.A. | Pergamon Press Inc., Maxwell House, Fairview Park, Elmsford, New York 10523, U.S.A. |
| CANADA | Pergamon of Canada, Suite 104, 150 Consumers Road, Willowdale, Ontario M2J 1P9, Canada |
| AUSTRALIA | Pergamon Press (Aust.) Pty. Ltd., P.O. Box 544, Potts Point, N.S.W. 2011, Australia |
| FRANCE | Pergamon Press SARL, 24 rue des Ecoles, 75240 Paris, Cedex 05, France |
| FEDERAL REPUBLIC OF GERMANY | Pergamon Press GmbH, 6242 Kronberg-Taunus, Hammerweg 6, Federal Republic of Germany |



First edition **1981**
EUR 7109

**British Library Cataloguing in Publication Data**

Houghton-Evans, W.
Solar houses in Europe. - (Pergamon international library).
1. Solar heating - Europe
2. Solar houses - Europe
I. Title II. Turrent, D.
III. Whittaker, Chris IV. Commission of the European Communities
697'.78'094 TH7413 80-49713

ISBN 0 08 026743 2 (Hardcover)
ISBN 0 08 026744 0 (Flexicover)

*In order to make this volume available as economically and as rapidly as possible the author's typescript has been reproduced in its original form. This method unfortunately has its typographical limitations but it is hoped that they in no way distract the reader.*

*Printed and bound in Great Britain by*
*William Clowes (Beccles) Limited, Beccles and London*

PERGAMON INTERNATIONAL LIBRARY
of Science, Technology, Engineering and Social Studies
*The 1000-volume original paperback library in aid of education, industrial training and the enjoyment of leisure*
Publisher: Robert Maxwell, M.C.

# SOLAR HOUSES IN EUROPE

## *How They Have Worked*

Published for the Commission of the European Communities,
Directorate General—Information Market and Innovation—Scientific
and Technical Communication, Luxembourg

# PREFACE

In the last few years, solar heating of dwellings and in particular solar water heating has attracted a lot of new interest. Though the number of currently existing solar houses is still very small when compared with the total number of houses in Europe, increasing oil prices and the need for energy conservation will certainly lead to a considerable extension of this market.

This book presents for the first time a comprehensive analysis of European solar houses. Monitored data from houses in various C.E.C. countries and climates have been compiled and this information serves to highlight the performance of typical solar heating systems.

The book was prepared within the framework of the European Communities Solar Energy R & D Programme. The cooperation of architects, building engineers, owners of solar houses and solar energy experts which made this work possible, is gratefully acknowledged.

I hope that the book will lead to a better understanding of solar heating and in particular to a more realistic assessment of its benefits.

G. SCHUSTER

Director General
for Research, Science and Education
Commission of the European Communities

# CONTRIBUTORS

This book was prepared for the Commission of the European Communities by Stephen George and Partners (i). The project was co-ordinated by Chris Whittaker, David Turrent (ii) and Ramiro Godoy (ii). Chapters 1 - 5 were written by W. Houghton-Evans. Editorial assistance was given by Peter Flack and Alec Gillies.

Individual solar-heated projects, which had been established by various national and other agencies, were investigated by Stephen George and Partners (within the United Kingdom) and by their sub-contractors, listed at Chapter 7 below, in other member countries of the Community, using the reporting format which was developed in the first stages of the study.

The monitoring of the projects was investigated by Ove Arup and Partners, Consulting Engineers (iii), who prepared and analysed a separate questionnaire and made recommendations.

The project was directed by the editors and by the Performance Monitoring Group of Directorate-General XII of the Commission.


Chapters 1 - 5 W. Houghton-Evans 5 Dryden Street, London W.C.2E 9NW. Telex 299533 AST.

Chapter 6 David Lush & Robert Aish 13 Fitzroy Street, London W.1.P 6BQ.

Chapter 7 C. Den Ouden Institute of Applied Physics, TNO, Stieltjesweg, Delft 2208, Netherlands.

Prof. G. Kuhn IUT Genie Thermique, 25 Rue Casimir-Brenier, 38031 Grenoble, France.

Poul Kristensen Thermal Insulation Laboratory, Technical University of Denmark, Building 119, DK 2800 Lyngby, Denmark.

W. Esposti CNR/ICITE, Viale Lombardia 49, S. Giuliano Milanese, 20098 Milano, Italy.

Prof. E. Hahne Institut für Thermodynamik und Wärmetechnik, University of Stuttgart, Seidenstrasse 36, 7000 Stuttgart 1, Federal Republic of Germany.


(i) 5 Dryden Street, London W.C.2E 9NW. Telex 299533 AST.
(ii) Energy Conscious Design, 44 Earlham Street, London W.C.2 9LA.
(iii) 13 Fitzroy Street, London W.1.P 6BQ.

## CONTENTS

# CHAPTER 1

# INTRODUCTION

Over two-thirds of a year's fuel requirements for a family dwelling may be replaced by energy gained locally from the sun.

This is the conclusion of a study of recent experience with solar heating throughout the EEC, from Italy in the south to Denmark and the UK in the north (Fig. 1). The proportion will, of course, vary with the elaboration and cost of the building and of the equipment used, but with moderate expense, solar energy can save one-third of the fuel needed to provide space and water heating. Fuel savings in the projects studied are indeed typically much more than this. Better thermal insulation and better control of ventilation and heating equipment are required to exploit solar gains to the full, and add greatly to the efficiency of the buildings studied. Compared with a normal post-war house, fuel savings could be as high as 90% - typically, 50% to 75%. If domestic consumption of fuel were to fall by 60%, total fuel consumption in the UK, for instance, would fall by 15%. Against this, of course, it would be necessary to place the additional cost and energy needed for improved insulation and equipment. But rising prices and limited fuel reserves make it increasingly worth-while - even in Northern Europe, where sunshine is less plentiful - to exploit solar energy to the full.

New techniques cannot be developed without 'teething troubles', and this report will reveal many of these. And a study based on a wide variety of experiences over varying periods in different countries inevitably provides information which is difficult to analyse and compare. Much more remains to be learnt, and it may be that future progress will discard some of the things which have been tried in favour of others yet to be developed. But the evidence from this first generation of solar houses, is that it is now possible to design for a predictable and worth-while performance, avoiding mistakes which have been made.

Sunshine varies in duration and intensity - especially with latitude, although clouds, pollution, screening and other circumstances can reduce local maxima by as much as 20%. Annual figures from the Italian examples show as much as 1500 kWh per sq. metre to be locally available, and from the UK, a minimum of 800 kWh per sq. metre (Fig. 2). Although, as Fig. 3 shows, demand varies also, annual supply throughout the more populous parts of Europe is everywhere enough to supply the highest demand. Figures 4 and 5 give an indication of feasibility.

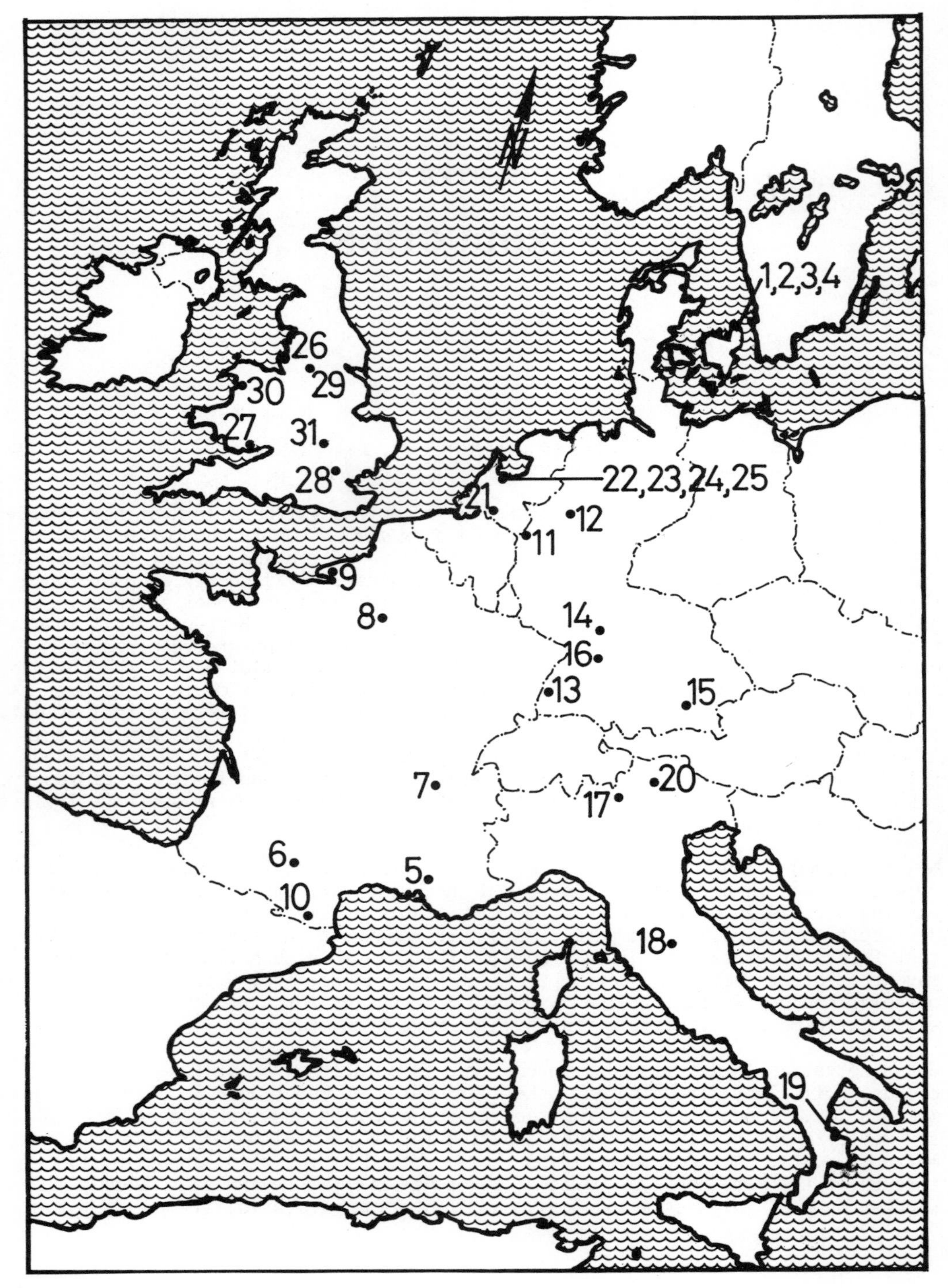


Fig. 1: the thirty-one projects

| No | Location | | passive system | dry system | heat recovery | long-term store | DHW only | warm air htg | h.w. rads. | u/fl htg. | Remarks |
|---|---|---|---|---|---|---|---|---|---|---|---|
| 1 | Blovstrod | (DK) | | | | | * | | | | |
| 2 | Gentofte | (DK) | | | | | | | * | | |
| 3 | Greve | (DK) | | | * | | | | * | * | |
| 4 | Lyngby | (DK) | | | * | * | | * | | | |
| 5 | Aramon | (F) | | | | | | * | | * | group of houses |
| 6 | Blagnac | (F) | | | | | | * | | | group of houses |
| 7 | Bourgoin | (F) | | | | | | * | | * | apartments |
| 8 | Dourdan | (F) | | * | | | | * | | | rock storage |
| 9 | Le Havre | (F) | | | | | | * | | | group of houses |
| 10 | Odeillo | (F) | * | * | | | | * | | | |
| 11 | Aachen | (D) | | | * | * | | | * | | |
| 12 | Essen | (D) | | | | | | | | * | |
| 13 | Freiburg | (D) | | | | | | | * | | apartments |
| 14 | Walldorf | (D) | | | | | | | | * | non-residential |
| 15 | Otterfing | (D) | | | | | | | | * | some soil storage |
| 16 | Wernau | (D) | | | * | | | | * | * | |
| 17 | Fiume Veneto | (I) | | | | | | * | * | | apartments |
| 18 | Firenze | (I) | | | | | | * | | | non-residential |
| 19 | Rossano Calabro | (I) | | | | | | * | | | apartments |
| 20 | Sequals | (I) | | | | | | * | | | group of houses |
| 21 | Eindhoven | (NL) | | | | | | * | | | |
| 22 | Zoetermeer 1 | (NL) | | | | | | | * | * | |
| 23 | Zoetermeer 2 | (NL) | | | | | | | * | | |
| 24 | Zoetermeer 3 | (NL) | | * | | | | * | | | |
| 25 | Zoetermeer 4 | (NL) | | | | | | * | | | |
| 26 | Bebbington | (UK) | * | * | | | | * | | | group of houses |
| 27 | Cardiff | (UK) | | | | | * | | | | apartments |
| 28 | London | (UK) | | | | | * | | | | group of houses |
| 29 | Macclesfield | (UK) | | | * | | | | * | | |
| 30 | Machynlleth | (UK) | | | | * | | | | * | non-residential |
| 31 | Milton Keynes | (UK) | | | | | | * | | | |

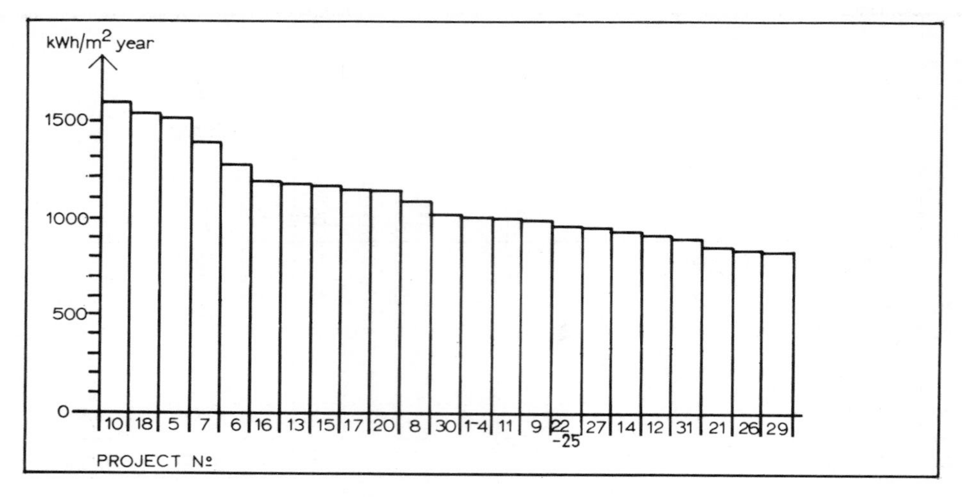


Fig. 2: global irradiation on the horizontal plane

This varies from 852-1605 kWh/m$^2$ per annum, and sites may be approximately classified as:

800 -1000 kWh/m$^2$ : Northern Europe (8 sites)
1000-1200 kWh/m$^2$ : Central Europe (10 sites)
1200-1600 kWh/m$^2$ : Southern Europe (5 sites)

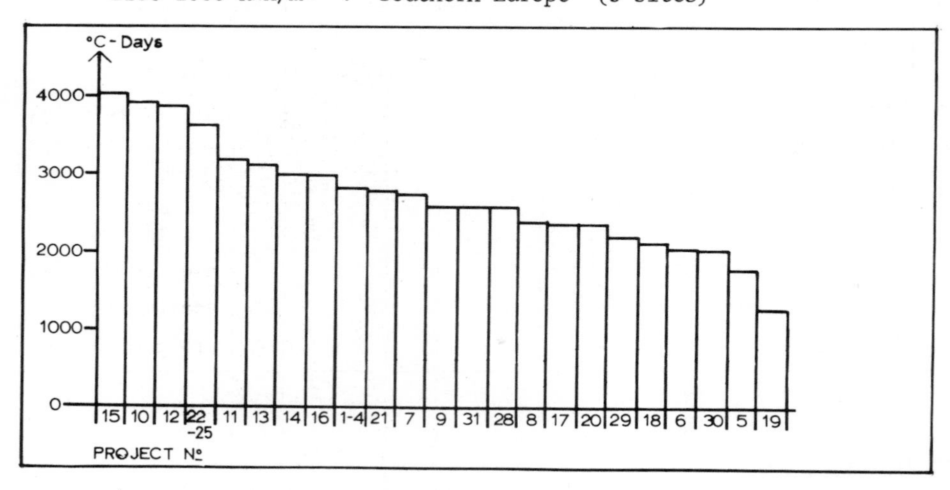


Fig. 3: annual degree days for 23 sites

Annual degree days are calculated from the product of the number of days per year that ambient temperature falls below a desirable 'base temperature', and the difference between base and ambient temperatures. The range shown is from 1782-4045°C days (a ratio of 2.2:1). Variation is partly due to differences in base temperature and on methods of calculation, for which no simple adjustment is possible, as correction would depend on local temperature fluctuations. A correction factor would vary from less than 200°C days in the south to over 300°C days in the north. There is a clear need for international uniformity.

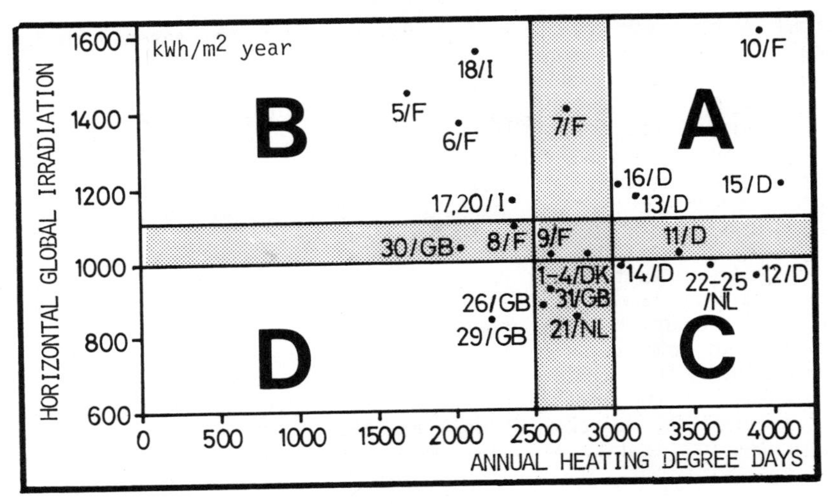


Fig. 4: Global irradiation and degree days

50% of sites fall within the shaded areas, and four zones (Fig. 5) can be discerned as follows:

zone A: consistently sunny; cold winters; e.g. Odeillo (10) Otterfing (15)
zone B: milder winters; sunny; e.g. Aramon (5) Firenze (18)
zone C: severer winters; low sunshine; e.g. Essen
zone D: milder winters; low sunshine; e.g. Macclesfield

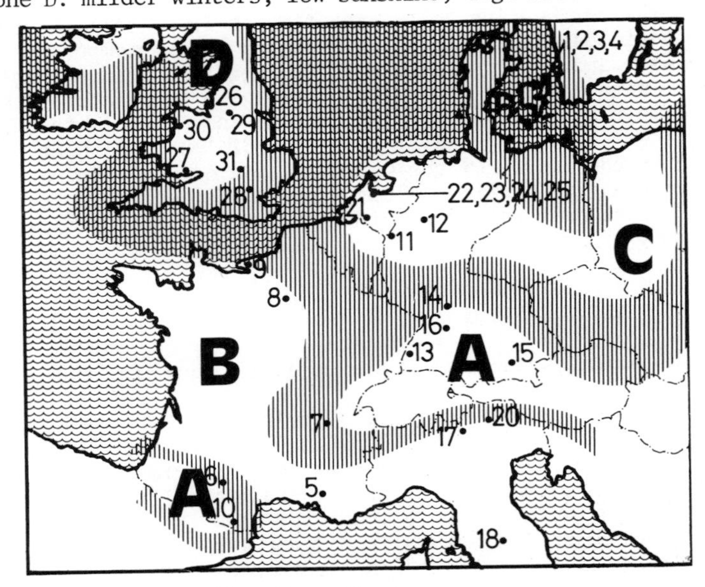


Fig. 5: climate zones

Zones A, B, C and D (Fig. 4) approximately mapped. The shaded boundary areas include much which is densely populated, and greater precision will require more data. Sites especially appropriate for solar heating are those in zone A (e.g. Odeillo (10)). The other climatic extreme (zone D) includes, for example, Macclesfield (29).

The principal problem for solar heating lies in daily and seasonal changes. Peak demand occurs in winter or at night, when little or no sunshine is available. The solution obviously lies in collecting heat from the sun when it is shining, and storing it for use later in the day or the year. Total reliance on solar energy implies long-term or seasonal storage. None of the projects studied were successful with this (typically this was not more than a week) and there is obvious room for improvement here. Short-term storage was more successful, and can shorten the heating season and reduce demand for fuel - especially in spring and autumn, and on chilly summer evenings.

Another major problem is temperature. For space heating, temperatures below 20°C are of little use, and domestic hot water needs to be no less than about 50°C. Figures from this study show the operating range for solar collection to be up to boiling point, but mostly no higher than 80°C. The higher temperatures, moreover, occur when they are least needed - in mid-summer. The storage problem is thus exacerbated (the hotter the store, the faster the heat loss) and in practice it is generally worthwhile only to collect at temperatures well over 5°C, thus reducing the utilization of solar energy to between 10% and 20% of the total available.

An inevitable consequence of these limitations has been the employment of an auxiliary system to 'top-up' or supplement solar heat. Among the projects studied, there is one optimistic attempt (4) to avoid this by retaining within the dwelling the 'free-heat' gains from occupants and electrical appliances. At the other extreme, those which confined themselves to domestic hot water heating, were able to use low temperatures to pre-heat cold water from the mains supply. With space heating, this poses difficulties - especially where hot water is used to distribute warmth through radiators. Although it is possible to use larger radiators or under-floor heating coils operating at lower temperatures than is customary in conventional central heating systems, return temperatures below 15°C are not practicable, and nothing can therefore be gained when the solar supply is as cool as this. Warmed air in principle permits low temperatures to be used, and was employed in twelve projects. The extreme case here is the French project (10) which relied on the 'Trombe Wall' principle of 'passive' heating, which uses heat accumulated in the fabric of the building, and natural convection. Thirteen projects used water as the heating medium - seven of them in under-floor coils. In general, additional heat was provided by gas, oil or electricity, usually with a boiler, sometimes (7 projects) a heat pump.

## SOLAR HEATING SYSTEMS

Two methods of collecting solar energy are possible - passive and active.

Passive systems have the longest history. They rely on a building's capacity to store heat in its fabric, and to admit sunshine through windows. Glazing, since it admits solar radiation and hinders re-emission, has greatly improved passive performance - especially in the recent past. Traditional practices in architecture of orientation and site selection have long exploited the principle, as has the use in vernacular building of thick walls and the like. Passive systems for domestic hot water are mainly confined to the tropics. Of the projects studied, only two (10, 26) used passive systems - both for space heating only. There is great dependence upon the thermal characteristics of the building and the vagaries of the weather, and overheating in summer may cause difficulties.

Active systems are so called because they use energy other than that derived locally from the sun to collect and distribute solar heat. They comprise essentially a collector, a store and heat emitters (Fig. 6). A circulating medium

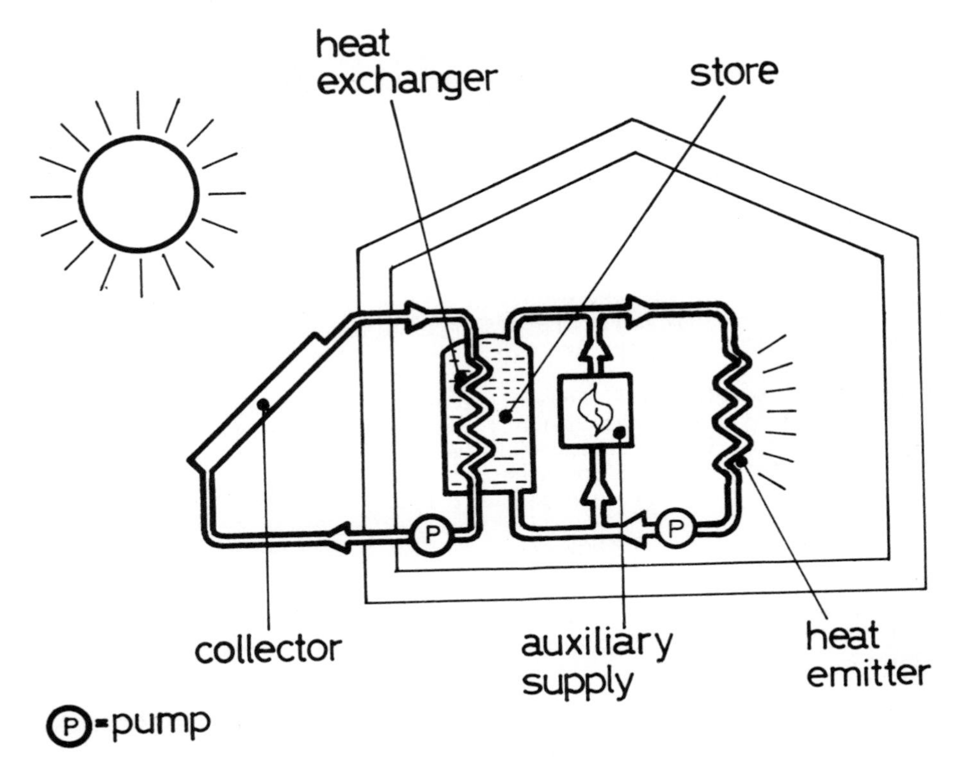


Fig. 6: a typical active system

(usually water, sometimes air) is used to transfer heat through the system, and this is propelled by electric pumps or fans. Both domestic hot water and space heating are possible, and they may be separated or combined. Equipment and controls need careful design and maintenance, and have caused difficulties in many of the projects studied. It is customary for frost protection to add glycol as an anti-freeze, and non-return valves are necessary where reverse circulation could occur. At Eindhoven (21) a unique system which incorporates glycol-filled expansion tubes is being tried.

DEMAND

Efficient design must begin with an adequate assessment of the demand for space heating (or cooling), ventilation and domestic hot water. Apart from effects due to climate, variations in the way buildings are built can cause wide differences (Fig. 7). The way people behave can also have a great effect, and it is known that in similar houses, one household may use up to five times as much heat as another of similar composition.

The way buildings are built will affect both heat loss and heat storage. Most of the projects had better insulation than is normal - some of them exceptionally so. With warm air systems it is very desirable to control air leakage and to ventilate artificially, at least during the heating season. Some projects (3, 4, 11, 29) incorporated devices to recover heat from stale air, and some (4, 11) from waste water. Many projects were concerned not only with solar heating, but with energy conservation in general, and included such devices as shutters and triple glazing.

Values of volumetric heat loss for nineteen projects are shown in Fig. 7. Lower values represent the better insulated buildings. The range is from 0.03 - 1.82 W/m$^3$.K, with a mean value of 1.06 W/m$^3$.K. It can be seen that the projects fall into three distinct groups:

| Vol Heat Loss W/m$^3$.K | Number |
|---|---|
| 0 - 0.5 (highly insulated) | 2 |
| 0.5 - 1.0 (well insulated | 8 |
| 1.0 - 2.0 (poorly insulated) | 9 |

Aachen (11) is an example of very highly insulated building with U-values of 0.17 W/m$^2$.K for the external walls and 0.23 W/m$^2$.K for the roof. Typical U-values for the well-insulated buildings are 0.3 - 0.4 W/m$^2$.K.

For the twenty-two single family houses, floor areas range from 79.5 - 220m$^2$, with a mean value of 144m$^2$. In Great Britain, the average size of a five-person single family house is 100m$^2$, so the experimental projects studied are, on average, nearly 50% larger than this. Floor areas for the twenty-one single family houses with solar space heating may be grouped as follows:

| Floor Area m$^2$ | Number |
|---|---|
| 0 - 100 | 3 |
| 101 - 150 | 13 |
| 151 - 200 | 4 |
| 201 - 250 | 1 |

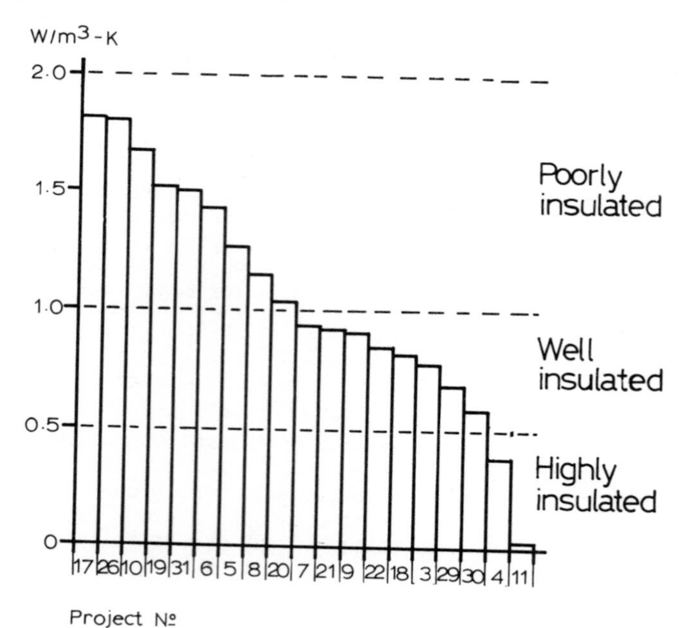


Fig. 7: volumetric heat loss (G)

Calculated heat losses per cubic metre of building volume for 19 projects

Fig. 8 illustrates the design loads for space and water heating for twenty projects. The range for single family houses (fifteen projects) is 4600 - 37900 kWh/year, with a mean value of 20307 kWh/year. Design loads for the single family houses can be summarised as follows:

| Load kWh/year | Number |
|---|---|
| 0 - 10,000 | 1 |
| 10,000 - 20,000 | 8 |
| 20,000 - | 6 |

The projects with higher loads tend to be the larger houses, for example, Dourdan (8), Otterfing (15). The one project with a total load below 10,000 kWh/year is Lyngby (4), which has less than the mean floor area and is very well insulated.

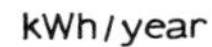


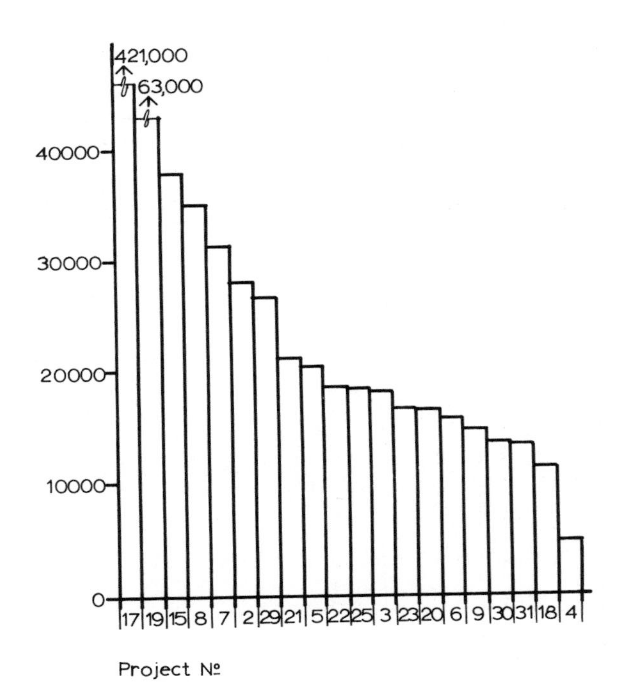


Fig. 8: annual design load

In Fig. 9, loads are expressed per unit floor area for the same projects. It can be seen that the range is from 38 $kWh/m^2$ - 261 $kWh/m^2$, with a much more even spread between the projects. The mean value is 134 $kWh/m^2$.

Thermal mass is also a factor affecting the thermal performance of a building. A lightweight building (for example, timber frame construction) will heat up more quickly, and therefore the space heating system can be switched on shortly before the building is to be occupied. This is an advantage where occupancy is limited to a few hours a day. In heavier buildings, constructed of brick, stone or

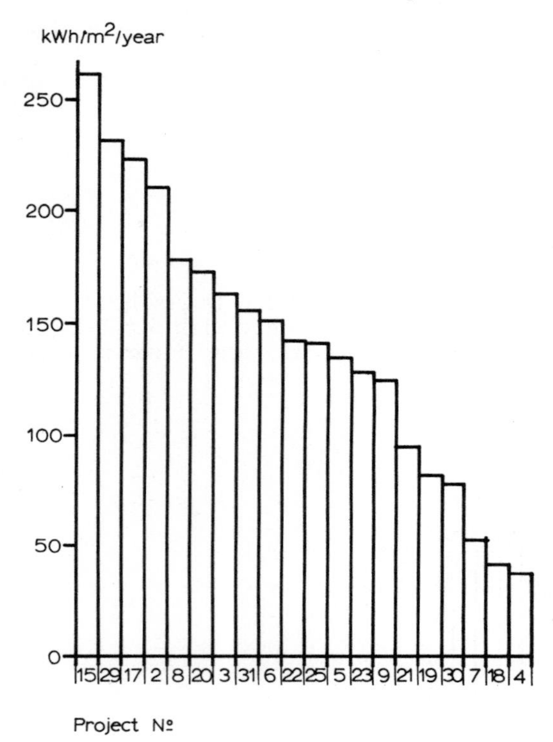


Fig. 9: design load per unit floor area

concrete, a proportion of the heat input will be used to raise the temperature of the structure, thus reducing the rate of increase in room temperature. Heavy-weight buildings are therefore more suitable for applications with a relatively continuous occupation, where the increased thermal inertia of the structure is useful in absorbing fluctuations in temperature.

Approximate figures for thermal mass are available for eighteen projects, and are summarised below.

| Thermal Mass kg/m$^3$ (building weight/volume) | Number |
|---|---|
| 0 - 150 | 7 |
| 150 - 250 | 5 |
| 250 - | 6 |

The construction of the buildings studied is diverse and includes prefabricated and traditional methods; lightweight and heavy construction; 2-storey detached houses and 3 or 4 storey apartment buildings. In a few cases (Lyngby (4), and Aachen (11)) they have been designed with special emphasis on minimum energy consumption. In the future, it is likely that new buildings will be designed from an 'energy conscious' point of view, incorporating high levels of thermal insulation and large windows on the south elevation to maximise passive solar gains, and the loads which the active solar heating system will have to satisfy are likely to be considerably less than the mean value of 20,000 kWh per annum found in the current sample of projects.

## COLLECTORS

All but one of the active systems used liquid cooled collectors, similar in principle to Fig. 10. In six cases, (13, 21-25) they incorporated a selective surface, which retains more of the solar radiation.

Three projects (11, 12, 13) included high-efficiency evacuation, which is claimed to double the energy collected at normal operating temperatures. One project (8) used an air-cooled collector in conjuction with rock storage, and one (25) combined an air-cooled collector with water storage.

Single-glazing was used in sixteen projects. Double-glazing was used in thirteen - mainly in the colder climates. Although there is some advantage in siting collectors so that they lie at right-angles to as much as possible of the useful incident sunshine, in the examples studied slopes varied from $20^{o}$ to $90^{o}$ to the horizontal. This variation bears little relation to latitude, and may be a consequence of roof slope or other design considerations.

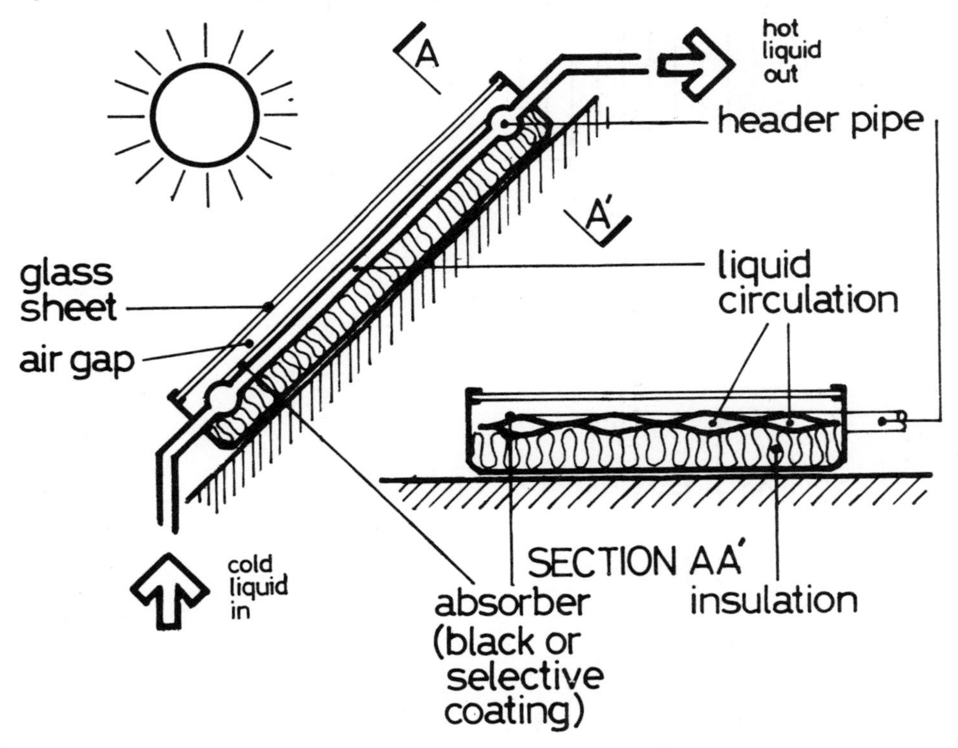


Fig. 10: a typical liquid cooled collector

Fig. 11 shows for seventeen projects the range of collector area employed: from 20.3 $m^2$ to 80 $m^2$; with a mean of 44.7 $m^2$. Fig. 12 shows this as a percentage of floor area: from 17 to 55%, with a mean value of 32%. Fig. 13 shows collector area as a proportion of the design load. The range here is from 109 kWh per $m^2$ to 1031 kWh per $m^2$, with a mean of 500 kWh per $m^2$. Most show values in the range 400 kWh per $m^2$ to 600 kWh per $m^2$. From thirteen of the projects it is possible to calculate the quantity of fluid circulating in the primary loop from collector to store: between 0.32 litres per $m^2$ and 16.7 litres per $m^2$, with a mean of 3.3 litres per $m^2$ of collector area.

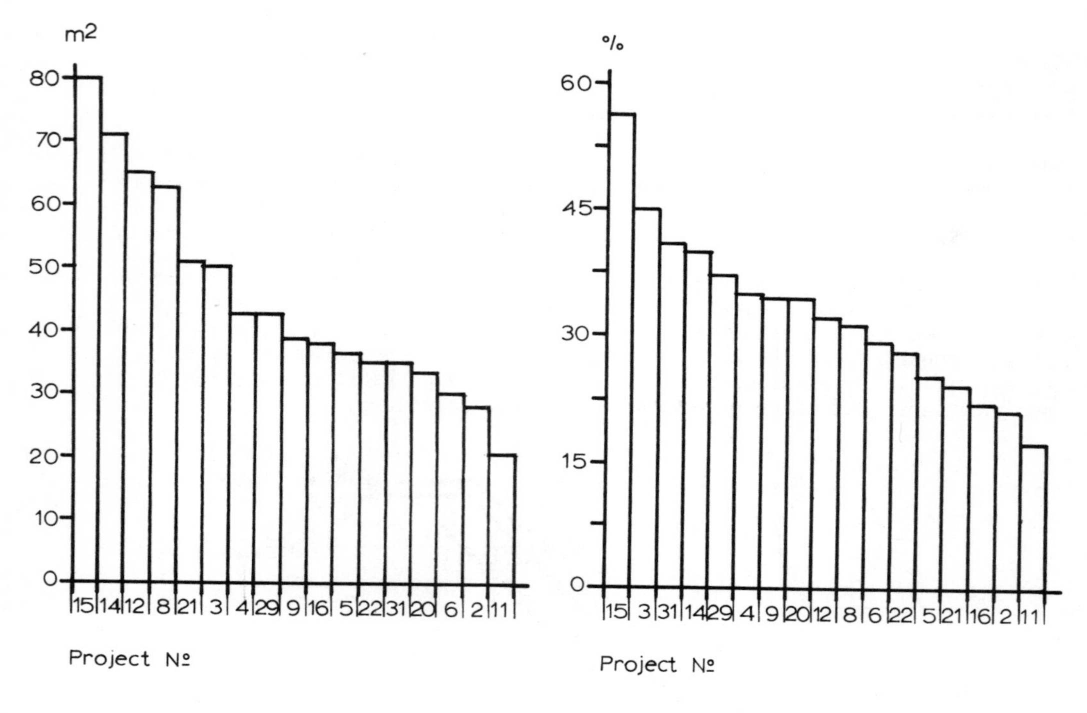


Fig. 11: collector area

Fig. 12: collector area per unit floor area

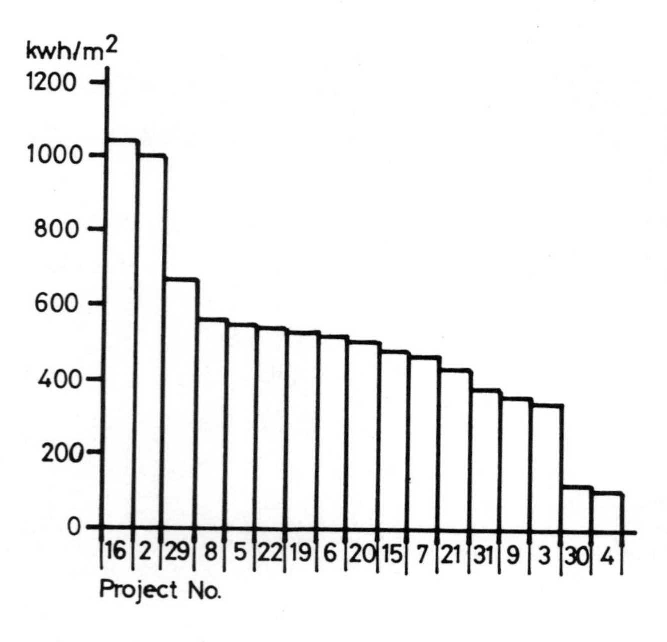


Fig. 13: design load per unit area of collector

## HEAT STORAGE

The house at Dourdan (8) in France is exceptional among the active systems in using air to collect heat. It is exceptional also in using a rock store, 40 $m^3$ (estimated as equivalent to 14 $m^3$ of water). At Otterfing (15) in Germany there was a disappointing experiment with soil storage. From Bourgoin (7) in France comes a report of an incomplete experiment with a 'phase-change' material, held in plastic containers within the store, designed to make use of latent heat.

All the other active systems relied exclusively on water as a storage medium, usually in single tanks, with volumes varying from 1 $m^3$ to 8 $m^3$ (mean 4.5 $m^3$). (Fig. 14). Multiple tanks are included in 7 projects, but in only 3 is there any attempt to exploit the possibility of grading heat storage into, for example, low, medium and high temperature compartments. The solar house at Eindhoven (21) is noteworthy in this respect. Most examples allow for only two to three days storage. Lyngby in Denmark (4) Aachen in Germany (11) and Machynlleth in Wales (30) attempt long-term storage with limited success.

Figure 15 relates storage volumes to the collector area for 17 single-family houses. The range is from 30 litres per $m^2$ to 145 litres per $m^2$, with a mean of 95.5 litres per $m^2$.

The size of store required has resulted in some cases in its being outside the building. Within the building, losses from it may be disadvantageous in summer, but are certainly useful in winter. Good insulation is in all cases most desirable.

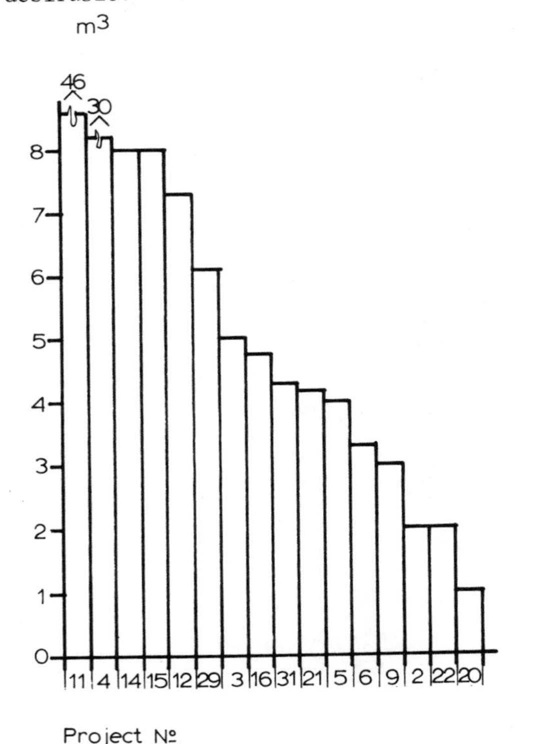


Fig. 14: storage volume

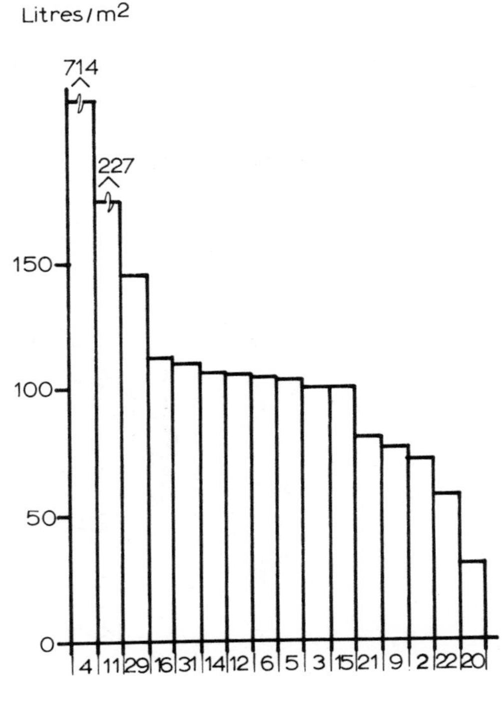


Fig. 15: storage volume per unit area of collector

CIRCULATION AND CONTROL

The primary circuits (collector to heat store) are essentially similar in all the active systems (Fig. 16). When the temperature of liquid in the collector is significantly higher (normally at least 2°C) than the lowest temperature in the store, a control switches on the primary pump, switching it off again when the temperatures are the same. In air systems, switch-on may be delayed until the temperature difference is higher than 5°C. Wet systems with 'drain down' facilities may require more elaborate control - as at Lyngby (4).

There is greater variety in the secondary circuits from store to heat emitters, and Fig. 17 summarises the main types. (Domestic hot water is omitted for simplicity). Some of them make provision for heat to be taken directly from collector to dwelling, using the store only when it is advantageous to do so.

Control of the secondary circulation begins with room temperature. When this falls below the desired level, the system operates in much the same way as conventional heating systems. When necessary, flow temperature is maintained above the design minimum by the auxiliary heating system. When temperature is high, performance is best controlled by reducing the rate of flow. Design flows in most of the systems were between 50°C and 80°C. In type D, inlet temperatures as low as 24°C were possible. In domestic hot water circuits, cold water from a heater tank is pre-heated via a heat-exchanger within the heat store, and then brought to the required temperature with an immersion heater or similar means.

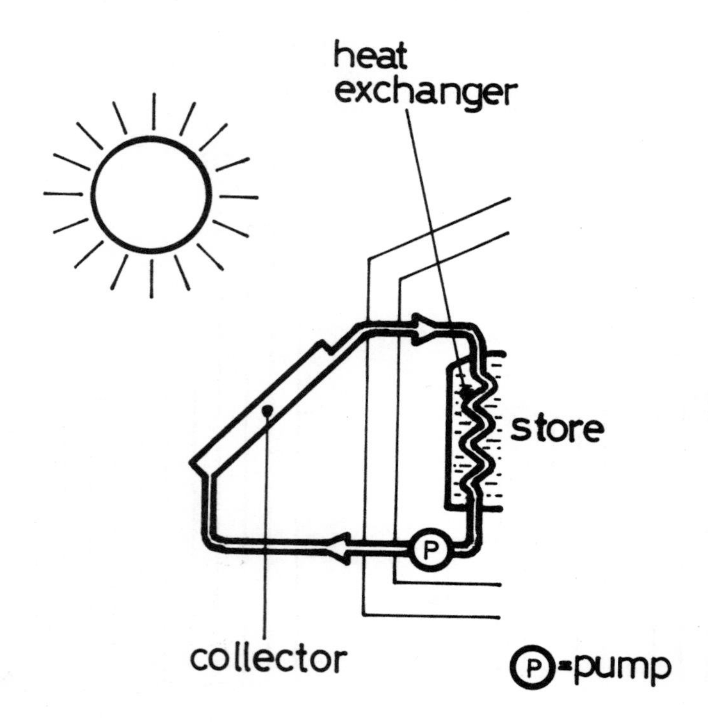


Fig. 16: a typical primary circuit for an active 'wet' system

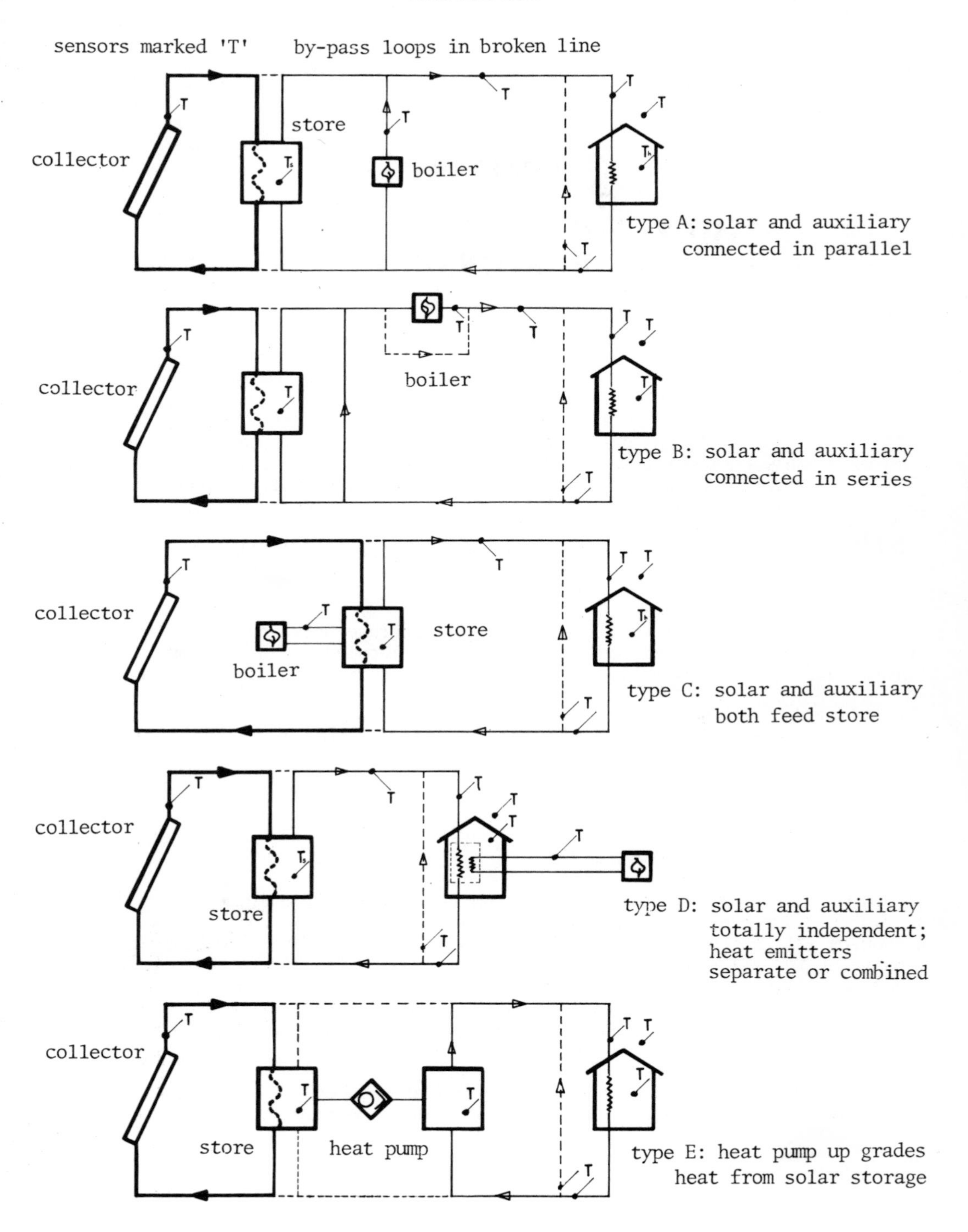


Fig. 17: typical active systems

# CHAPTER 2

# HOW THE BUILDINGS HAVE PERFORMED

Data were analysed for two separate groups of projects, one consisting of thirteen projects with similar characteristics, and the other of all the twenty-three projects for which data are available.

GROUP I (13 projects): Projects No (2), (3), (7), (9), (12), (14), (16), (21), (22), (23), (24), (25), and (31)

These are all active systems, with short-term storage only, providing space heating and pre-heating of domestic hot water. All are located in Northern and Central Europe, and with the exception of project 7 (Bourgoin), are single family houses.

GROUP II (23 Projects)
All projects except (8), (13), (17), (19), (20), (26), (27), (28)

## ANNUAL PERFORMANCE

It can be seen (Fig. 18) that total annual loads range from 8,000 - 30,000 kWh/year. The four projects with loads greater than 20,000 kWh/year - Wernau (16) Eindhoven (21) Essen (12) and Walldorf (14) - have floor areas 20% larger than the mean value for all projects. Eindhoven, Essen and Walldorf also have collector areas 40% larger than the mean for all projects, resulting in markedly higher solar contribution. The figure also shows the significant difference in annual load in the terraced houses at Zoetermeer between the ends (25, 22) and the middle (23, 24).

Very well insulated houses such as Lyngby (4) and Aachen (11) (not shown) have much lower loads, viz 9,600 kWh and 5,600 kWh respectively.

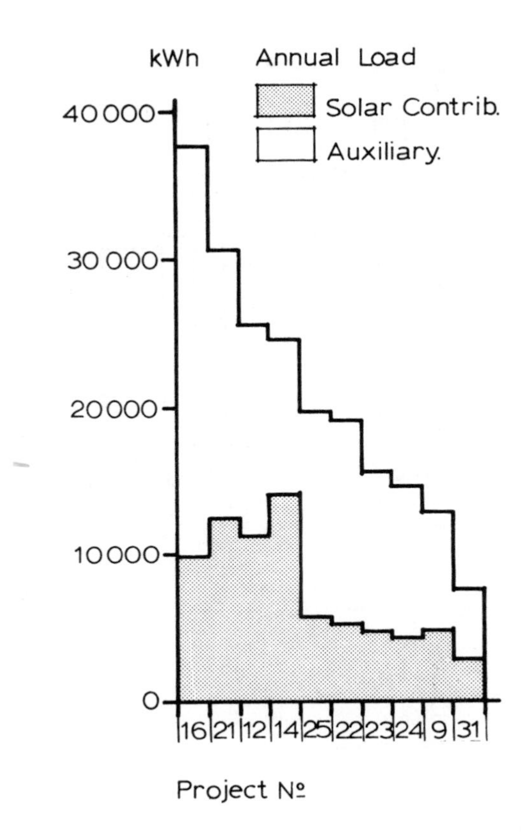


Fig. 18: measured annual energy consumption for group I projects, excluding apartment buildings

Figure 19 shows the proportion of solar to total annual consumption for Group I projects. Figures range from 16 - 58%, with a mean value of 32%. This compares with a range of estimates from 38 - 61% for the same group of projects (mean value 43%).

For the Group II projects (not shown), Firenze (18) has the highest solar proportion with 72%, although this figure includes a contribution to summer cooling load. The passive system at Odeillo (10) has a 70% solar contribution to space heating due to exceptional climate conditions. In the well-insulated houses with long-term storage, solar contribution to space and water heating loads was 35% for Lyngby (4) and 50% for Aachen (11).

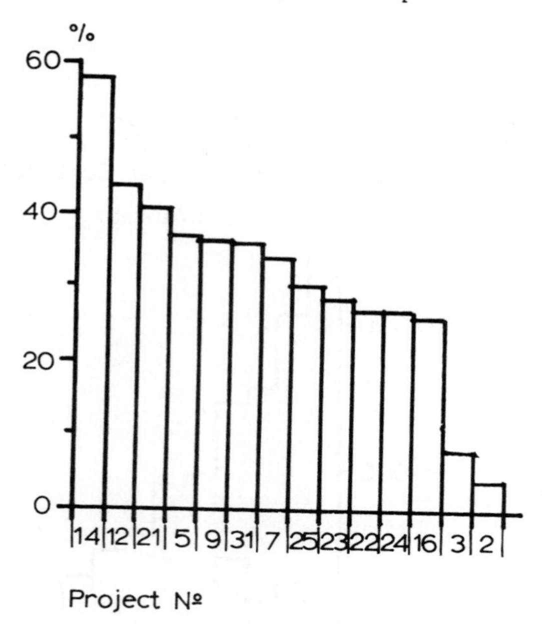


Fig. 19: solar contribution to total annual consumption

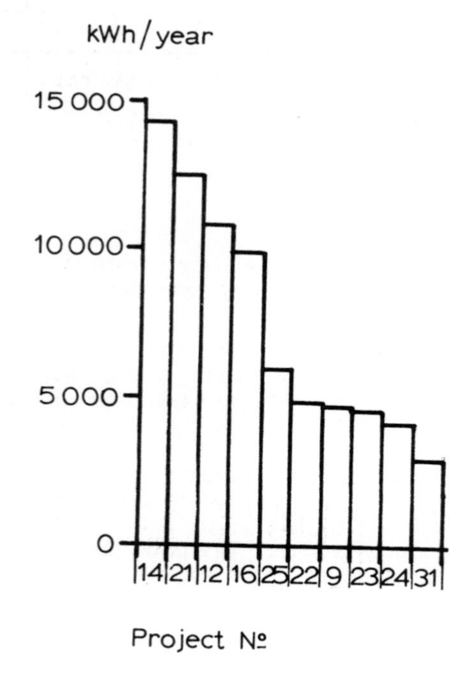


Fig. 20: solar energy used: group I

Figure 20 shows the amount of solar energy used in Group I projects. The range is 3,000 - 14,000 kWh. Again, two distinct sub-groupings can be seen. The higher range of 10,000 - 14,000 kWh/year corresponds to the four projects with higher annual loads. The remaining six projects with a range of 3,000 - 6,000 kWh are very similar systems in comparable geographical locations.

Solar energy used in some projects not shown increases the range. Thus, Bourgoin (7) with 23,400 kWh, Odeillo (10) with 19,000 kWh and Firenze (18) with 14,100 kWh/year show markedly higher solar energy usage.

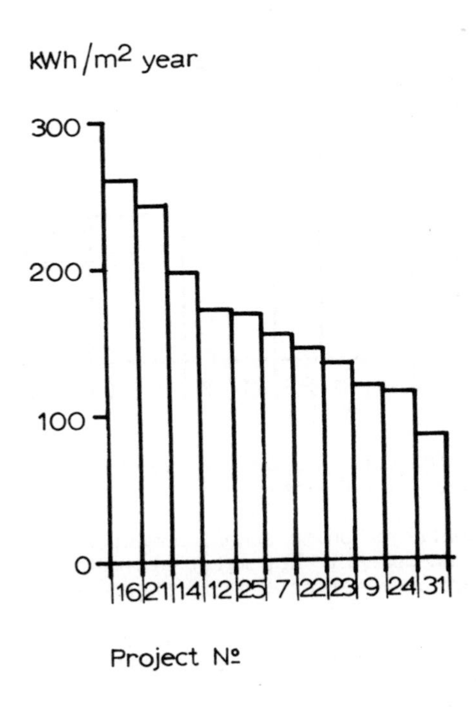


Fig. 21: solar energy used per unit area of collector

In order to compare the different projects, Fig. 21 shows annual solar energy used per unit collector area. For Group 1 projects, the range is 83 - 260 kWh/m$^2$, with a more even spread between projects. With the exception of Eindhoven (21) the four highest performances are from heat pump assisted systems. With respect to this, it is perhaps important to note that Essen (12) and Eindhoven (21) have summer solar energy usage substantially higher than all other projects.

From the Group II projects, Odeillo (10) has the highest output, at 347 kWh/m$^2$. Mean and standard deviations for annual solar energy used per unit collector area for both groups are:-

Group I 162 ± 52 kWh/m$^2$ year

Group II 152 ± 69 kWh/m$^2$ year.

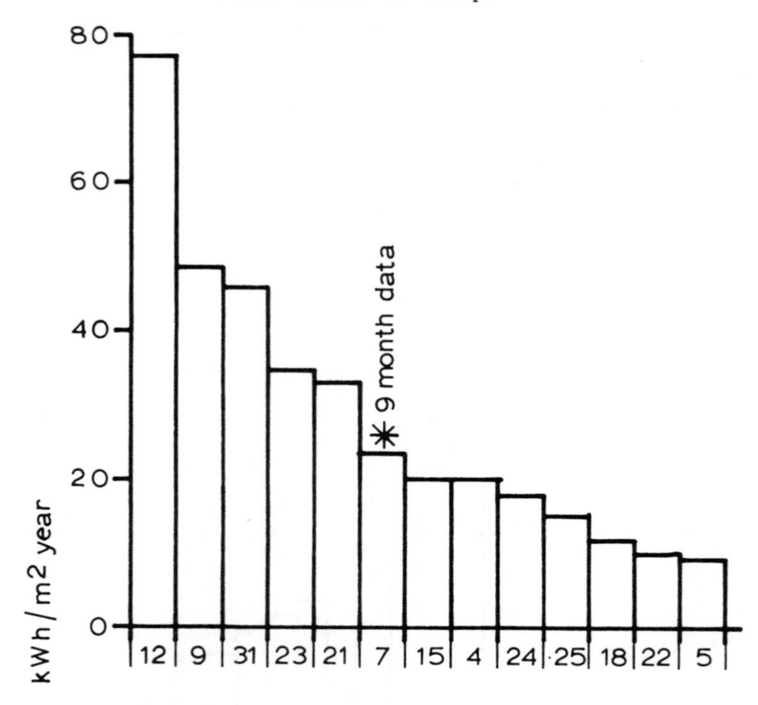


Fig. 22: solar contribution to domestic hot water per unit area of collector

Figure 22 shows solar contribution to annual domestic hot water load for the 13 combined space heating and domestic hot water projects with data available. Output per unit of collector area varies from 10 kWh/m$^2$ year to 77 kWh/m$^2$ year. A large number of factors have affected the domestic hot water performance. Thus, the highest output corresponds to Essen (12), where priority is given to domestic hot water, while in a number of projects (e.g. Zoetermeer) priority was given to space heating. In some systems, poor heat transfer efficiency of hot water preheat tanks resulted in poor solar contribution to DHW (e.g. Aramon (5)). The four houses in Zoetermeer (22) - (25) give contributions to DHW ranging from 10 to 35 kWh/m$^2$ year. By comparison, the DHW only system, (Blovstrod (1)), gives over 90 kWh/m$^2$ year.

The solar contribution to domestic hot water represents between 5% to 55% of the total annual energy used.

Figure 23 summarises annual system efficiency for Group II projects. System efficiency is defined as the ratio of solar energy used to incident energy on the collector, and varies from 7 to 30%. The main factors affecting system efficiency are seasonal variations in the load pattern and in incident energy, and the relative size of the system (Fig. 23). Efficiency of between 10 - 20% may be considered as the norm. As can be seen, there are four projects with efficiencies higher than 20%, these being Bourgoin (7), Odeillo (10), Eindhoven (21) and Wernau (16). It should be noted however that high annual system efficiency may result from abnormally high summer loads. The mean and standard deviation for annual system efficiency is, for Group I, 16.3 ± 6.3%; and for Group II, 15.3 ± 7.0%.

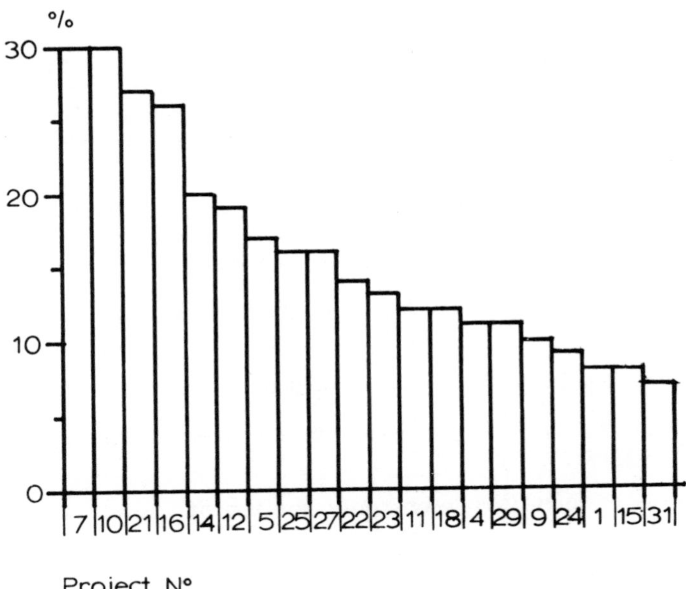


Fig. 23: annual system efficiency: group II

There are 4 projects with system efficiencies over 20%. They are:

Odeillo, France (10) - 30%
Bourgoin, France (7) - 30%
Eindhoven, Netherlands (21) - 27%
Wernau, W. Germany (16) - 26%

ODEILLO: A passive space-heating system, based on the Trombe-wall principle, and located high in the Pyrenees. The site is ideal for solar heating applications, with 2460 annual sunshine hours and nearly 4000 heating degree days. More importantly, 200 hours of sunshine are typical in January, compared to 320 in July. The vertical collection surface maximises winter solar collection, when demand for heating is greatest, but occasional heat demand on summer evenings also serves to improve overall system efficiency.

BOURGOIN: An active wet system designed to provide both space and water heating. Again, collectors are vertical to maximise winter collection. The available data cover nine months from September 1978 to May 1979. During this period, heating demand is high and incident energy low, so that most of the energy collected can be used. Consequently, a high system efficiency is obtained. This will be reduced when lower efficiencies obtained during the summer months are included. There is however another important factor: the system serves a group of 12 apartments, and performance is thus less affected by individual patterns of occupancy. Demand is more consistent and this permits the system to operate more efficiently.

EINDHOVEN: An active wet system combined with a warm-air heating system. Good annual performance figures can be attributed partly to careful design, and partly to higher than normal summer loads. The collector area is 51 $m^2$ (large for a single family house), and selective surface absorbers have been used. The storage tank is designed to ensure good stratification by:-

- bringing water from the collectors into the store through an inlet which is designed to float at the level where storage temperature is the same as that of the incoming water;
- preventing turbulence within the store by keeping circulation below 0.1 $m.s^{-1}$;
- pre-heating incoming fresh air around the base of the store;
- pre-heating domestic hot water in a heat-exchanger rising the full height of the store;
- heating by auxiliary means only the top of the store;
- taking heat for space heating from that level within the store which is at the required temperature.

WERNAU: An active wet system, incorporating a heat pump, air heat recovery and elaborate controls. Fourteen operating modes are possible. These are controlled so that the most efficient can be selected automatically. The heat pump can use very low storage temperatures. The collector area is relatively small (38$m^2$), and heat demand is high, although the house is not occupied.

The following 4 projects have lower than average system efficiencies:

Blovstrod, Denmark (1)
Otterfing, W. Germany (15)
Gentofte, Denmark (2)
Macclesfield, Great Britain (29)

BLOVSTROD: An active wet system designed to pre-heat domestic hot water supply. Heat losses from the store were much greater than anticipated, due to imperfect insulation of the pipework and reverse circulation due to syphonic action. The oil-fired post-heater tank was maintained at 65$^oC$, compared with the delivery temperature of 50$^oC$. Mixing with cold water was therefore necessary, and consequently only some of the hot water used was pre-heated by the sun. The boiler was however switched off for 2 months in the summer, indirectly saving in addition an estimated 1175 kWh.

OTTERFING: An active wet system for space and water heating, incorporating a very large (80$m^2$) collector, 8$m^3$ heat storage plus some heat storage in soil. Poor results are mainly due to three factors: because soil storage temperature was too low, only 40% of the collected energy could be used; the floor heating system used the low-grade heat which was available inefficiently; and there were technical problems with components which reduced the efficiency of the system.

GENTOFTE: An active wet system for space and water heating, fitted in an existing house. The system is relatively small, and the existing standard-size radiators were retained for heat distribution. As a result, the return temperatures were too high (60°C) for the system to gain any benefit from the heat store. Provisional figures indicate a 17% solar contribution to space heating, and a slightly better performance with domestic hot water.

MACCLESFIELD: An active wet system with open trickle collectors, multiple storage tanks, heat pump, and mechanical ventilation recovering heat from stale air. The primary pump is switched on at 15 min. intervals to check on the availability of energy, and is left on or off, depending on the temperature rise across the collector. This system is potentially wasteful of stored energy, and also results in increased power consumption by the pump. For two months in winter, exposed parts of the primary circuit became frozen. It is also important to note that (due to a limited monitoring budget) some of the performance data is estimated, probably on the conservative side. For example, the air heat recovery system absorbs a fraction of the standing losses from the heat storage tanks, but such gains are not included in the figures.

In general, the results illustrate the significance of annual demand in assessing efficiency. Higher demand usually results in a higher solar contribution. Size of collector is also significant: large collector areas at Eindhoven, Essen and Walldorf gave good results. Some efficient projects also make use of more elaborate technical devices such as selective surface collectors, heat pumps, and stratified thermal storage.

It should be noted that with very well insulated houses there are smaller loads and a shorter heating season. This has encouraged a strategy based on long-term inter-seasonal storage, such as has been attempted with limited success at Lyngby and Aachen.

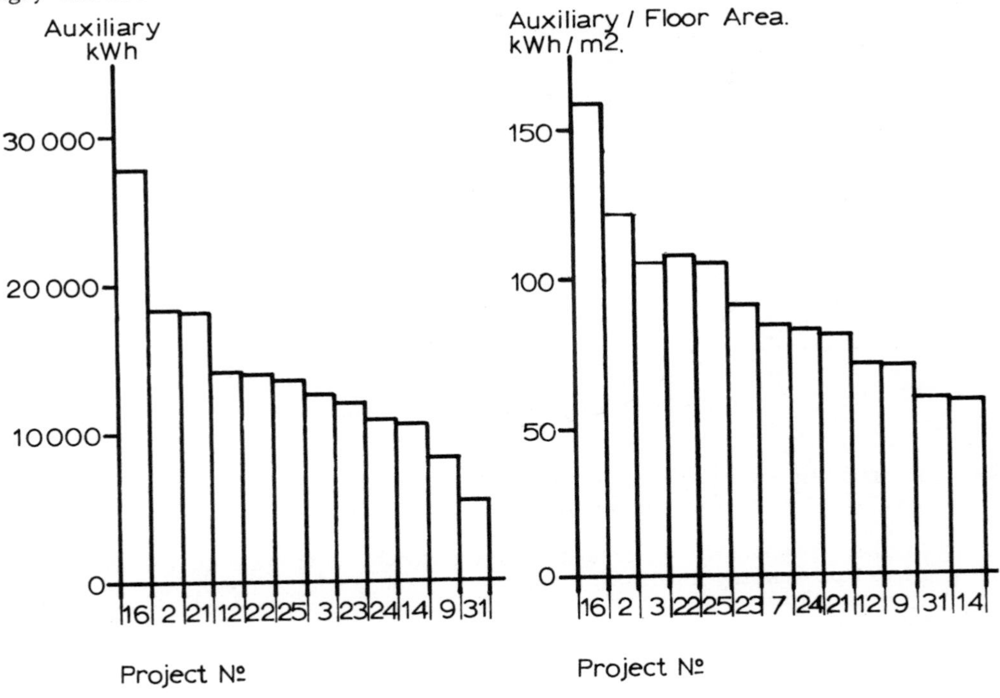


Fig. 24: auxiliary consumption

The main objective of solar heated houses is to reduce the consumption of auxiliary energy. Fig. 24 shows auxiliary energy consumption for Group I projects. The range is from 5,000 to 28,000 kWh, with a mean value of 13,000 kWh. These figures are also expressed per unit floor area. The range is then from 60 to 160 kWh/$m^2$ with a mean value of 90 kWh/$m^2$.

Within the Group II Projects, Aachen (11) and Lyngby (4) have attempted to reduce auxiliary energy consumption to a minimum. Their annual loads are respectively, 5,500 kWh and 9,620 kWh. Both projects have very high levels of thermal insulation, some long term (seasonal) heat storage and heat recovery from waste water and air ventilation system.

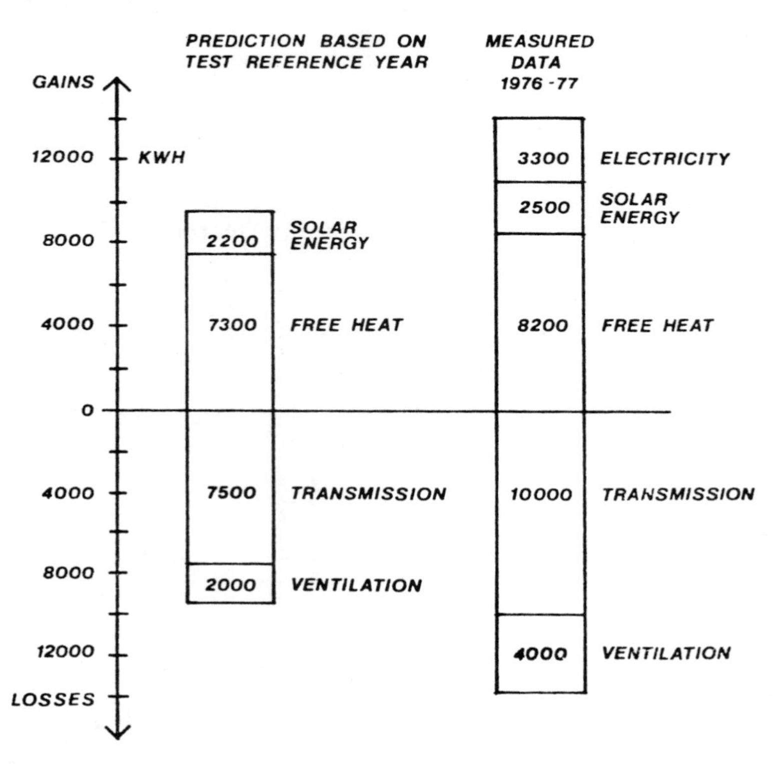


Fig. 25: energy balance for Lyngby (4)

Figure 25 shows the energy balance for Lyngby, derived from 1976/77 measured data. This figure illustrates four important points:

- the high proportion of free heat gains which contribute to reducing the space heating load;

- the higher than anticipated heat losses resulting in an additional 3,300 kWh electricity required;

- the close agreement between predicted and measured solar contribution.

- a reduction on the percentage solar due to increased total load

ZOETERMEER 2 (23)

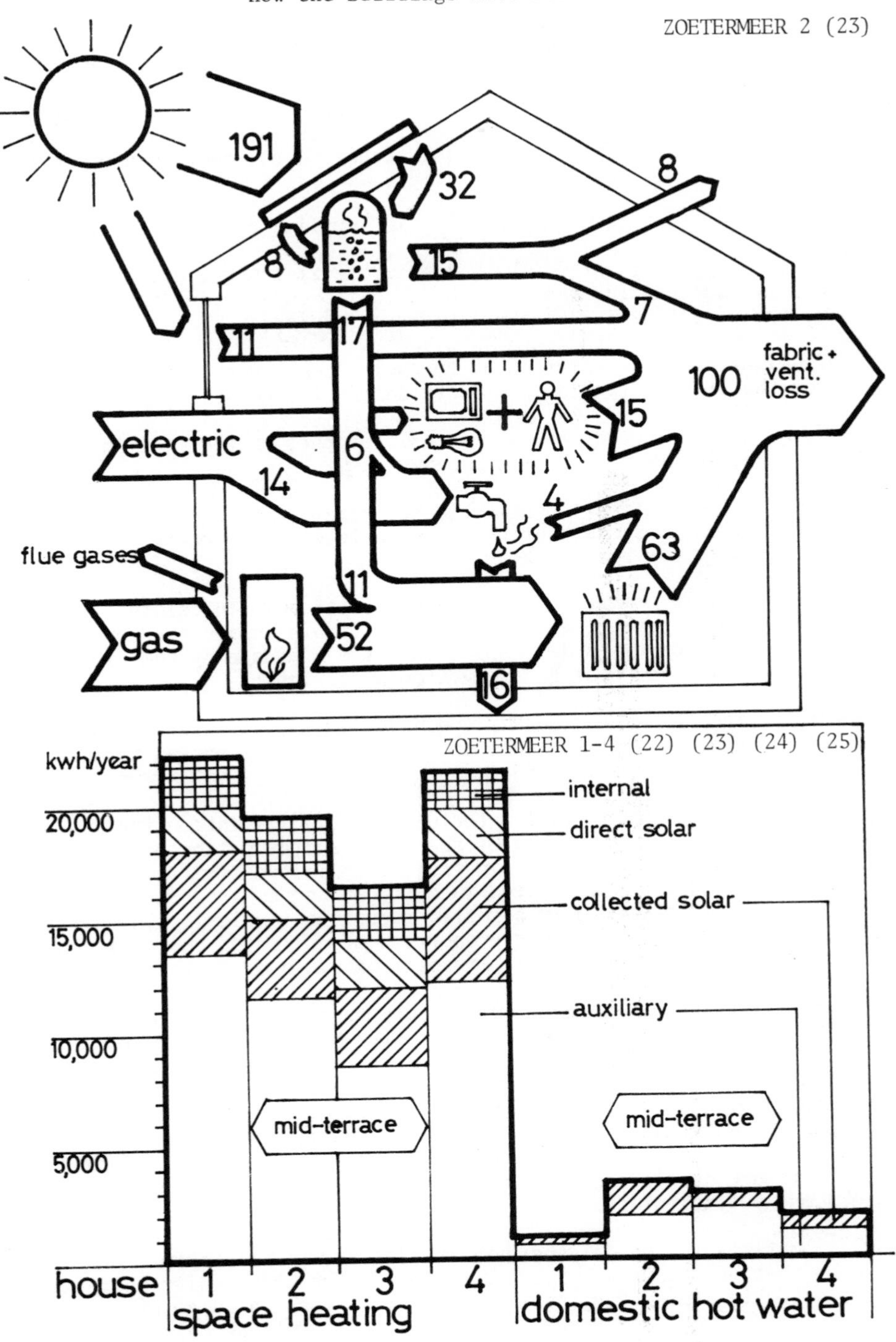


Fig. 26: annual energy flow for Zoetermeer
gross annual space heating load = 100%
figures approximate to ± 10%

This study has revealed a considerable lack of knowledge about the thermal performance of buildings. Unfortunately few of the projects gave an adequate account of annual energy flow for the building and evaluation has consequently had to concentrate on energy flows in the heating system alone. However, Fig. 26 attempts to show all energy flows in one of the Zoetermeer houses. Quantities are expressed as a percentage of the annual load required to maintain a comfortable internal environment. It can be seen that:

- 20 - 30% of the space heating load is supplied by free heat gains;
- total energy incident on the collector is 2½ times greater than the annual space heating load. Much of this occurs during the summer, but the potential for seasonal storage is clear;
- of all useful heat supplied by the solar heating system, 30% are losses from the storage tank and pipework inside the house;
- 20% of all collected energy was rejected through the over-load heat exchanger to prevent boiling in storage;
- standing losses from the domestic hot water tank make a significant contribution to space heating, comparable with the energy supplied by the solar pre-heat tank;
- energy used in domestic hot water consumption is 16% of the space-heating load; some of this energy will remain inside the house, before waste-water is discharged into the drainage system; this shows the potential for heat recovery from waste water;
- much of the total heat losses (probably 40 - 50%) occur through natural ventilation; and some of this could be recovered through a heat-exchange system.

Within a solar heating system itself, the efficiency of solar energy conversion and storage is of prime importance. Of the total energy collected, it is the proportion used which matters. To illustrate this, information from projects with appropriate data (all of them located in Northern and Central Europe) was analysed (Fig. 27). Although only a small proportion (typically 26%) of incident energy is stored, a relatively large proportion of collected energy is eventually used, especially when the useful heat losses from storage are taken into account.

Figure 28 shows the annual energy flow in four projects. The figures for useful losses reflect some design features of the system but it must be stressed that although thermal losses from storage tanks and primary circuit can be measured or calculated fairly accurately, the proportion of these losses contributing to space heating of the living area may be difficult to evaluate. Calculating how much of these losses are actually reducing demand for auxiliary heating is even more difficult, particularly during mid-heating-season, when it is likely that at least part of the storage losses occur when the house is unoccupied, or when free heat gains are alone sufficient to cope with the heating demand.

Figure 29 summarises the annual energy flow diagram for all Group I projects. The figure shows that only a very small percentage of the total solar energy incident on the collector is stored: typically 26%. By contrast, a large proportion of the collected energy is used: over 60% if useful losses are included.

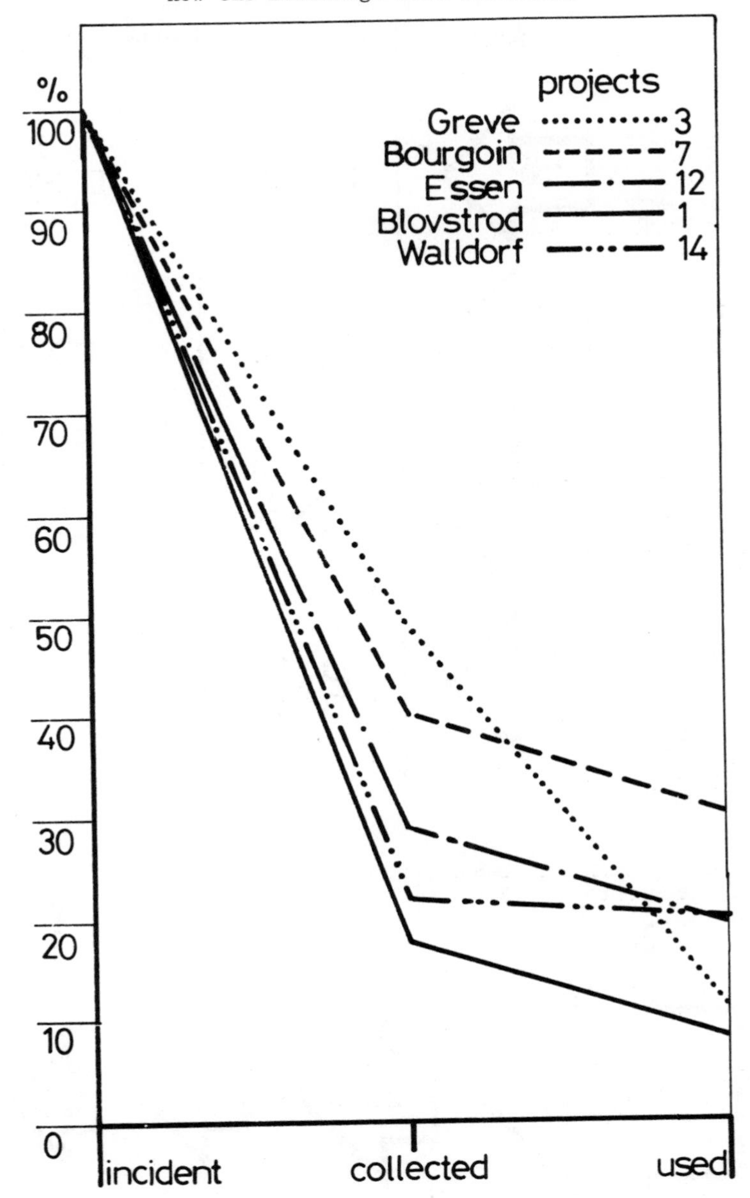


Fig. 27: annual energy flows

When the storage is located outside the heated volume of the house, and no losses contribute to space heating, the figure for usage efficiency is inevitably lower (e.g. 22% for Le Havre (9)). Long term storage tanks also have lower usage efficiency figures (Lyngby: 48%, Aachen: 33%) although this may be offset by higher collection efficiency.

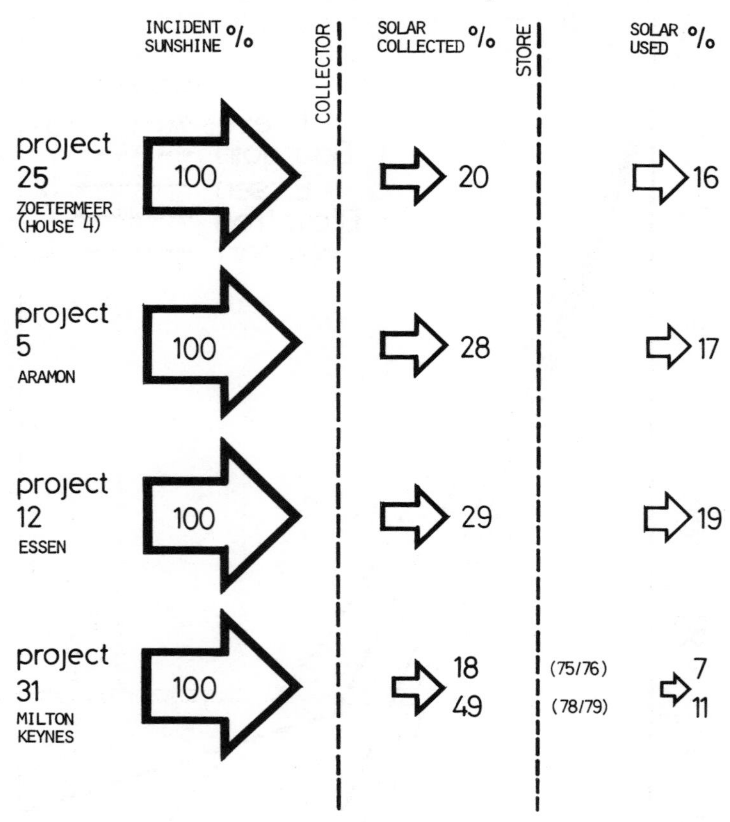


Fig. 28: annual energy flow: 4 projects

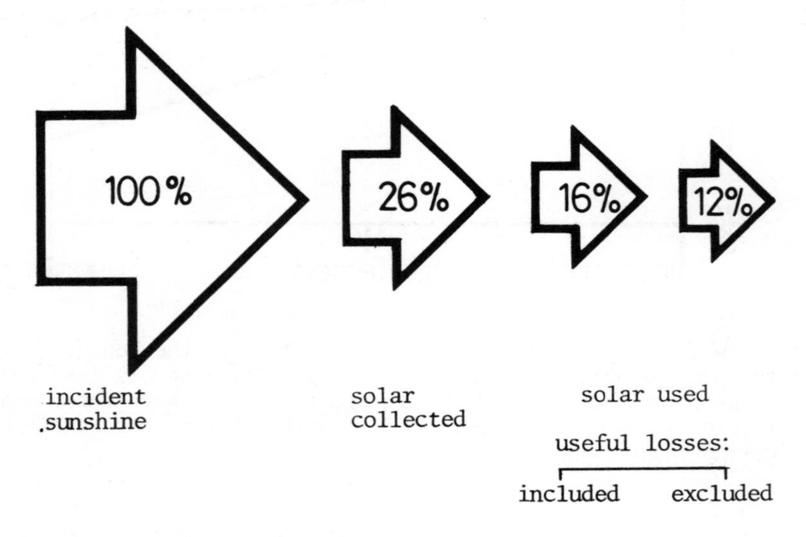


Fig. 29: annual energy flow: group I

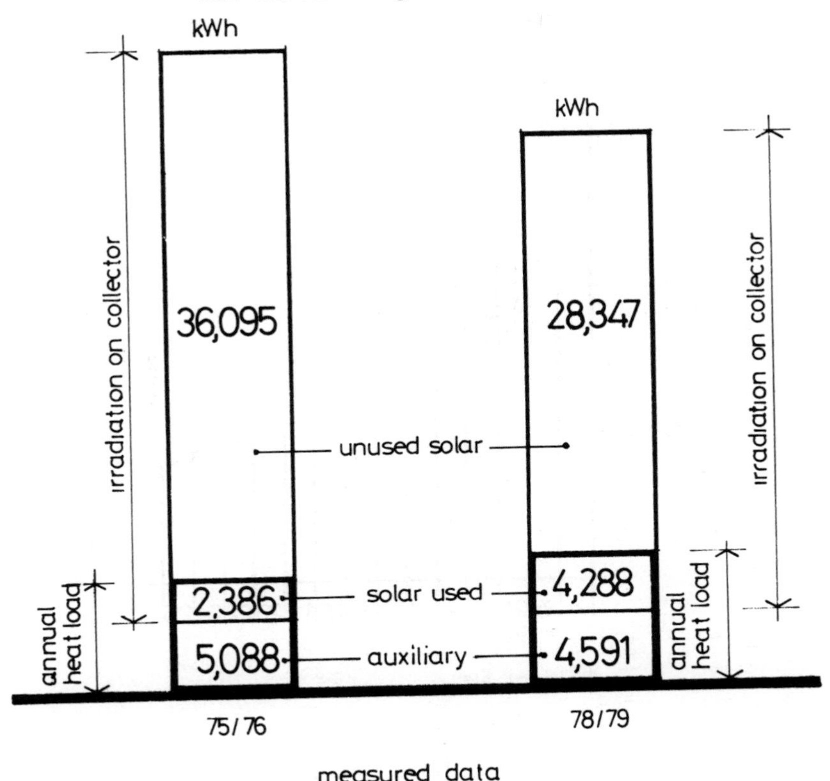


Fig. 30: annual performance: Milton Keynes (31)

The project at Milton Keynes (31) is particularly interesting, as it shows (Fig. 30) a significant improvement in performance over a three year period of monitoring - itself a striking vindication of the value of efficient monitoring. The improvements were a consequence of successive technical modifications, namely:

- spraying of anodised collectors with matt black paint for improved absorptivity (0.82 to 0.98).

- change in control system to allow separate operation of solar and auxiliary. The system operates now on different modes according to storage temperature. For storage temperatures above or below the set temperatures, either only solar or auxiliary is used. Between these two temperatures, auxiliary and solar are alternated, solar being used merely to reduce the cooling rate of the building. The system permits full solar heating at storage temperatures well below 40°C in mild weather and storage temperatures generally as low as 24°C.

- reduction in DHW set temperature from 65°C to 50°C.

The relative importance of these modifications can be fully evaluated only with a computer, but their combined effect has shown a marked increase of system output. It should be added that in the heating season following the modifications, global irradiation was lower. Although there were changes in occupancy and lower ambient temperatures, the proportion of collected energy used rose by 67%.

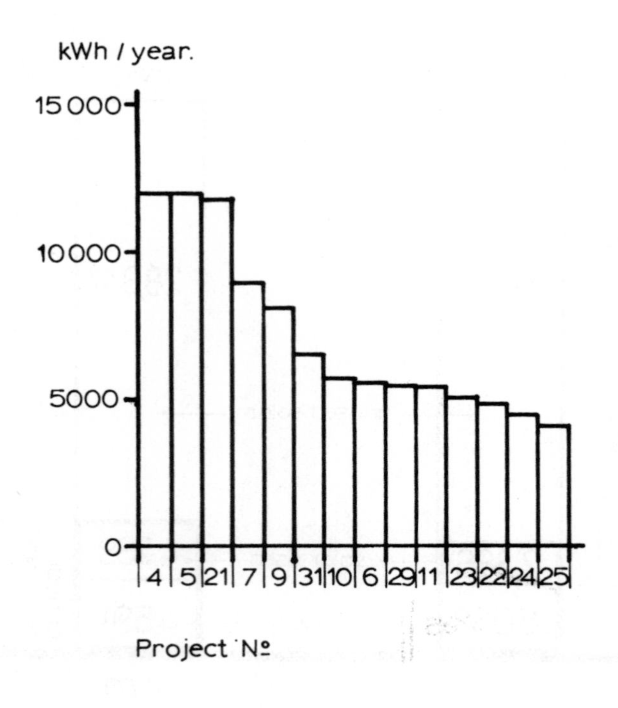


Fig. 31: free heat gains

Figure 31 shows the quantity of annual free heat gains for fourteen projects with data available. The figures represent the energy contribution to the annual energy balance of the building from people (body heat), appliances and direct solar gain through windows. The proportion of free heat gains contributing to space heating is difficult to establish; part of the internal gains are exhausted immediately through increased air ventilation rate, waste water etc. and part of the direct solar gain occurs during the summer, and may result in overheating.

The various estimates for useful free heat gains range from 4,000 kWh/year to 9,000 kWh/year, with a mean and standard deviation of 5,900 ± 1,400 kWh/year.

Figure 32 summarises the annual space heating load for four typical projects, and shows the relative proportions of solar energy (active system contribution), auxiliary energy, and free heat gains. Useful heat losses from storage and pipework are also shown. It can be seen that the solar contribution is typically 10 - 20%, free heat gains 25 - 50% and auxiliary energy 20 - 60%.

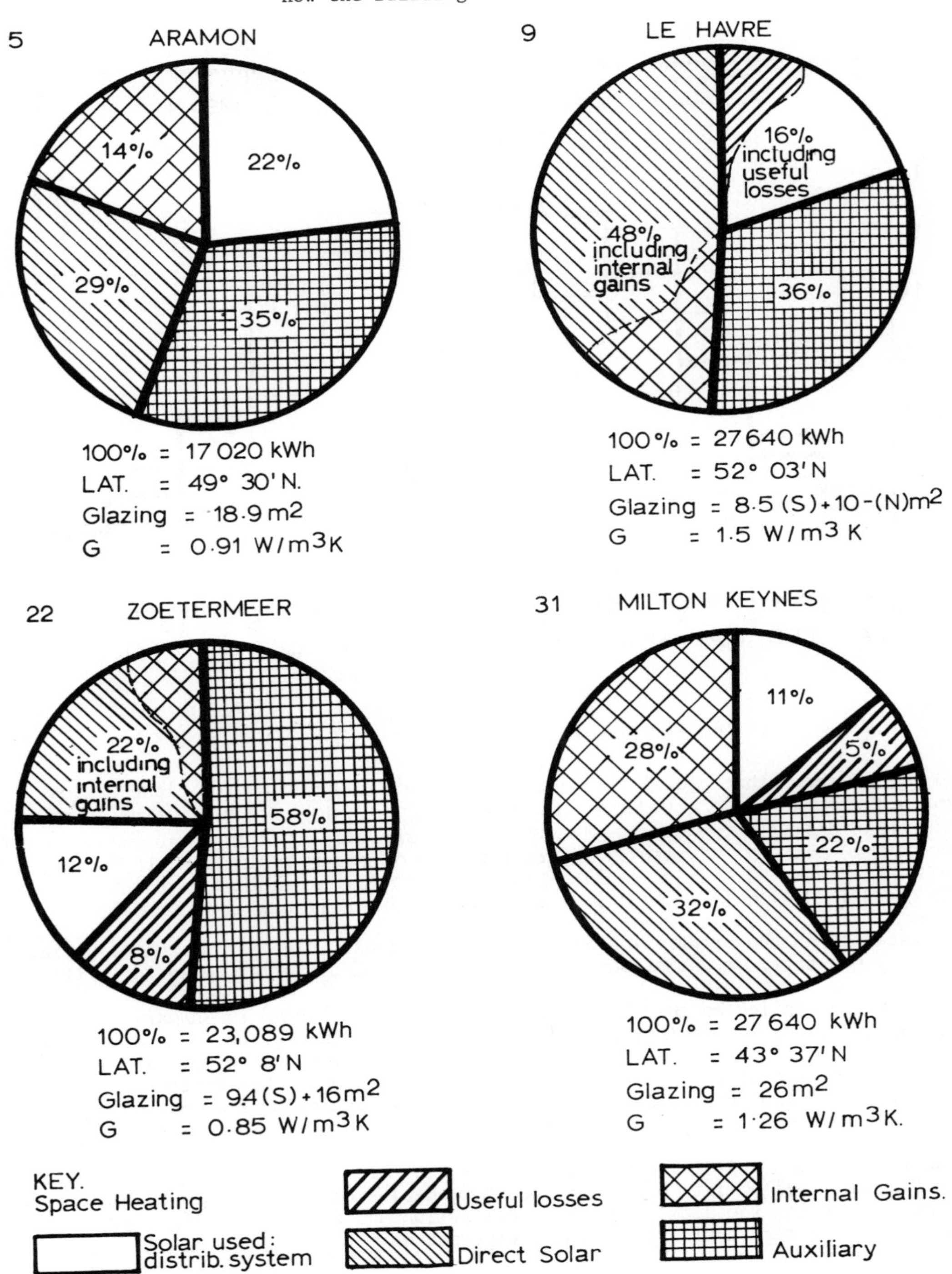


Fig. 32: annual space heating load for 4 projects

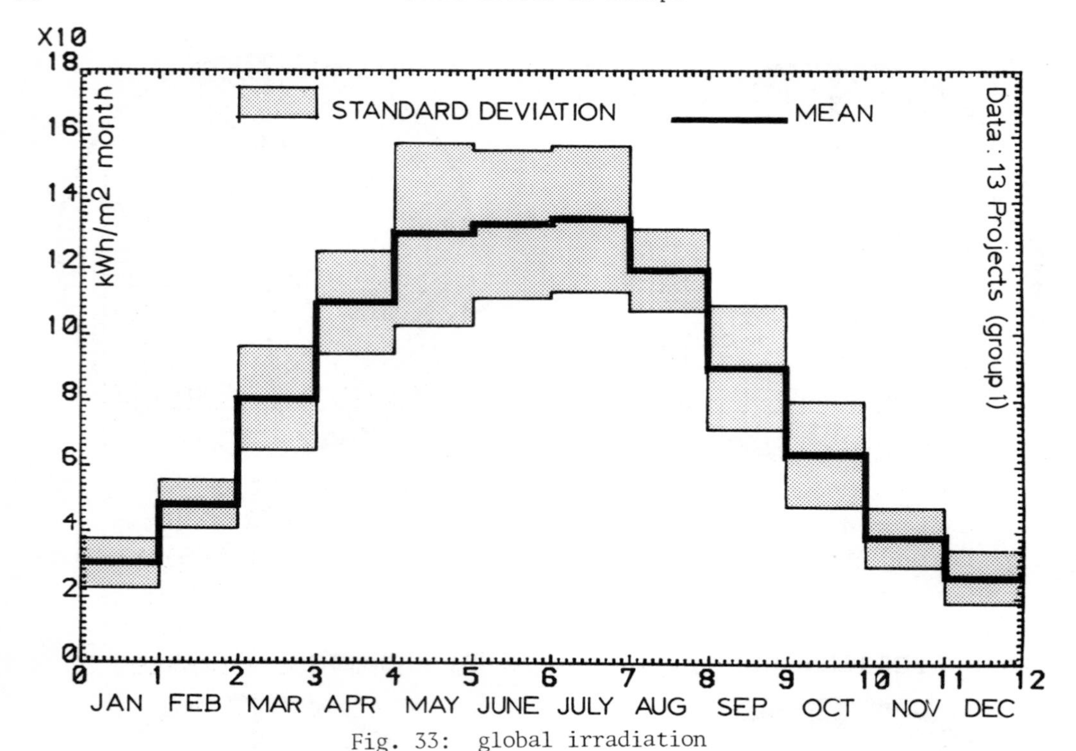


Fig. 33: global irradiation

MONTHLY PERFORMANCE

Figure 33 shows that global irradiation measured on the collector plane varies between 30 kWh/m$^2$. month in winter, increasing to 130 kWh/m$^2$. month in summer (Group I projects). The values are similar for Group II projects, but the scatter of the data (typically 20% either side of the mean for Group I) increases to $\pm$ 30% for Group II. This results, from the inclusion in Group II, of projects such as Odeillo (10) or Lyngby (4) with respectively high and low irradiation.

The scatter of the irradiation data is only marginally higher than annual variations for any one site, or variations due to the effect of different collector tilts.

Solar energy collected increases from a minimum of 10 kWh/m$^2$ in January to 35 kWh/m$^2$ at the end of the heating season, as shown in Fig. 34. There are large variations in these figures from project to project, the scatter of data being typically of the order of $\pm$ 40% either side of the mean. This scatter is particularly significant during the summer and at the beginning of the heating season, when geographical location and system design are relatively more significant.

The figure shows that collected energy peaks in May. This is due to a combination of factors: availability of energy, weather, depletion of the store and demand for heating. During the summer, hours of collection and collection efficiency decrease as a result of high storage temperatures.

Figure 35 shows solar energy used for space heating and domestic hot water preheating for Group I projects. Monthly figures vary between 8 kWh/m$^2$. month in winter, up to 28 kWh/m$^2$. month in mid-season, gradually declining during the

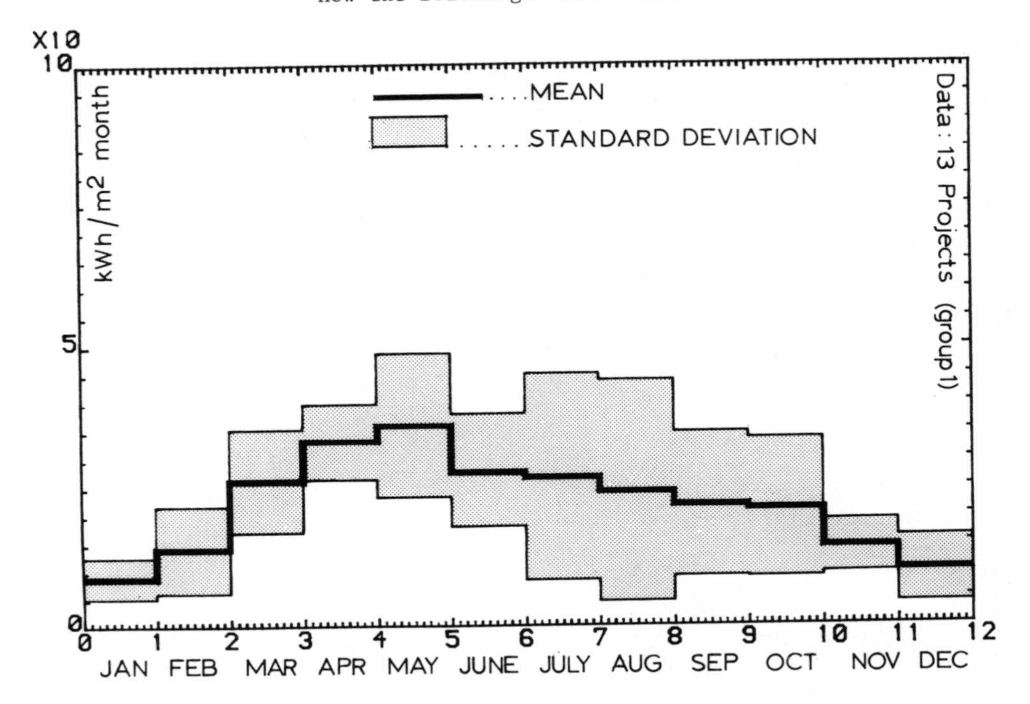


Fig. 34: solar energy collected

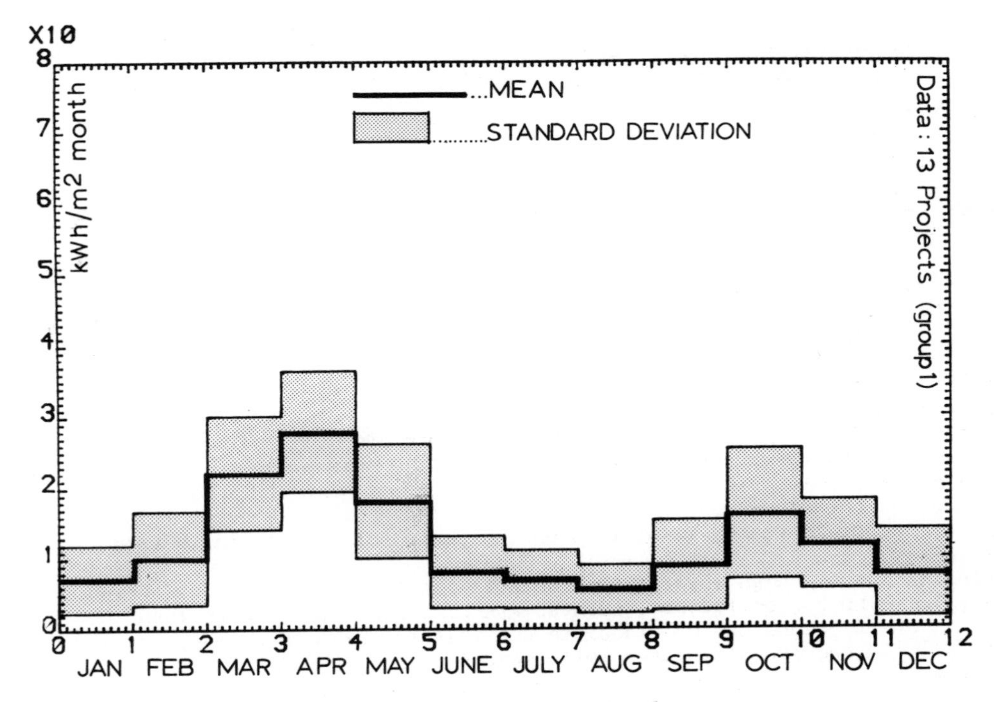


Fig. 35: solar energy used

summer to 5 kWh/$m^2$. month. Separate monthly figures (Fig. 36) present very large variations from project to project, with monthly figures varying typically between 30% and 60% either side of the mean. This variation is due to many factors, including geographical location, system design, load and occupancy. In the winter, system output is strongly dependent on system design and control. In the summer, it is dependent on the load and occupancy characteristics.

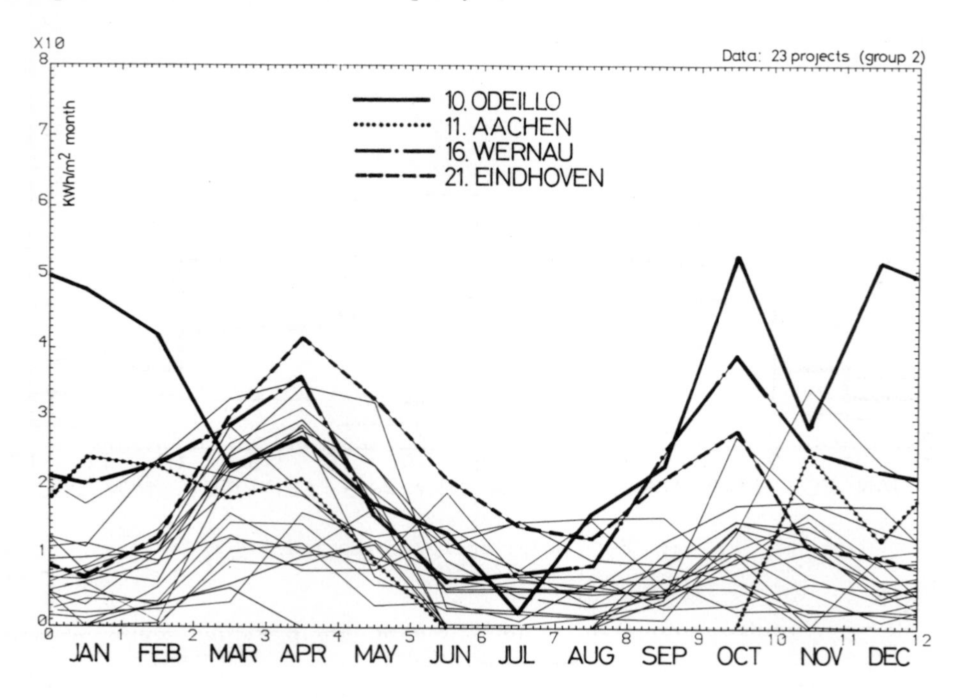


Fig. 36: solar energy used per unit area of collector

Figure 36 compares monthly performance data for all projects in Group II, with four projects selected to illustrate special features:

ODEILLO: this project shows high system output, with figures of 40 to 50 kWh/$m^2$. month between October and February. This is due mainly to exceptional global irradiation, since system efficiency during that period is not significantly higher than for other projects (33% compared to 25%: Fig. 21);

EINDHOVEN: shows a high system output from March to July. This is partly due to the large total load sustained during this period;

AACHEN: data shows no solar heating required until November. This is the effect of increased insulation on the length of the heating season;

WERNAU: shows high solar system outputs throughout the heating season. A large number of factors, such as use of a heat pump, large loads per collector area and use of elaborate controls to maximise output, combine to give good overall system performance.

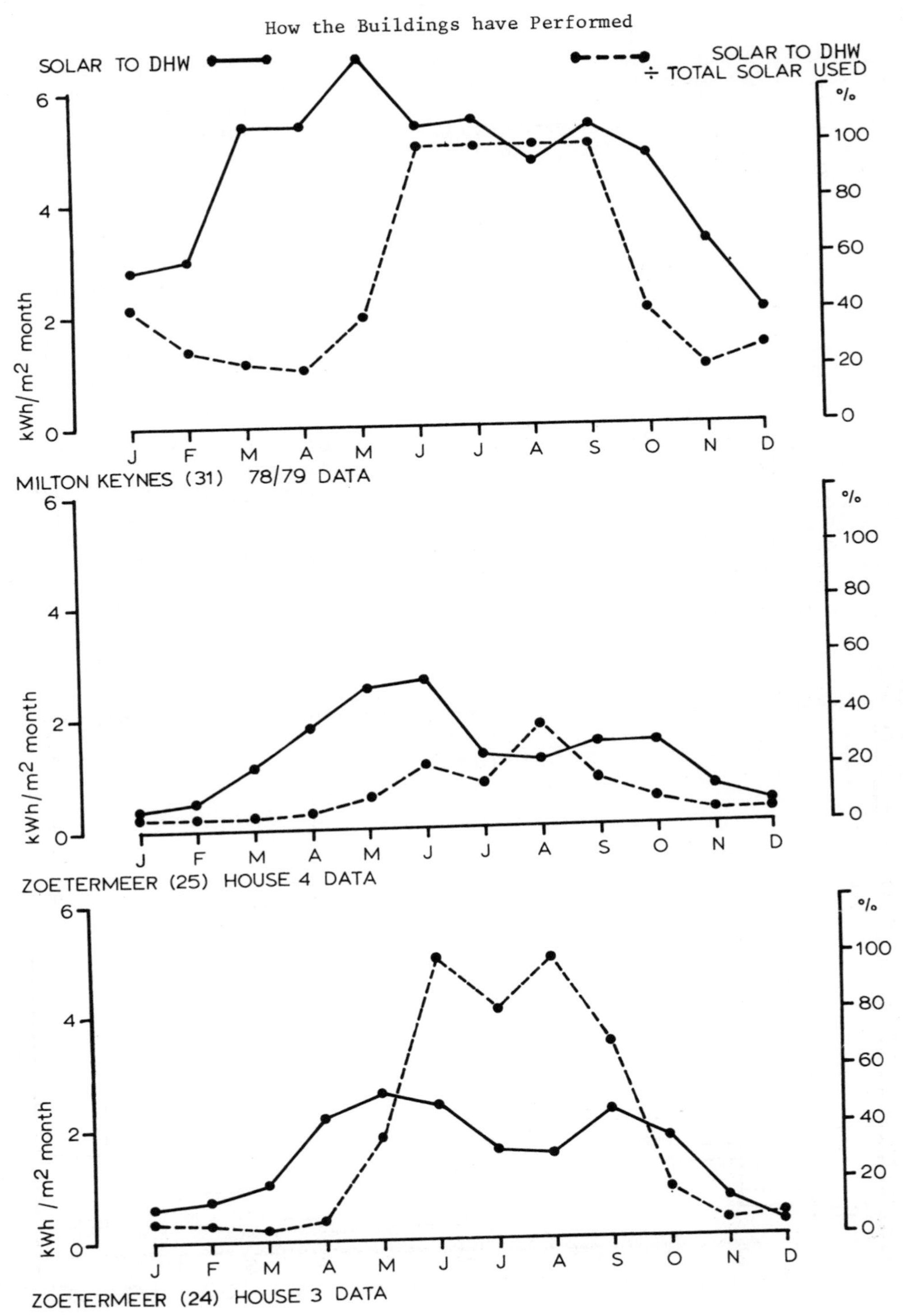


Fig. 37: solar contribution to domestic hot water DHW per unit area of collector

Analysis of the solar contribution to domestic hot water pre-heat shows a very wide range of performance (Fig. 37). Output is minimum during the winter (mean for group I: 1.4 $kWh/m^2$. month), and increases to a maximum of 5 $kWh/m^2$. month at the beginning of the summer, when tap water temperature is low and irradiation high. Scatter of data is large, typically ± 50% either side of the mean, but the general trend can be seen.

Milton Keynes (31) shows the additional feature, present in a number of other projects, of a small increase in the proportion of solar energy used which contributes to hot water during mid-heating season. This is due to low storage temperatures, incompatible with space heating, but acceptable with the lower temperatures required for hot water pre-heat.

Zoetermeer shows lower outputs due to the higher priority given to space heating, and very significant variations of solar contributions to domestic hot water among the four houses.

Figure 38 shows the relative importance of mean values of global irradiation, solar energy collected and solar energy used for 13 projects (group I). It can be seen from this that only a very small fraction of the solar energy available is actually collected: on average 26% in winter and 20% during the summer. The two projects with high efficiency collectors - Aachen (11) and Essen (12) - show improved figures of the order of 35% for winter and 26% for summer collection efficiency. The figure also shows that the three curves are out of phase by about a month. This is due to the natural phase lag between outside temperature and global irradiation, and to the additional effect introduced by tank storage capacity.

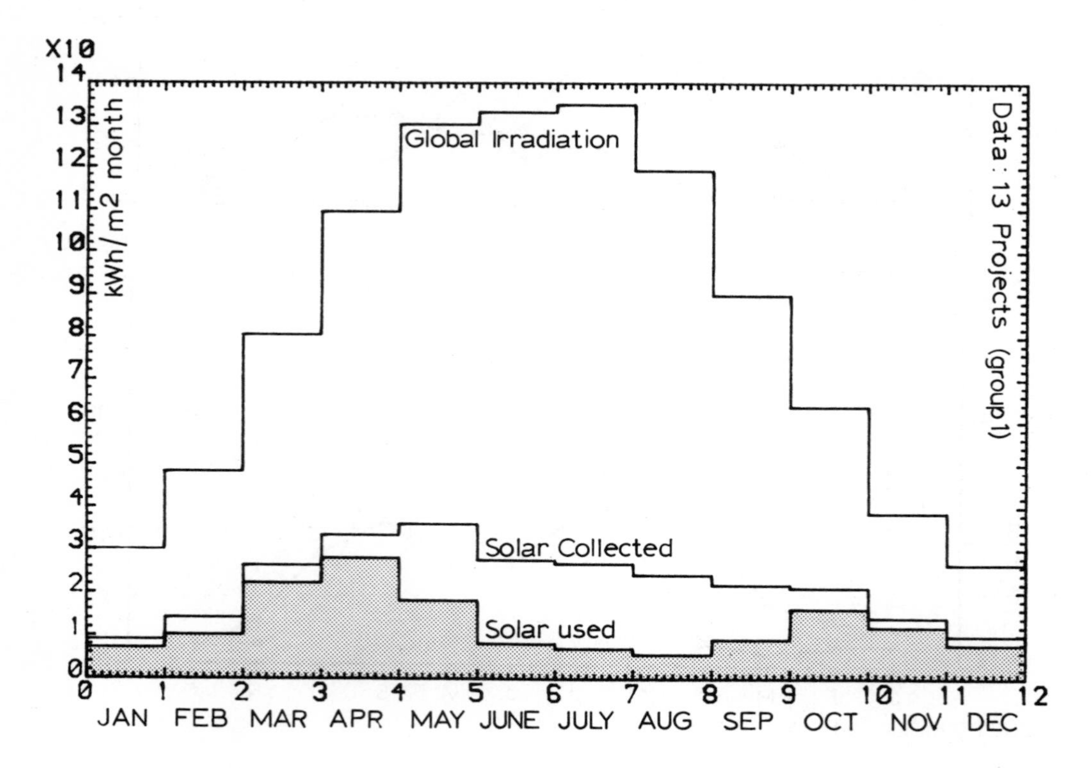


Fig. 38: mean monthly values

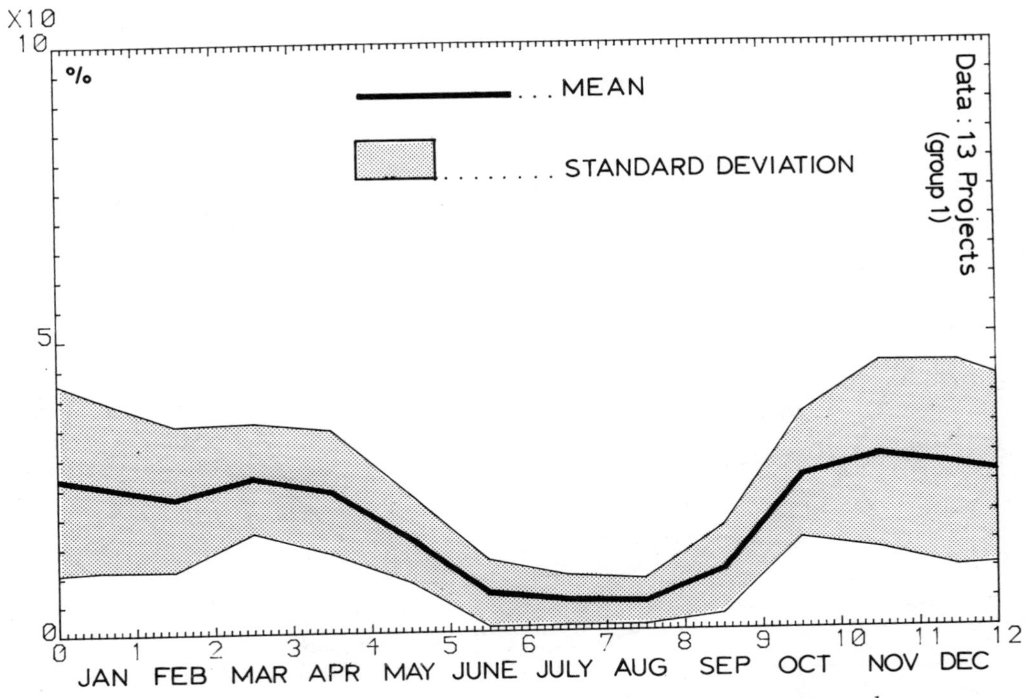


Fig. 39: mean proportions of incident solar energy used

Figure 39 shows monthly figures for system efficiency (the proportion of solar energy used to total solar energy incident on the collector), ranging from a typical maximum of 25% in winter, gradually decreasing to a minimum of 5% in summer. The variation of performance around the mean is significant. Many variables affect system performance, but in general those associated with system design are more important during the winter, while occupancy and climate become more important at the end of the heating season and during the summer, when system output is load dependent.

Figure 40 shows monthly system efficiency data for all projects, with five individual cases identified:

AACHEN (11) data show high system efficiency during the heating season. This is the effect of the seasonal storage contribution to total load. From June to October, free heat gains and waste heat recovery are sufficient to achieve internal comfort conditions;

WERNAU (16) AND BOURGOIN (7) are two projects with good overall monthly system efficiency;

EINDHOVEN (21) AND ODEILLO (10) show high system efficiency during the summer. During the winter both systems show fairly similar system efficiency, although climatic characteristics are totally different.

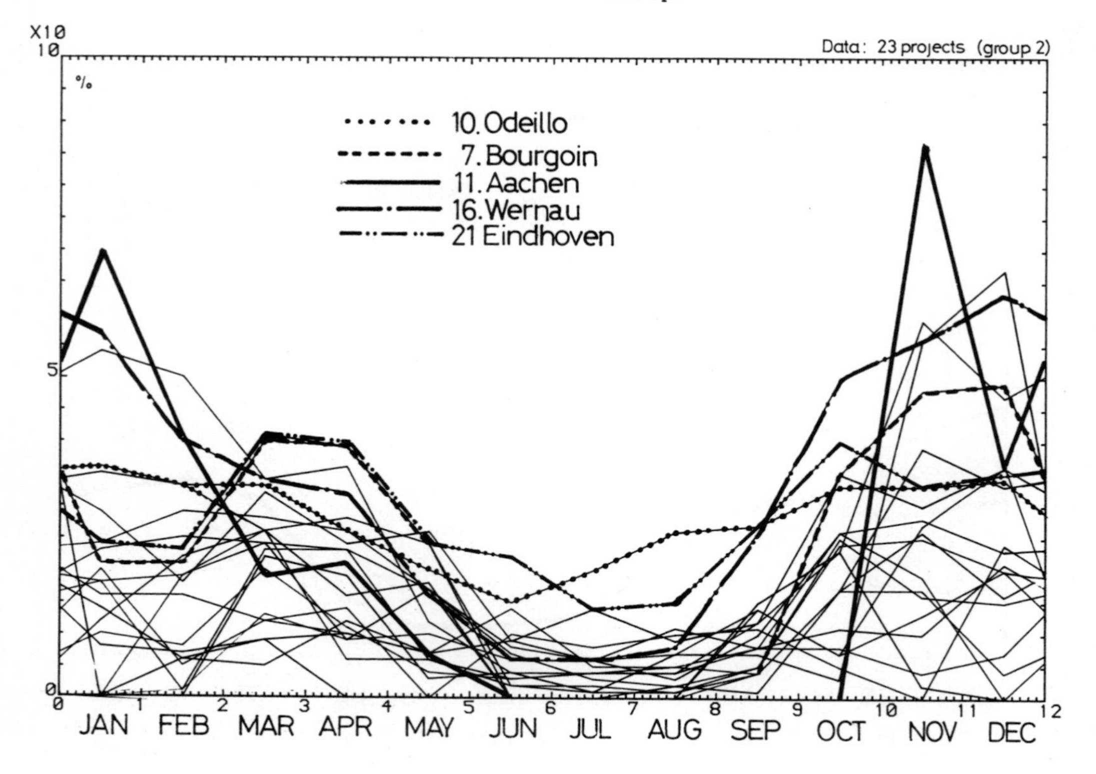


Fig. 40: proportions of incident solar energy used

The main conclusions to be drawn from an analysis of monthly performance data, are that systems generally perform best at either end of the heating season, and that system efficiency is very much higher in winter than in summer. System output typically peaks in April/May and October/November. Over 50% of the total annual system output is used in these months. Collected energy reaches a peak in May, after which it declines consistently through the summer due to high storage temperatures, and only begins to increase again in February. The length of the heating season therefore has important implications for the design of the system.

System efficiency reaches a maximum of 25% in winter and drops sharply to 5% in summer when there is only a hot water load to satisfy.

From the point of view of monthly performance, the following projects are expecially interesting:

ODEILLO: This project, above all the others, shows high system output, with figures of 40 - 50 kWh/m2. month between October and February. This is due mainly to exceptional global irradiation since system efficiency during that period is not significantly higher than for other projects (33% compared to 25%).

EINDHOVEN: High performance from March until January is partly due to the large total load sustained during this period. These high mid-season and summer loads have contributed to high annual figures.

WERNAU: Consistently high system output throughout the heating season. A number of factors, such as the use of a heat pump, high loads per collector area, and elaborate controls combine to give good overall performance.

AACHEN: Shows the effect of good thermal insulation. The heating season has been reduced to 5½ months. From June to October, free heat gains and waste heat recovery are sufficient to achieve internal comfort conditions.

ZOETERMEER: Comparing monthly data for the four houses, there is a fairly consistent pattern. Allowing for a technical problem in February (House 2), system output is similar for all houses during the heating season, with perhaps a slightly higher output from the end terrace houses. During the summer there is a wider scatter: solar energy, used for water heating only, varies by a factor of 4. There is also very little correlation with the number of occupants. House 4 with one to two people has twice the load of House 1 with two people.

CARDIFF: The energy balance for this water heating system was examined over a period of six days. It is interesting to see that 38% of the incident energy corresponds to heat losses during collection, and another 38% could not be collected because the temperature from the collectors was less than the temperature in the storage. System efficiency for the same period was 20%, compared to a long-term figure of 17%.

In those systems which combine pre-heating of DHW with space-heating, there is a very wide range of performance, but maximum monthly output is at the beginning of the summer when cold water supply temperature is low and irradiation high. Some projects, (such as Milton Keynes) show a small increase in the proportion of used solar energy which contributes to hot water pre-heat in winter. This is due to lower storage temperatures, insufficient for space-heating but acceptable for water-heating.

## DAILY PERFORMANCE

There are insufficient data for a detailed analysis of daily performance. Figure 41 shows hourly values of total energy incident on collectors, collected energy, and indoor, outdoor and storage temperatures on two consecutive days in August for one of the houses in Zoetermeer. There was no demand at this time.

Comparison between incident and collected energy shows the importance of collector losses. In the evening, and (especially) the morning, losses were greater than incident energy, and no collection took place. The high collector losses are due to high storage temperatures, while the apparent asymmetry between morning and evening losses is partly due to greater differences in the morning between storage and outdoor temperatures.

Storage temperatures reach a maximum at the end of the collection period, and then gradually diminish until next day. Drop in temperature is due to losses from the storage tank, and possibly also to some overnight convection within the initially stratified storage tank.

The figure also indicates a lag of a few hours between fluctuations in indoor and outdoor temperatures.

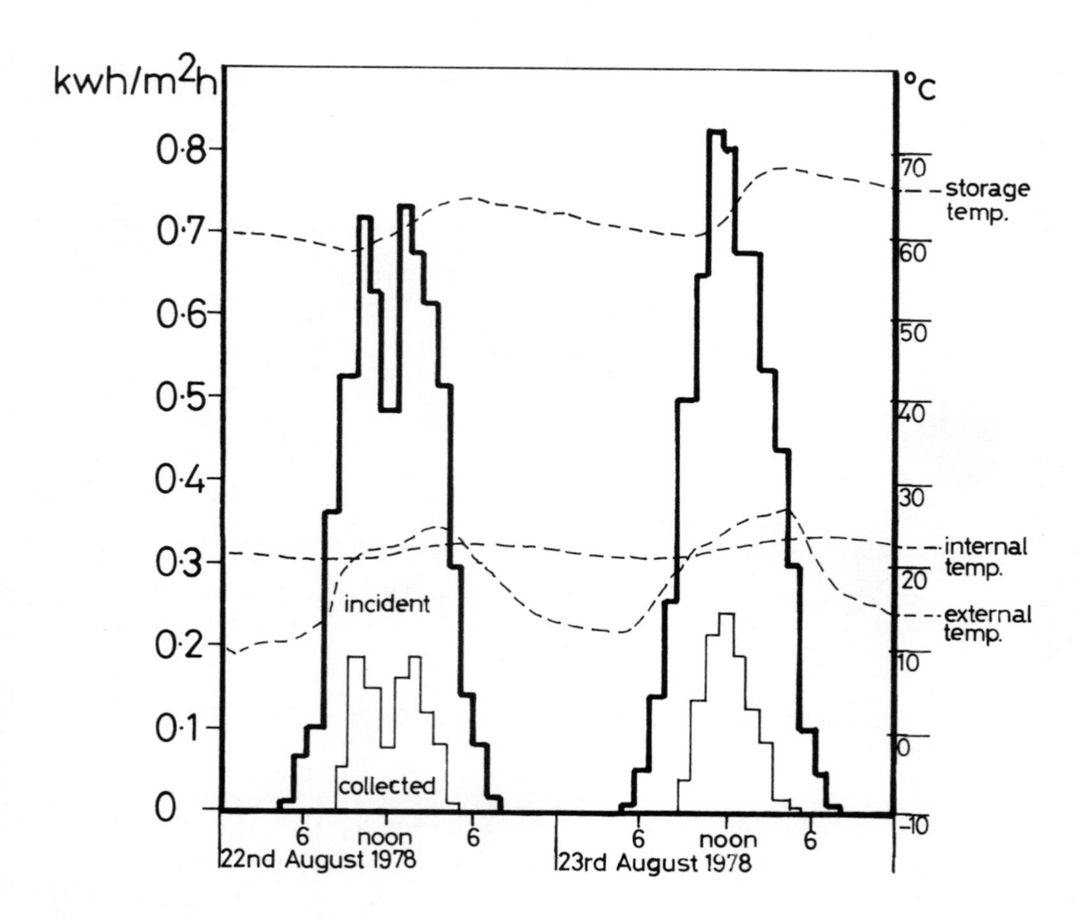


Fig. 41: daily performance, Zoetermeer 1 (22)

# CHAPTER 3

# LESSONS FROM THE SURVEY

In Northern Europe, the main purpose of installing a heating system in any building is to maintain internal temperatures above outside air temperatures for at least six months of the year. Solar energy can be used to assist this process when it is available, and when this availability coincides with a demand for heating. As we have seen, it is possible to store heat for a period of a few days, but since the store becomes depleted and solar radiation is sometimes low, it is necessary to have an auxiliary supply of energy which is adequate to maintain comfort conditions under the worst winter conditions. Consequently, the efficiency of the solar heating system is strongly dependent on the relationship between the supply of energy (solar radiation) and the demand for heating (the house load). System performance is therefore time-dependent, and highly sensitive to a whole range of factors: geographical location, system design, component efficiencies, controls, size of heating load, and occupancy.

Figure 42 illustrates the effect of improving the level of thermal insulation in a house. Heat losses through the fabric are significantly reduced, and the space heating load is therefore considerably less. At the same time, the length of the heating season is reduced by several weeks because incidental gains become sufficient to maintain comfort conditions for a longer period. The relatively larger amounts of solar energy that are available at the beginning and end of the heating season, cannot therefore be used. It is interesting to note that in well-insulated houses, the space heating load can be similar to the domestic hot water load. Ventilation heat losses also become relatively more important, and account for 40 - 45% of the space heating load.

However system performance is assessed, the short-term balance between solar energy supply and demand is the major factor. Other factors will tend to influence system performance, but only indirectly, and for the most part only as they affect this balance.

## CLIMATE

It is, however, with these other factors that design is mainly concerned, and their influence may be summarised as follows.

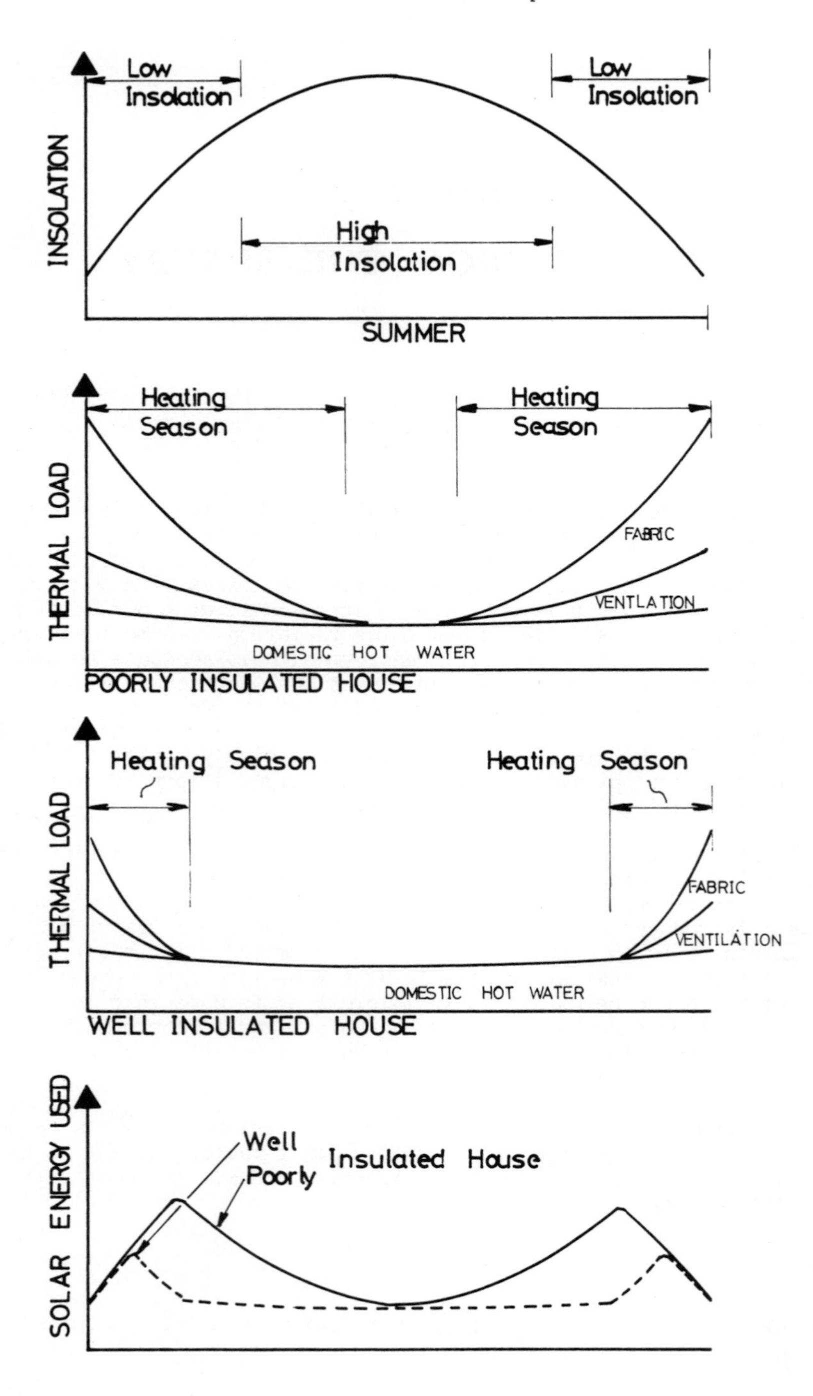


Fig. 42: effect of thermal insulation

Incident energy (solar radiation) determines the pattern of supply of energy and outside air temperature determines the pattern of demand for heating. Wind speed and direction affect collector efficiency, whilst precipitation, wind speed and relative humidity can affect the thermal properties of the building, and hence, the demand.

Climatic conditions will also tend to influence the behaviour of the occupants, affecting their social habits, clothing etc., thus affecting the heat demand and the amount of free heat gains.

Ideal climatic conditions for high solar contribution are *sunny and cold*. Solar heating systems perform best when solar energy incident on the collectors is available at times when heating is required; this is so because storage is not very practicable beyond three to seven days capacity. Because sunshine and cold days do not usually coincide,

1. during the summer, when there is more sun than can be used, any increase in load results in increased solar output;

2. in winter, when thermal load is much greater than solar energy incident on the collectors, any increase in load results in increased auxiliary consumption;

3. use of solar energy 'peaks' at the end of the heating season.

## BUILDING

Building thermal load directly affects energy demand. The space heating load is conditioned by the size and contruction of the building, levels of thermal insulation, areas of glazing and natural ventilation, as follows:

direct solar gains through south facing windows, energy consumption by appliances (refrigerators, lamps, etc) and body heat provide much energy which can contribute to internal comfort conditions;

direct solar gains are susceptible to control through adequate design of glazing areas and the south-face of the building;

the size of the building and its degree of exposure affects the load;

the thermal mass of the building moderates internal temperatures and affects warming up time;

unintentional ventilation through cracks, etc. is usually much larger than predicted;

thermal insulation can reduce heating load very significantly and is therefore most desirable.

## OCCUPANTS

The occupants probably have the largest single influence on the performance of the heating system. Comfort conditions are highly subjective and average house internal temperatures can range from 16°C - 22°C. The rate of natural ventilation is also strongly dependent on the way people behave. For example, opening and closing of windows and doors, even internal ones, during the heating season will significantly alter the load.

In addition, all activities of the occupants in the building generate some heat - estimated between 2,000 kWh/year to 8,000 kWh/year for single family houses - some of which can contribute to the maintenance of internal comfort conditions. Finally, occupants operate (or don't!) curtains or insulated shutters and may interfere with the controls. As a result records of household energy consumption show that for identical buildings loads can vary by a factor of over 4:1.

The efficiency of a solar heating system is system dependent in winter and occupancy dependent in summer.

'Free' heat gains can make a contribution comparable to the amount gained from an active solar system.

## CONTROLS

The control system plays an important role in determining the amount of collected energy that can be used, and hence affects the overall system efficiency. It must be designed to ensure that solar energy, when available, has priority for satisfying space-heating demand, either directly from the collector or indirectly via the heat store.

Experience has shown that careful 'tuning' of controls is necessary to achieve optimum working conditions. Adequate controls are essential to ensure that:

1. collection takes place whenever the amount of collected energy is greater than the energy consumption of the pumps;

2. solar energy is given first priority whenever possible;

3. the auxiliary does not prevent use of warm solar heated water, and does not transfer energy into the storage;

Inlet temperatures to heat emitters (radiators, warm air unit etc,) and domestic hot water set temperature should be kept as low as possible. Solar heated water should not be allowed to mix with cooler water, and should be used down to as low a temperature as possible. This will ensure that the collectors also operate at as low a temperature as possible and therefore at maximum efficiency, and that high grade heat is conserved. It may be achieved by incorporating a heat pump, providing large heat exchangers and heat emitters, and good control to separate hot and cold water.

## COMPONENT AND SYSTEM DESIGN

Designs vary considerably and very few have yet been technically optimised. Nevertheless, performance monitoring has provided some useful information on the behaviour of individual components and their influence on overall system performance. The size and type of collector, its surface treatment, number of glazing sheets and so on, affect the amount of energy collected. Similarly the configuration and location of the heat store, and the type of heating system used, affects the amount of heat that can be extracted from the system.

COLLECTORS

Under optimum working conditions, solar collectors can convert over 70% of incident energy into low grade heat. The efficiency of collection reduces however, as the temperature difference between inlet and outlet increases; as the angle of incidence increases; or as (in winter, for example) irradiation levels decrease. Annualy, collection efficiencies for the arrays of collectors and interconnecting pipes are typically in the order of 20 - 40%.

The thermal inertia of the collector is an important factor, especially during periods of intermittent irradiation, when warming-up time may be too long for effective collection. The tilt of the collector should have regard to the sun's position when solar energy is to be collected. Collection efficiency of the collector array and interconnecting pipework is lower than laboratory measurements, but good design of collectors and shorter runs of well-insulated pipes improve efficiency. The predicted demand, storage capacity and collector area need to be well matched if the system is to run efficiently.

HEAT STORAGE

Heat losses from the storage tank can be very significant, even when well-insulated and protected from damp. Under optimum conditions, up to 60 - 70% of the annual energy stored can be recovered, but this figure may drop to as little as 20 - 30% when the thermal insulation is imperfectly positioned or if it is crossed by cold bridges. A proportion of these heat losses can be considered as a useful contribution to space heating, if the store is located within the house. The size of the heat store affects the rate of heat loss, not only because of surface area, but also because its size relative to the collector area and heating load will affect temperatures within the store.

High storage temperatures, likely during the summer in installations with a small store, result in poor system efficiency.

Low storage temperatures, likely during the winter in systems with a large heat store, may hinder the use of solar energy for space heating.

Good temperature stratification in the store will promote good year-round performance.

Uncontrolled heat losses from the store, primary circuit and domestic hot water tank are large and, to a certain extent, unavoidable. Good design should, therefore, use as much as possible of these losses for space heating, giving due consideration to the possibility of over-heating during the summer.

# CHAPTER 4

# RECOMMENDATIONS FOR DESIGN

The results show that measured system performance tends to fall short of design predictions. There are many reasons for this. Design is necessarily based on many assumptions regarding occupancy, ventilation rates, heat losses, and so on. The weather also varies from year to year. A 20% variation in solar radiation between one year and another is not uncommon. Another reason may be that collector efficiencies are lower than the figures obtained under laboratory conditions, because of condensation, dirt and heat losses from connections. Storage efficiencies have also been found to be lower than expected, typical efficiency being no more than 40%. Wrong assumptions are often made about the final configuration and insulation of pipework, and the heat losses in practice may well be higher than assumed in design.

The following can significantly affect performance, and every assumption about them should be checked as rigorously as possible:

- solar radiation
- outside air temperature
- collector efficiency
- internal temperature
- DHW consumption
- DHW set temperature
- pipework losses
- storage losses
- ventilation losses
- fabric losses
- free heat gains
- control settings
- occupancy

Any assessment of the annual contribution from the solar heating system should include:

- total amount of solar energy used for both domestic hot water preheat and for space heating;

- the contribution to space heating from uncontrolled heat losses from the storage tank and the primary circuit;

- hidden savings in auxiliary energy resulting from disuse of the boiler at times when its operation at low loads would have resulted in poor efficiency, and unnecessary standing losses from the boiler and/or hot water tank;

- the additional bonus of having more abundant warm water (especially in summer).

Future designs should integrate the design of the house and the heating system -

including thermal insulation, use of free heat gains, and heat recovery with the solar heating system -and should be optimised in terms of long-term performance.

Good system performance depends on keeping design temperatures as low as comfort conditions will allow. This implies using solar heated water (or air) down to the lowest possible temperature, and preventing hot fluids from mixing with cold ones. This can be achieved by setting a low temperature for domestic hot water; by stratified storage with correctly sized heat-exchangers; and by using larger quantities of cooler air or water for space heating. It was found that reducing the set temperature to 50°C gave improved performance without causing any inconvenience to the occupants. For a given capital investment, it is not yet possible to say whether large radiators, underfloor heating or warm air systems are best.

Good performance for a solar heating system depends on continuous, reliable operation. In the projects studied, this has rarely been achieved. In practice, component failure or malfunction has resulted in stoppage of the system for days, (sometimes weeks) with a resulting loss in performance.

## COLLECTORS

Breakage of the inner pane of glass in double-glazed collectors is a common problem, generally caused by high temperatures. It is important that the detailing of the glazing - especially the lower pane -allows some thermal movement. When prefabricated 'box' collectors are used, it has been found that the weatherproofing around each unit is complicated and installation time is increased. Incorporation of the collector in the roof construction, will improve both weathering and appearance. In most cases, glazing is in ordinary 6 mm float glass, sometimes with low iron content to increase transmission. Planning authorities sometimes have required the use of non-reflective glass to prevent glare. Condensation on the inner surface of the glazing is a common problem. This tends to clear itself by mid-morning, but can become persistent, reducing transmission and affecting collector efficiency. The solution is a small hole in the bottom of the collector to create some movement of air. Insulation behind the collector should be adequate to prevent losses, be able to withstand temperatures of 200°C and be well sealed.

Dirt is another problem affecting collector efficiency. On the outer surface of the glass, this is usually cleared by rainfall, but it can also occur on the inside in the form of deposits from the sealing or insulation materials. In the case of the air collector at Zoetermeer (24), a white deposit was formed (caused by condension on the inner glass surface) reducing the transmission coefficient from 0.84 to 0.76. Another problem with air collectors, reported from Dourdan (8) is an unpleasant smell, caused by the action of hot air on surrounding timber. Many people have experienced leakage from collectors caused either by poor plumbing, or (more usually) jubilee clips on flexible piping loosened by heat. The latter is such a commonly reported problem that the use of rubber connectors is not recommended. Absorber plates are generally either steel or aluminium. In the case of aluminium there is considerable concern about durability, as degradation, in the form of pin-holes, has been noticed. Finally, it has been found that with time, insulating material can move away from the back of the collector, thus increasing convection losses. It is recommended that measures are taken to ensure that good, permanent contact is established.

## PRIMARY CIRCUIT

For maximum efficiency, it is best to minimise heat losses by reducing the length of flow and return pipework to the collectors. Here, one of the major problems has been depletion of the heat store by reverse circulation at night - something which has affected at least ten of the projects. One project also experienced reverse circulation through a mixing valve in the domestic hot water circuit. A non-return valve is essential, but the type of valve is important as it must be resistant to the water-glycol solution. The 'rubber bellows' type appears to fail in this respect, and the 'nylon-chamber' type is preferred. It is essential that all pipework be properly insulated before the final adjustment of sensors and valves, to avoid subsequent disturbance. The primary pump is a key component. Standard central heating pumps are generally used, although they are sometimes down-rated. The main problems reported were noise, and insufficient flow rate. The malfunction of 3-way valves is also a commonly reported problem: mainly slow response and leakage to the by-pass. As a final point, it is important to ensure that safety valve and automatic air vent (in sealed systems) are incorporated, and that the drain-down facility is easily accessible.

### • HEAT STORAGE

As heat losses tend to be high, it is recommended that the store should be within the house, so that losses during the heating season will contribute to space heating. In practice, this will place a limitation on the size of the store, which may conflict with design optimisation of the whole system. In new houses, it may imply designing the house around the store, and structural loading may require special attention. In existing houses, it is generally only possible to incorporate a relatively small store within the house unless (as was the case in Gentofte (2)), there is basement space available and the store can be made up on site. Larger stores tend to be positioned outside the house and below ground. However, considerable problems have been experienced with this approach (tracing and remedying leaks for instance) and it is not recommended. Insulation of the store is obviously of prime importance. However, in practice, actual heat losses can be higher than design values due to the effects of cold bridges either at the bottom or at the sides of the store, and these must be eliminated.

Stratification within the heat store is always beneficial, and it may be achieved by:

- using several tank in series
- using tall slim storage tanks
- pre-heating domestic hot water by circulating upwards through the full height of the store
- using low rates of flow and/or internal baffles to reduce turbulence inside the storage
- returning water from the collectors near to the top of the storage, and flow to the collectors near the base; in indirect systems, the heat exchanger in the primary circuit should be vertical
- positioning cold returns from the distribution circuit also near the base.

### SYSTEM CONTROLS

Control is the key feature of the whole system. In practice, it has been found that considerable care must be taken in the adjustment of all system controls to achieve proper operation. To achieve optimum performance, the system has to be 'tuned' over the first few weeks of operation and this involves re-calibration of sensors on site, checks on wiring and re-adjustment of thermostats. Operating temperature differences are very small, so the accuracy of sensors is an important consideration. One of the main problems is the primary pump switching on and off in rapid succession - the 'hunting effect' - and the control system must be designed to overcome this. In some cases, temperature sensors have become loose, and good thermal contact between sensors and pipe work must be ensured, with a firm permanent fixing.

# CHAPTER 5

# STATISTICAL SUMMARIES

(Typical figures excluding exceptional extremes)

| | |
|---|---|
| Solar irradiation on horizontal surface: | |
| Northern Europe | 800 - 1000 kWh/annum |
| Central Europe | 1000 - 1200 kWh/annum |
| Southern Europe | 1200 - 1600 kWh/annum |
| Degree days: | 2000 - 4000 $^{o}$C.days |
| Solar energy used/collector area | 70 - 245 kwh/annum.m$^2$ |
| Solar energy collected/irradiation on collector | 20 - 30 % |
| Solar energy used/collected | 45 - 70 % |
| Solar energy used/total thermal load | 16 - 72 % |
| Total solar energy used per project | 4000 - 9000 kWh |
| Total auxiliary heat used per project | 6000 - 18000 kWh |
| Total 'free heat' gains per project | 5000 - 10000 kWh |
| Volumetric heat loss | 0.03 - 1.82 W/m$^3$.K |
| | mean 1.06 W/m$^3$.K |
| Design loads | 4600 - 37900 kWh/year |
| | mean 20307 kWh/year |
| Design load/unit floor area | 38 - 261 kWh/m$^2$ |
| | mean 134 kWh/m$^2$ |
| Collector area | 20.3 - 80.0 m$^2$ |
| | mean 44.7 m$^2$ |
| Collector area/unit floor area | 17 - 55 % |
| | mean 32 % |
| Design load/unit area of collector | 109 - 1031 kWh/m$^2$ |
| | mean 500 kWh/m$^2$ |
| Storage volume/unit area of collector | 30 - 145 litres/m$^2$ |
| | mean 95.5 litres/m$^2$ |
| Annual load | 8000 - 30000 kWh/year |

The tabulations which follow have been abstracted from the reports on the 31 projects. Some figures are based on more recent information which has become available since the reports were originally submitted.

TABLE 1 Climate

| | Annual Global Irradiation on Horizontal Plane | Annual Sunshine Hours | Degree Days | Base Temperature (°C) |
|---|---|---|---|---|
| 1 Blovstrod | 1024 | 1580 | 2829 | 17 |
| 2 Gentofte | 1024 | 1580 | 2829 | 17 |
| 3 Greve | 1024 | 1580 | 2829 | 17 |
| 4 Lyngby | 1024 | 1580 | 2829 | 17 |
| 5 Aramon | 1525 | 2700 | 1788 | 18 |
| 6 Blagnac | 1285 | 2050 | 2070 | 18 |
| 7 Bourgoin | 1402 | 2050 | 2780 | 18 |
| 8 Dourdan | 1132 | 1707 | 2917 | 18 |
| 9 Le Havre | 1012 | 1603 | 2332 | 18 |
| 10 Odeillo | 1605 | 2460 | 3942 | 18 |
| 11 Aachen | 1109 | 1510 | 3231 | 15 |
| 12 Essen | 932 | 1257 | 3881 | 15 |
| 13 Freiburg | 1180 | | 3147 | 12 |
| 14 Walldorf | 952 | 1540 | 3050 | 12 |
| 15 Otterfing | 1093 | 1567 | 4045 | 20 |
| 16 Wernau | 962 | 1763 | 3100 | 12 |
| 17 Fiume Veneto | 1159 | 2012 | 2386 | 19 |
| 18 Firenze | 1539 | 2213 | 2144 | 19 |
| 19 Rossano Calabro | 1382 | 2200 | 1297 | 19 |
| 20 Sequals | 1159 | 2007 | 2386 | 19 |
| 21 Eindhoven | 1284 | 1400 | 2785 | 18 |
| 22 Zoetermeer Ho 1 | 976 | 1400 | 3644 | 18 |
| 23 Ho 2 | 976 | 1400 | 3644 | 18 |
| 24 Ho 3 | 976 | 1400 | 3644 | 18 |
| 25 Ho 4 | 976 | 1400 | 3644 | 18 |
| 26 Bebbington | 835 | 1451 | 2531 | 18 |
| 27 Cardiff | 978 | 1571 | | |
| 28 London | 911 | 1489 | 2600 | 18 |
| 29 Macclesfield | 874 | 1241 | 2211 | 15.5 |
| 30 Machynlleth | 1030 | 1300 | 2053 | 15.5 |
| 31 Milton Keynes | 911 | 1489 | 2022 | 18 |
| Notes | kWh/m² | hrs. | °C Days | |
| Mean: | 1108 | 1683 | 2836 | |
| 2nd max : 2nd min | 1.9:1 | 2.1:1 | 2:2.1 | |
| S.D. | 230 | 358 | 667 | |

TABLE 2 Building Parameters

| | Floor Area | Heated Volume | Thermal Mass | 'G' |
|---|---|---|---|---|
| 1 Blovstrod | 160 | | | |
| 2 Gentofte | 133 | 390 | 100 | |
| 3 Greve | 110 | 264 | 60 | 0.78 |
| 4 Lyngby | 120 | 288 | 100 | 0.41 |
| 5 Aramon | 150 | 393 | 650 | 1.26 |
| 6 Blagnac | 103 | 277 | (150) | 1.43 |
| 7 Bourgoin | * 95.5 | * 229 | medium | 0.93 |
| 8 Dourdan | 197 | 534 | 340 | 1.15 |
| 9 Le Havre | 116 | 290 | 125 | 0.91 |
| 10 Odeillo | 79.5 | 300 | 300 | 1.67 |
| 11 Aachen | 116 | 290 | | 0.03 |
| 12 Essen | 200 | 552 | (brick) | |
| 13 Freiburg | * 107 | * 714 | | |
| 14 Walldorf | 176 | | | |
| 15 Otterfing | 145 | 900 | | |
| 16 Wernau | 174 | ** 1400 | | |
| 17 Fiume Veneto | * 84 | | | 1.82 |
| 18 Firenze | 260 | 750 | light | 0.82 |
| 19 Rossano Calabro | * 130 | * 550 | 500 | 1.51 |
| 20 Sequals | 95 | 332 | 380 | 1.04 |
| 21 Eindhoven | 220 | 650 | 294 | 0.92 |
| 22 Zoetermeer Ho 1 | 130 | 375 | 200 | 0.85 |
| 23 Ho 2 | 130 | 375 | 200 | 0.85 |
| 24 Ho 3 | 130 | 375 | 200 | 0.85 |
| 25 Ho 4 | 130 | 375 | 200 | 0.85 |
| 26 Bebbington | 50 | 118 | 2200 | 1.9 |
| 27 Cardiff | | | | |
| 28 London | 107 | | | |
| 29 Macclesfield | 114 | 360 | | 0.69 |
| 30 Machynlleth | 170 | 460 | 60 | 0.59 |
| 31 Milton Keynes | 85 | 234 | 93 | 1.5 |
| Notes | $m^2$ | $m^3$ | $kg/m^3$ | $W/m^2 {}^oC$ |
| Mean | 142 | 402 | 342 | 1.02 |
| 2nd max : 2nd min | 3.3:1 | 3.3:1 | 7:1 | 3.5:1 |

* ... per flat ** incl. basement and attic

'G' = volumetric heat loss in $W/m^3$

TABLE 3 System Parameters

| | COLLECTOR | | | | litres | STORAGE | | |
|---|---|---|---|---|---|---|---|---|
| | Glazing | Absorb | Tilt | $m^2$ | $m^2$ | $m^3$ | Medium | No. Tanks |
| 1 Blovstrod | S | B | $20^{\circ}$ | 10 | 50 | 0.5 | W | 1 |
| 2 Gentofte | D | B | $45^{\circ}$ | 28 | 71 | 2. | W | 1 |
| 3 Greve | D | B | $38^{\circ}$ | 50 | 100 | 5. | W | 1 |
| 4 Lyngby | D | B | $90^{\circ}$ | 42 | 714 | 30. | W | 1 |
| 5 Aramon | D | B | $90^{\circ}$ | 37.6 | 106 | 4. | W | 1 |
| 6 Blagnac | D plast. | Trickle | $75^{\circ}$ | 30 | 107 | 3.2 | W | (1) |
| 7 Bourgoin | S | B | $90^{\circ}$ | 25.3* | 198 | 5* | W(pc) | 6 |
| 8 Dourdan | S | B | $70^{\circ}$ | 62.5 | | 40 | Rocks | 1 |
| 9 Le Havre | D | B | $45^{\circ}$ | 39.5 | 76 | 3 | W | 1 |
| 10 Odeillo | D | B | $90^{\circ}$ | (55) | passive | 33 | Concrete | 1 |
| 11 Aachen | High eff. | S1 | $48^{\circ}$ | 20.3 | 227 | 46 | W | 2 |
| 12 Essen | D | B | $48^{\circ}$ | 65 | 111 | 7.2 | W | 7 |
| 13 Freiburg | High eff. | | $55^{\circ}$ | 9.1* | 412 | 3.8* | W | |
| 14 Walldorf | S | B | | 71.5 | 112 | 8 | W(+2pc) | 1 |
| 15 Otterfing | D | B | $30^{\circ}$ | 80 | 100 | 8 | W(+105s) | 1 |
| 16 Wernau | D | B | | 38.4 | 122 | 4.7 | W(+s ) | 1 |
| 17 Fiume Veneto | S | B | $60^{\circ}$ | 10.8* | 28 | 0.25* | W | 1 |
| 18 Firenze | D | B | $48^{\circ}$ | 110 | 150 | 16.5 | W | 3 |
| 19 Rossano Calabro | S | B | $25^{\circ}+90^{\circ}$ | 20* | 60 | 1.2 | W | 1 |
| 20 Sequals | S | B | $45^{\circ}+90^{\circ}$ | 33 | 30 | 1. | W | |
| 21 Eindhoven | S | S1 | $42^{\circ}$ | 51 | 80 | 4.1 | W | 1 |
| 22 Zoetermeer Ho 1 | S | S1 | $50^{\circ}$ | 35 | 57 | 2 | W | 1 |
| 23 Ho 2 | S | S1 | $50^{\circ}$ | 35 | 57 | 2 | W | 1 |
| 24 Ho 3 | S | S1 | $50^{\circ}$ | 35 | 57 | 2 | W | 1 |
| 25 Ho 4 | S | S1 | $50^{\circ}$ | 35 | 57 | 2 | W | 1 |
| 26 Bebbington | D | B | $90^{\circ}$ | 19 | Passive | | Concrete | 1 |
| 27 Cardiff | S | B S1 | $28^{\circ}$ | 6.2* | 32 | 0.2* | W | 1 |
| 28 London | | | $30^{\circ}$ | 5 | 100 | 0.5 | W | 1 |
| 29 Macclesfield | S | TrickleB | $34^{\circ}$ | 42 | 145 | 6.1 | W | 4 |
| 30 Machynlleth | D | TrickleB | $34^{\circ}+55^{\circ}$ | 100 | 1000 | 102 | W | 2 |
| 31 Milton Keynes | S | B | $30^{\circ}$ | 35 | 120 | 4.2 | W | 2 |
| Notes ( ): Secondary *: per flat | | | | | | | | |
| S: Single D: Double pc: phase change | | | | 39.7 | 160 | 11.6 | Mean | |
| W: Water A: Air s: soil | | | | 25.3 | 211 | 20.6 | S.D | |
| B: Black S1: selective | | | | 10:1 | 22:1 | 153:1 | 2nd max÷2ndmin | |

TABLE 4 Heating System

| | Heat Rec. | Heat Pump | BOILER oil | gas | solid | + Electric | HEAT DISTRIBUTION Air | Radiators | Underfloor |
|---|---|---|---|---|---|---|---|---|---|
| 1 Blovstrod 0 | | | X | | | | 0 | 0 | 0 |
| 2 Gentofte | | | X | | | | | X | |
| 3 Greve | X | | X | | | | | X | (X) |
| 4 Lyngby | X | | | | | X | X | | |
| 5 Aramon | | | | | | X | Aux | | Solar |
| 6 Blagnac | | | | X | | | X | | |
| 7 Bourgoin | | | | | | X | Aux | | Solar |
| 8 Dourdan | | X | | | | | X | | |
| 9 Le Havre | | | | | | X | X | | |
| 10 Odeillo | | | | | | X | X | | |
| 11 Aachen | X | X | | | | | X | X | |
| 12 Essen | | X | | | | | | | X |
| 13 Freiburg | (X) | | X | | | | | X | |
| 14 Walldorf | | X | X | | | | | | X |
| 15 Otterfing | | | X | | | | | | X |
| 16 Wernau | X | X | | X | | | | X | X |
| 17 Fiume Veneto | | | | X | | | X | X | |
| 18 Firenze | | | | X | | X | X | | |
| 19 Rossano Calabro | | | | | | X | X | | |
| 20 Sequals | | (X) | | X | | | X | | |
| 21 Eindhoven | (X) | | | X | | | X | | |
| 22 Zoetermeer Ho 1 | | | | X | | | | X | X |
| 23 Ho 2 | | | | X | | | | X | |
| 24 Ho 3 | | | | X | | | X | | |
| 25 Ho 4 | | | | X | | | X | | |
| 26 Bebbington | | | | | | X | X | | |
| 27 Cardiff 0 | | | | | | X | 0 | 0 | 0 |
| 28 London 0 | | | | | | X | 0 | 0 | 0 |
| 29 Macclesfield | X | X | | | X | | (X) | X | |
| 30 Machynlleth | | | | | X | | | | X |
| 31 Milton Keynes | | | | X | | | X | | |
| Notes Total | 5 | 6 | 6 | 11 | 2 | 10 | 17 | 8 | 9 |

secondary system: (X) not included in total

DHW only .........(0)

(+) Immersion heaters excluded, except for DHW only projects

TABLE 5 Annual Performance

| | INTERNAL GAINS | DIRECT SOLAR | SOLAR USED | AUXILIARY USED |
|---|---|---|---|---|
| 1 Blovstrod O | | | 915 | 1023 |
| 2 Gentofte | | | 3601 | 19151 |
| 3 Greve | | | 4106 | 14321 |
| 4 Lyngby | 8300 | 3700 | 3330 | 6290 |
| 5 Aramon | 12000 | | 6408 | 10679 |
| 6 Blagnac | 5625 | | 4360 | 13766 |
| 7 Bourgoin M | 9675 | | 24172 | 49355 |
| 8 Dourdan | | | | |
| 9 Le Havre | 8150 | | 4731 | 8202 |
| 10 Odeillo | 5800 | | 19094 | 8050 |
| 11 Aachen | 5460 | | 2718 | 2730 |
| 12 Essen | | | 11222 | 14517 |
| 13 Freiburg M | | | | |
| 14 Walldorf | | | 14233 | 10510 |
| 15 Otterfing | | | 7450 | 25150 |
| 16 Wernau | | | 9990 | 27760 |
| 17 Fiume Veneto M | | | | |
| 18 Firenze | | | 14121 | 5397 |
| 19 Rossano Calabro M | | | | |
| 20 Sequals | | | | |
| 21 Eindhoven | 7300 | 4700 | 12492 | 18013 |
| 22 Zoetermeer Ho 1 | 2760 | 2220 | 5069 | 14057 |
| 23 Ho 2 | 2960 | 2130 | 4673 | 12000 |
| 24 Ho 3 | 2560 | 1930 | 4019 | 10808 |
| 25 Ho 4 | 1860 | 2170 | 5966 | 13676 |
| 26 Bebbington | | | | |
| 27 Cardiff O | | | 1943 | 4758 |
| 28 London O | | | | |
| 29 Macclesfield | 3660 | 4640 | 2967 | 7782 |
| 30 Machynlleth | | | | |
| 31 Milton Keynes ** | 5100 | 5900 | 2386-4288 | 5088-4591 |
| | kWh | kWh | kWh | kWh |

Notes **:2 years data

DHW ——————(O) 9 Month data ——(2) 7 month data house un-occupied—(3)

Prov. Data——————(2,3) 4 Month data ——(27) 9 month data———(7)

Multi-occupancy—— (M) Inc Summer cooling(18)

TABLE 6 Annual Performance

| | Solar Used (kWh/m²) | | % Solar | % System Efficiency |
|---|---|---|---|---|
| | Excl. losses | Inc. losses | | |
| 1 Blovstrod | 92 | 92 | 47 | 8 |
| 2 Gentofte | 88 | 127 | 16 | 14 |
| 3 Greve | 82 | 82 | 22 | 8 |
| 4 Lyngby | (79) | 79 | 35 | 11 |
| 5 Aramon | 170 | (170) | 37 | 17 |
| 6 Blagnac | | 145 | 24 | |
| 7 Bourgoin | 154 | 154 | 33 | 27 |
| 8 Dourdan | | | | |
| 9 Le Havre | 120 | (120) | 36 | 10 |
| 10 Odeillo | 347 | | 70 | 30 |
| 11 Aachen | (134) | | 50 | 12 |
| 12 Essen | | 173 | 43 | 19 |
| 13 Freiburg | | | | |
| 14 Walldorf | | 199 | 58 | 20 |
| 15 Otterfing | 93 | | 23 | 8 |
| 16 Wernau | | 260 | 26 | 26 |
| 17 Fiume Veneto | | | | |
| 18 Firenze | | 128 | 72 | 12 |
| 19 Rossano Calabro | | | | |
| 20 Sequals | | | | |
| 21 Eindhoven | 206 | 245 | 41 | 27 |
| 22 Zoetermeer Ho 1 | 93 | 145 | 27 | 14 |
| 23 Ho 2 | 96 | 134 | 28 | 13 |
| 24 Ho 3 | 58 | 115 | 27 | 11 |
| 25 Ho 4 | 134 | 194 | 30 | 16 |
| 26 Bebbington | | | | |
| 27 Cardiff | 78 | 78 | 29 | 17 |
| 28 London | | | | |
| 29 Macclesfield | 71 | | 24 | 9 |
| 30 Machynlleth | | | | |
| 31 Milton Keynes** | 62-112 | 68-122 | 32-48 | 6-12 |

Notes **: 2 years data

DHW——(0) 9 month data—(2) 7 month data house un-occupied ——(3)

prov. data——(2,3) 4 month data—(27) 9 month data ——(7)

multi-occupancy—(M) including summer cooling—(18)

TABLE 7 Annual Solar Contribution

| | Total consumption (kWh) | | Solar proportion (%) | |
|---|---|---|---|---|
| | Predicted | Measured | Predicted | Measured |
| 1 Blovstrod | 1938 | 1938 | 49*** | 47 |
| 2 Gentofte | 28000 | 22752* (22584) | 30 | 16* |
| 3 Greve | 18000 | 18437* (18847) | 50 | 22* |
| 4 Lyngby | 4600 | 9620 | 100 | 35 |
| 5 Aramon | 20372 | 17087 | 47 | 38 |
| 6 Blagnac | 15495 | 18130 | 66 | 24 |
| 7 Bourgoin | 70287 | 72527 | 38 | 33 |
| 8 Dourdan | 35000 | | 50 | |
| 9 Le Havre | 14500 | 12934 | 49 | 36 |
| 10 Odeillo | | 24144 | | 70 |
| 11 Aachen | | 5448 | | 50 |
| 12 Essen | | 25739 | | 43 |
| 13 Freiburg | | | | |
| 14 Walldorf | | 24743 | | 58 |
| 15 Otterfing | 37900 | 32600 | 35 | 23 |
| 16 Wernau | 39600 | 37750 | 38 | 26 |
| 17 Fiume Veneto | 420998 | | | |
| 18 Firenze | 11040 + | 19518 + | | 72+ |
| 19 Rossano Calabro | 63400 | | 53 | |
| 20 Sequals | 16513 | | 42 | |
| 21 Eindhoven | 21000 | 30505 | 61 | 41 |
| 22 Zoetermeer Ho 1 | 18541 | 19126 | 29 | 27 |
| 23 Ho 2 | 16519 | 16673 | 37 | 28 |
| 24 Ho 3 | | 14827 | | 27 |
| 25 Ho 4 | 18383 | 19642 | 32 | 30 |
| 26 Bebbington | | | | |
| 27 Cardiff | 4000 | 6701 | 30-35 | 29 |
| 28 London | 4320 | | 45 | |
| 29 Macclesfield | 26500 | 12450* | 13 | 24* |
| 30 Machynlleth | 13300 | | 100 | |
| 31 Milton Keynes ** | 13250 | 7474-8879 | 59 | 32-48 |

** 2 years data

* estimate from incomplete data (corrected figures)

+ includes summer cooling

***revised predictions using measured weather date and consumption.

# CHAPTER 6

# MONITORING

All of the projects had a research and development objective, and for this some monitoring of performance is required. Monitoring during the early stages of operation is essential if the system is to be 'tuned' to its best performance. It is desirable over a year or more if efficiency is to be properly assessed and subsequent design improved. Some continuing monitoring is necessary if the efficiency of the system is to be maintained and faulty components readily identified.

A 'performance monitoring format' was used to collate information, and as a result of the experience gained, it is now possible to suggest for future use a low-cost data acquisition procedure, and this will be described later.

Accurate local weather data are an essential prerequisite for good design. In o[illegible] ny of the projects which provided information, weather measurements were made on site. Information like this will be invaluable in extending existing knowledge and in improving design.

## DESIGN

To monitor successfully, it is first of all necessary to have clear objectives. These may be reduced to a list of simple questions:

- how much fuel does the system save?
- how closely do predicted and measured performances agree?
- is it possible to improve the design?
- is the system working as economically as possible?
- how long will the system last?
- how is performance influenced by the building and its occupants?

Answers to these questions will indicate the appropriate monitoring strategy. Possible strategies are tabulated in Table 8, from a minimum level (I), to a maximum (III) which would embrace all three.

The design of a monitoring system is best done at the same time as the design of the installation which is to be monitored. Only in this way will it be possible to ensure that instruments are easily accessible for reading and adjustment,

TABLE 8 Three Levels of Monitoring

| | I | II | III |
|---|---|---|---|
| CLIMATE | | | |
| Insolation | Total irradiation on collector plane. | Total irradiance on:<br>-collector plane<br>-horizontal plane<br>-vertical plane | Direct and diffuse irradiance |
| Temperature | Dry bulb temperature | | Wet bulb temperature |
| Miscellaneous | | Wind speed | Precipitation<br>Wind direction<br>Atmospheric pressure |
| BUILDING | | | |
| Space | Temperature | Mean radiant temperature | |
| Ventilation | | Tracer gas measurement | Door & window monitoring |
| Fabric | | | Heat fluxes |
| Free Heat Gains | | Cooking<br>Direct solar gains<br>Heat storage losses | Light & small power<br>Pumps, fans,<br>flue losses |
| HEATING SYSTEM | | | |
| Collector | | Heat flow (or inlet, outlet temperatures and flow)<br>Inlet temperature | Surface temperature<br>Temperature at the rear of insulation |
| Storage SH and/or DHW | Heat flows (or inlet, outlet temperature & flows) | Draw off temperature<br>Set temperature | |
| Auxiliary Heating (SH and/or DHW) | Heat flows | | Auxiliary heating efficiency |
| Miscellaneous | | Heat storage: mean temperature<br>Cold water usage | Duty times<br>Storage: strata temperature<br>Surface temperature of DHW/SH storage insulation |

Level I: Column I
Level II: Columns I & II
Level III: Columns I, II & III

DHW: domestic hot water
SH: space heating

that power is available where required, and that the necessary straight pipe runs, isolating valves etc. are provided.

In design, the first step is to decide on the necessary minimum information to be obtained, the frequency of readings, and the maximum acceptable error. Instrumentation must be appropriate for the energy flows to be measured, which may include all those illustrated diagrammatically in Fig. 43. There is no point in obtaining more information than can be used, and all similar instruments should measure with comparable accuracy.

Having identified in principle the appropriate measuring points, instruments should be selected with the required degree of accuracy and an adequate range of measurement. They should also be robust enough to withstand such adverse conditions as may occur (corrosion, dust, vibration etc.).

Instruments should be located so that they perform accurately and are accessible for reading, calibration and maintenance. It is important to remember that accuracy is ultimately dependent upon the accuracy of the least accurate instrument. Small changes in temperature are difficult to measure, and sensors should be positioned with this in mind. Flow meters should be in straight runs of pipe which are preferably in lengths of not less than 15 pipe diameters. For temperature measurement it is necessary to ensure that:

- for air temperature, the sensor is shielded from the influence of surrounding walls, etc;
- for wall surface temperature, there is good contact with the wall; heat conducting compound should be used;
- for immersion temperatures, there is adequate protection from corrosion, etc. and good thermal contact between liquid, protecting sheath and sensors.

Where possible, design should allow for measurements to be made over at least one year and for monthly 'energy audits' to be calculated.

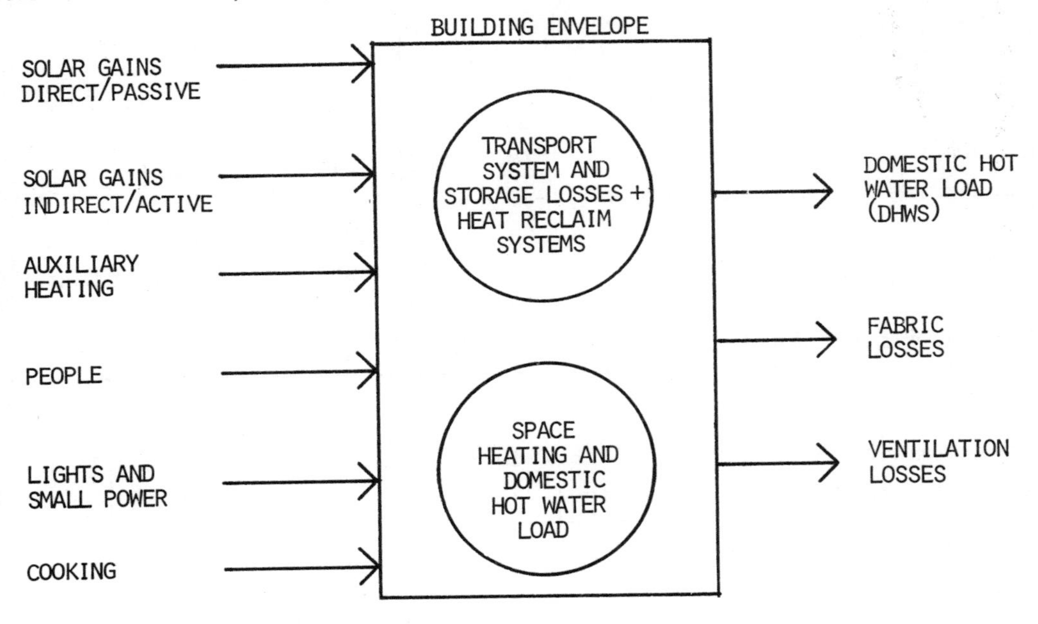


Fig. 43: energy flow

TABLE 9. Typical Measurement Instrumentation

| INSTRUMENT TYPE | MANUFACTURER | RANGE | ACCURACY | NO. OF PRO-JECTS USED ON | RE-CALIB. PERIOD | SAMPLE RATE |
|---|---|---|---|---|---|---|
| IRRADIATION | | | | | | |
| Solarimeter CM5 | Kipp & Zonen | 0-1500 W/m$^2$ | 0.75% | 9 | | |
| Thermopile | " " | 1 W/cm$^2$ | 3% | 1 | | |
| Pyranometer | " " | | 3% | 4 | | |
| WIND | | | | | | |
| Anemometer | Precitechnique | 10cm/s | 6% | 1 | | |
| " | Pen-Lann | 0-48cm/s. | 0.3m/s | 1 | | 10 min. |
| " | Jules Richard | 725cm/s | 5% | 3 | 1 yr. | |
| " | Schilt Knecht | | ± 20% | 1 | 1 yr. | 1 min. |
| Wind Vane | Chauvin-Arnoux | 360$^o$ | 20% | 2 | | |
| " | SIAP | 0-355$^o$ | ± 3% | 2 | 1 yr. | 15 min. |
| TEMPERATURE (Outside) | | | | | | |
| Thermocouple (Typet) | Thermo Electric | 203, 8$^o$C<br>389, 9$^o$C | ± 0.5% | 3 | ½ yr. | 20 min. |
| Thermafilm PRT | Matthey | -10 - 100$^o$C | ± 0.25% | 1 | ½ yr. | |
| Platinum Res | Brion-Leroux | -40 - 60$^o$C | 0-15% | 1 | | |
| " " | Degussa | | 0.2% | 1 | | 10 min. |
| " " | Tinsley | 0 - 100$^o$C | ±0.1$^o$C | | | Cont. |
| PT 100 | SIAP | -20 - +50$^o$C | | 2 | 1 yr. | 15 min. |
| NTC Thermometer | " | 0 - 100$^o$C | ± 3% | 1 | 1 yr. | 15 min. |
| OTHERS | | | | | | |
| HairHydrometer | SIAP | 0-100$^o$C | ± 3% | 2 | 1 yr. | 15 min. |
| Degree Days Meter | E.I.E. Society | | | 1 | 1 yr. | |

TABLE 9 (cont.) Typical Measurement Instrumentation

| | | | | | | |
|---|---|---|---|---|---|---|
| SYSTEM TEMPERATURES | | | | | | |
| PT-100 | Fischers & Sohne | | | 6 | ½ yr. | 30 min. |
| Thermafilm DRT | Matthey | -10 - 100°C | ± .25% | 1 | ½ yr. | |
| Silicon Diodes | Texac | 0 - 100°C | ± 0.2% | 1 | | |
| Platinum Resistance | Schlumberger | 100°C | 0.15% | 1 | | |
| " " | Degussa | | 0.2% | 2 | | 10 min. |
| MASS FLOWS | | | | | | |
| Ring Pistometer Type | Aquametro VZTH | 1 pulse = 1 litre | | | | |
| VZTH or VZFM, | VZFM | 1 pulse = 0.4ℓ | 1% | 3 | | |
| Vanewheel type SDT | SDT | 1 pulse = 0.4ℓ | | | | |
| Diaphragm | Schlumberger | $15 m^3/h$ | 7% | 1 | | |
| Woltman | Conteuro | $0-7 m^3/h$ | 2% | 2 | ½ yr. | 15 min. |
| Turbine | " | $0-3 m^3/h$ | 2% | 2 | ½ yr. | 15 min. |
| Flow-meter | Pen-Lann | 0-24m/s | 0.2m/s | 1 | | 10 min. |
| " | Sappel | | | 1 | | |
| Magnetic Flow Meter | Echardt | | ± 1% | 1 | 1 yr. | |
| " " " | | | | | | |
| Type DMA 460 | Andress & Hauser | 0.600ℓ/s | 5% | 1 | | Cont. |
| Turbine | Flow Technology | | ± 1% | 1 | 1 yr. | |
| (Litremeter Rotameter) | | 0.1-30 ℓ/m | 1 % | 2 | 2 wk. | 1 min. |
| HEAT QUANTITY | | | | | | |
| Calec - Ceh | Aquametro | | 1%ΔT@15°C<br>2%ΔT@7.5°C<br>5%ΔT@2.5°C | 2 | ½ yr. | |
| Heat Meter | Zanust | $0.3 m^3/h$ | | | ½ yr. | |

INSTRUMENTATION

Table 9 shows the variety in instrument and data acquisition systems used in the projects.

The accurate measurement of solar irradiation is crucial to the proper evaluation of system performance. Generally, measurements are taken in the plane of the collector. The instrument most commonly used is the Kipp and Zonen Solarimeter. Readings must be corrected for variations caused by: ambient temperature, wind speed, declination angle, convection due to inclination angle of solarimeter, speed of response and sun position. Measurements can be taken at intervals of 15 - 60 seconds, or continuously integrated, and an acceptable accuracy is considered to be ± 5%. There is also now on the market a cheap solarimeter which has to be read manually but gives a direct reading of total global irradiation in kWh.

There are three types of thermometer: space, contact and immersion. Commonly used instruments are platinum resistance elements and thermocouples. The former are more expensive but are highly accurate (± 0.2°C). They are generally used for differential temperature measurements where small temperature differences, common in solar assisted systems, require particular attention. Thermocouples cost less, and have an accuracy of 1°C. Thermopiles have also been used by some organisations. These give a higher signal output per unit temperature difference than thermocouples and can be made with an accurace of ±0.1°C. The degree of accuracy required is related to application, but the following accuracy standards seem necessary:

absolute temperature: ± 0.25°C

differential temperature: ± 0.1°C for $\Delta T \leq 15°C$
± 0.25°C for $\Delta T > 15°C$

Mass flow measurements are necessary for the computation of heat flows. The type of instruments used include: positive displacement, differential pressure, turbine, magnetic and ultra-sound, although the most popular seems to be the positive displacement type. The main difficulty lies in maintaining accuracy over long periods, especially at low flow rates. The most commonly used instruments - Acqua Metro, Pollux and Fischer and Porter - all gave acceptable results and have been checked for calibration individually. One particular instrument, the ring-piston meter proved to be too noisy, and problems have also been experienced with the Ultra-sound meter. With individual calibration and correction for flow rate and temperature, accuracies of between ± 2% to ± 5% can be obtained. Costs vary greatly. Flow meters often provide a digital signal which is then integrated by means of pulse counters read at regular intervals by the data loggers. There is a comprehensive range of integrated heat meters available which are suitable for either direct or remote reading. However, there is considerable difficulty in obtaining accurate and inexpensive instruments. At design flow and large temperature differentials, accuracies of ± 2% can be achieved, but differential temperatures below 10°C, common in solar assisted systems, may produce a disproportionate increase in error and so can flow rates below 50% of design. There is the additional question of how much the accuracy of these heat meters may be affected by the presence of additives.

The measurement of heat flow through the building fabric is important for checking the design values of thermal transmittance figures and the effect of varying climatic conditions. A thermopile instrument for surface mounting has been designed by TNO Delft. It has an accuracy of ± 5%.

Heat flow caused by air infiltration into a building is difficult to assess, unless mechanical ventilation systems are used. The standard method for measuring ventilation rate is by using a tracer gas, but there is no known method for continuous measurement in occupied houses. An alternative is to monitor the opening and closing of windows and doors to check the effect of occupancy on ventilation losses.

The measurement of auxiliary fuel consumption is also crucial. Standard kilowatt-hour meters with repeaters if necessary, can be used for monitoring electricity consumption. Where other fuels are used, the fuel input must be measured. Energy consumption for lighting and cooking should be measured separately to assist in the estimate of the resulting 'free heat gains'.

Finally, the duty times of the solar and auxiliary equipment needs to be recorded. This is done usually by means of a digital signal initiated by an interface relay.

## DATA ACQUISITION SYSTEMS

The information provided by the different measuring instruments, typically in the form of an output in mV or pulses, has to be recorded at regular intervals and converted into engineering units for further analysis. These two separate processes, recording and conversion, can be achieved in a number of ways.

### Manually

The instruments are connected to units providing a visual display (e.g. a digital voltmeter, an integrating meter etc.) and read manually. This procedure, the lowest in capital cost, is unreliable, very time-consuming and totally inappropriate for variables changing rapidly with time, such as incident solar radiation. However, regular readings of integrating instruments such as kilowatt-hour meters can give a useful impression of energy consumption over a given period of time.

### Chart Recorders

Here, measuring instruments are hard-wired to a chart recorder, thus providing a continuous visual display of the output of the various instruments, or of the state of the system (e.g. whether circulating pumps are on or off).

A multi-channel chart recorder is robust and reasonably cheap and gives a good visual impression of the condition of the system, a particularly useful feature during the early stages of commissioning to check for faults. However, the analysis of the resulting rolls of charts is tedious and extremely time consuming, It may well keep a person employed full time converting data from the charts into useful performance figures.

It is important to note that both manual readings and chart recording would usually require the assumption of a linear relationship between output from the instrument and the value of the variable recorded. The inaccuracies introduced by such an assumption would depend on the particular instrument used and on the range of the variable, but is unlikely to exceed 5 - 10%.

Table 10 shows a suggested range of accuracies for the measurment for the main variables, before they are transferred to the data logger.

TABLE 10 Proposed Performance Criteria for Instruments measuring typical Variables in Solar Heating Monitoring System

| No. Off | Variable | Measurement Interval or Rate | Recording Rate | Range of Instrument | Accuracy | Output from Transducers | Type of Instrument used as example | Interface required by Data Acquisition System |
|---|---|---|---|---|---|---|---|---|
| 1 | Solar Radiation | 15 secs | 1 hour | 0 to 1300 $W/m^2$ | ±5% | 0-12 mV<br>0-30 mV<br>0-60 mV | Kipp and Zonen | High gain amplifier |
| 5 to 10 | Space temperature | 30 minutes | 1 hour | 0 to 25°C | ±0.25°C | | Resistance thermometer or thermocouple | Bridge circuit |
| 1 | Outside temperature | 30 minutes | 1 hour | -15 to 30°C | " | | | Amplifier and reference junction is used for absolute temperature measurement |
| Up to 6 | Storage temperature | 5 to 10 mins | 1 hour | 10 to 90°C | " | | | |
| 3 | Differential temperature | 1 to 15 sec | | 10 to 90°C | ±0.1°C for $\Delta T \leq 15°C$<br>±0.25°C for $\Delta T > 15°C$ | | | |
| 3 | Mass flow | 1 to 15 sec | | | ±2% | Pulses | | Contact closure counters |
| Up to 6 | Heat flux | 15 sec | 1 hour | 1-30 watts | ±2% | .002-6mv | TNO Delft, heat flow meters | Input voltage integrators |
| 1 | kWh | | 1 hour | 0-15 | ±1% | Pulses | Landis and Gyr | Contact closure counters |
| 1 | Auxiliary heating | | 1 hour | 0-10 | ±5% | Pulses | Effectively a heat metering device | Contact closure counters |
| 1 | Pump/fan start stop | | 1 hour | 0-60 mins/hr | | Contact closures | | Binary inputs |

## Data Loggers

Data loggers provide the facility for regular scanning of all the variables being measured, and their storage, in digital form. These systems vary enormously in sophistication and cost and may be sub-divided into three categories:

1. Non programmable data loggers. At the cheaper end of the scale data loggers scan automatically all variables at regular intervals, and store them in digital form, usually a magnetic tape cassette. These systems are usually physically small and often come with rechargeable batteries and weather-proof casing and are therefore extremely easy and unobtrusive to use. However, the system does not provide facilities for self-adjustment of the scanning rate and is therefore potentially wasteful of storage space. Additionally, the number of channels are limited (typically 10 to 15) and a very significant amount of further work is required to translate the tapes.

2. Semi-programmable data loggers. Here a certain amount of conditional logging is allowed. The scanning rate is reduced when variables are not changing significantly (e.g. at night) and increased other times. In addition, some integrated pulse counters are more likely to be incorporated for items such as flow-meters, and it may be possible to adjust instrument readings to calibration curves.

3. Micro-processor controlled data loggers. A fully programmable data logger gives considerable flexibility in recording. It can be programmed to scan any channel in any order, suitable calibration curves for individual instruments can be stored for correcting instrument readings, and some calculations can be done between cycles, thus greatly reducing the volume of data recorded and the amount of further analysis required. Such a data logger would probably include a real time clock (an invaluable aid) a large number of channels for digital and analogue inputs, and a better storage medium, possibly with read/write facilities (e.g. cartridge). Packages such as the HP or the Solartron/DEC 11 belong to this category, but the benefits of increased flexibility in use, should be balanced against the cost of developing appropriate software and the high purchase cost.

The projects surveyed have shown that the increasing sophistication of the data acquisition and instrumentation systems used, has resulted in an increasing number of very specialised problems requiring a well qualified staff to solve them. Table 11 shows some of these problems.

## PROPOSAL FOR A LOW COST DATA ACQUISITION SYSTEM

A survey of the data acquisition systems commercially available, showed that no low cost system suitable for monitoring solar heated houses was available. It was therefore decided to propose a specially designed system, on the basis that future government or CEC sponsored programmes would result in a large number of solar assisted heating systems installed within the CEC. If such programmes are coordinated, the proposed monitoring scheme should result in lower total costs - including development costs. It would provide a common approach to monitoring and an easy means for communicating data between the different research groups involved.

The proposal includes a realistic appraisal of the cost implication. Personnel costs associated with different alternatives have been considerd. A marginally more costly data acquisition system which implies a potential saving in personnel costs may provide a lower total cost for data gathering than an apparently cheaper data acquisition system.

TABLE 11 Data Acquisition Systems - Main Problems

| PROJECT | MAJOR DATA ACQUISITION EQUIPMENT TYPES/SUPPLIERS | COMMENTS ON EXPERIENCE IN USE |
|---|---|---|
| GENTOFTE | HP 3051A with data logging system and HP 3455 DVM HP 9815A calculator | Voltage peaks from mains meant a stabilizer required |
| BOURGOIN-JALLIEU | 64 Channel Data CRTBT (CNRS) | Integrator for Meteorological data sensitive to power fluctuations. Paper tape badly folded or torn |
| DOURDAN | 100 input monitoring system. Digital cassette. | Parasitic pulses |
| FREIBURG | Solartron/DEC 11 Computer based data acquisition system | DEC 11 exchanged Disk drive exchanged |
| WALLDORF | Brown Boveri metrowatt instrumentation system | Only initial problems. Some core storage problems |
| FIUME VENETO | IBM data logger managed by Microcomputer | Failures in recorder |
| SEQUALS | Programmable data logging computer | Failures in power supply |
| ZOETERMEER | Doric Multiplexer 09, Intel MCS-8008 CPU Kennedy Type 9000 Formatter Type 9230C Texas Inst. Silent 745 Modem Sat. | Initial software problems |
| CARDIFF | Dual cassette data logging system Transferred to PDP 11-40 computer | Recording tape tangled. Delay due to maintenance on PDP 11-40. |

Although the objectives for the future projects are quite clear, there are still many unknowns. Therefore, these proposals provide a broad specification for data acquisition systems covering a number of possible alternatives. The possible variations may be due to the following factors:

1. The type of system - active or passive
2. The number of variables measured in each dwelling
3. The total number of dwellings in the trial
4. The number of separate sites for the dwellings
5. The geographical dispersal of dwellings on the different sites
6. Other cost and engineering considerations

The recommended system specification is based on the following criteria:

1. Low cost
2. Modular elements
3. Flexibility in terms of number of imputs and recording frequency
4. Suitable for individual houses or groups
5. No moving parts

The proposal has the following features:

1. A programmable micro-processor based data acquisition system located in each dwelling, capable of accepting data from appropriate transducers and processing and storing a week's data.

2. A data collection system using either:

   i) replaceable data storage devices such as tape cartridge, or,

   ii) a modem/telephone based system.

3. Regional data collection centres where data from a number of monitored dwellings in the same national territory or climatic zone will be collected, processed and stored on nine track magnetic tape.

Figure 44 describes a block diagram of the type of system envisaged. The system is based on a microprocessor. The read-only memory (ROM) contains the control program. This program will reflect the various options which relate to a particular project. There is the possibility of having a number of digital input modules. Similarly, there are readily available devices for analogue inputs which can be grouped to receive the required number of analogue signals. Both options are considered. Fast random access memory (RAM) will be used for storing immediate values and for calculating hourly means etc. When storing data for one day or one week, mass storage devices with longer access times can be used. The output interface for the data acquisition system of an individual dwelling in a remote location will be a modem linked to the domestic telephone.

The system is modular since different input interface cards (required for different combinations of transducers) can be combined with different output interface cards. Programming for this system is achieved by writing a modular suite of programs where there is a specific program module for each type of input or output interface card.

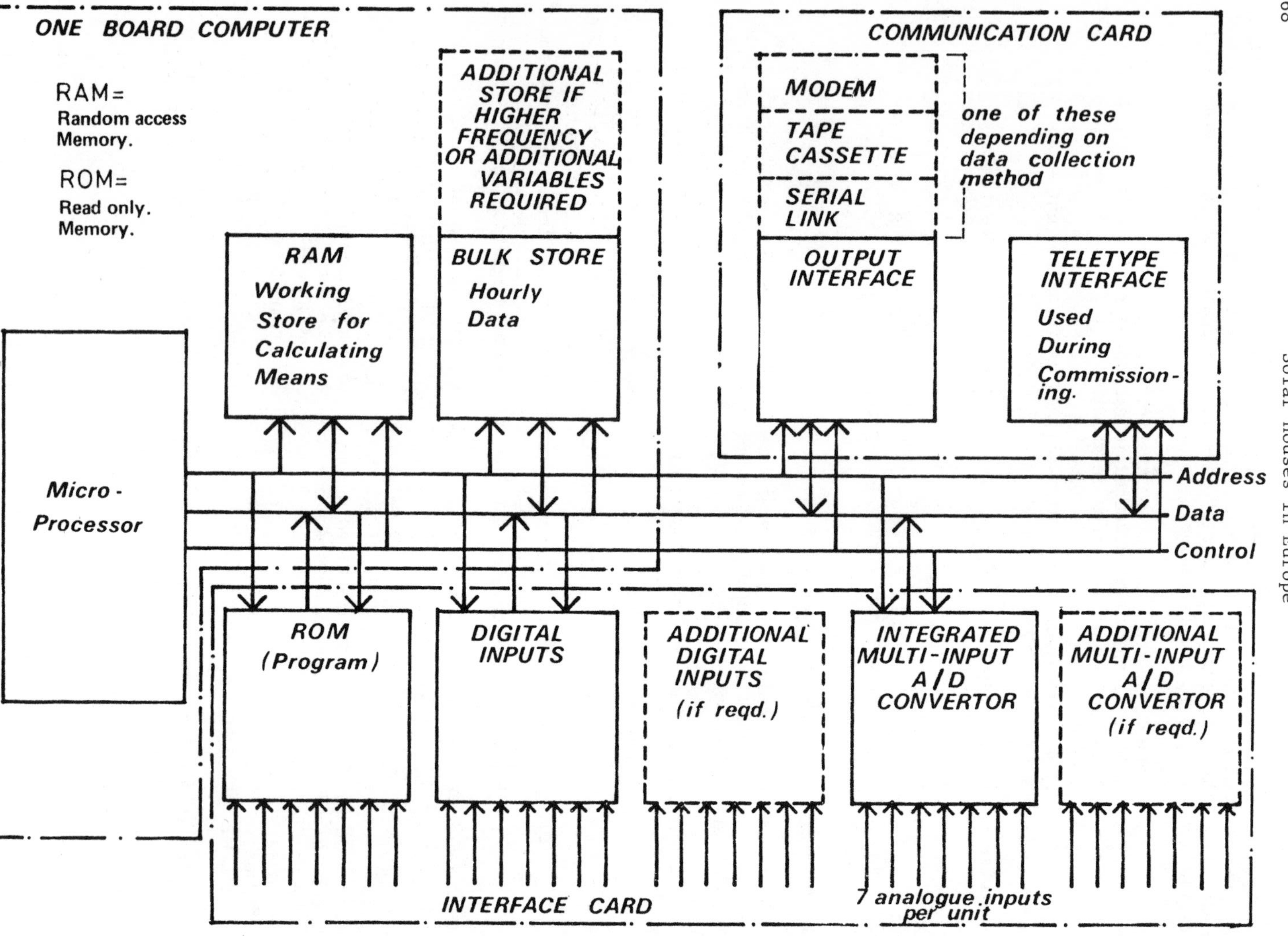


Fig. 44: block diagram of proposed microprocessor-based data acquisition system

The general view on the programming of micro-processors is that programs written in assembly languages are usually only understandable by the authors. One of the objectives of the future projects is to achieve a similar performance of monitoring and data acquisition systems so that the data is at least readable to the co-ordinating organisation. This implies that all software is developed in high level language on development systems and that each supplier of data acquisition systems has access to such development systems and provides fully documented listings of the programs required by the data acquisition systems.

## Regional Data Collection Centres

Associated with a group of such isolated sites will be a computer system acting as a regional data collection system which would use standard telecommunication techniques to interrogate each site to gather daily data.

Regional data collection centres do not need to be dedicated computers since they will only be used for a few hours each night. Therefore there is no need to develop or purchase special computer systems.

Alternatively, if a number of dwellings are grouped on the same site then it may be more effective to have a data gathering computer on that site. In this option the data acquisition system in each dwelling would have a direct serial data link to the data gathering computer. Another alternative can be considered if only a few variables are being measured in each dwelling and a number of dwellings (say up to 5) are grouped closely together, as in a terrace. In this arrangement it will be possible to use one data acquisition system for the group of dwellings thus considerably reducing the number and cost of data acquisition systems.

A teletype interface should also be provided so that a terminal device can be connected to the data acquisition system. This terminal would only be used during commissioning and is not a permanent feature of the data acquisition system.

# PERFORMANCE CRITERIA FOR THE PROPOSED DATA ACQUISITION SYSTEM

## Machine Size and Accuracy

It may be argued that 8 bit resolution is adequate for the analogue to digital conversion. This effectively gives a range of numbers from 0 - 256 which could be used to represent temperatures in the range of 0 - 100°C at 0.5°C accuracy and temperature differences in the range of 0 - 10°C at 0.05°C accuracy. However, to achieve this accuracy the input voltages into the analogue to digital converter should be accurately scaled to the voltage range of the converter. If the scale of the measured variable only uses half the voltage range of the converter then accuracy is also halved. It is suggested that the difference in cost between the 16 and 8 bit micro-processors is insignificant, compared with the savings made by the easier programming of the 16 bit machine. Therefore, it is recommended that 16 bit micro-processors are used and that analogue to digital conversion is performed at 12 bit accuracy with the corresponding greater resolution of analogue to digital conversion.

## Memory Capacity Requirements

The typical maximum storage capability of a 16 bit micro-processor of 64 K x 16 bit words is sufficient for storing one week's data.

For example, on one monitoring scheme which was considered during the study, there were sixty-five variables. Each could require 2 x 16 bit words - seven days a week.

65 x 2 x 24 x 7 = 21,840 words.

This leaves 42 K words for program and intermediate data storage.

The minimum number of variables required to be measured in each dwelling is thirteen. If five such dwellings were grouped closely together as in a terrace of houses then one data acquisition system would have the capacity to store one week's data for all five houses.

Data collection. The choice in data collection is between a replaceable physical device for holding data (such as a tape cartridge) or a modem/telephone-based data collection system. These alternative methods of data collection were evaluated for each of five main points for Data Acquisition Systems:

1. The short nature of the project and the need to minimise expenditure.

2. The data collection system should be capable of being used in other applications after the end of the initial monitoring project.

3. Uniformity of performance should be achieved and maintained if different project groups implemented or purchased their own data acquisition systems.

4. The unknown factors of geographical location and number of dwellings per site.

5. Frequency and level of maintenance and other routine visits by personnel from the monitoring organisations to each dwelling.

For the preferred basic systems certain cost parameters were considered:

1. Replaceable data storage solution

2. Modem/telephone data collection system.

Replaceable data storage option. In this option a number of days' data is stored within the data acquisition system using some replaceable data storage device, such as:

1. Paper tape
2. Tape cassette
3. Tape cartridge
4. Disk
5. Magnetic tape (½", 9 track)
6. Removable memory module.

The data is then transferred to a main frame computer for analysis. This option has cost implications in addition to the cost of purchasing recording units and the recording devices and relates to the personnel time which will be required;

(i) To physically collect the storage devices from the monitored dwellings

(ii) To supervise the transfer of data from the data storage devices to long term storage on 9 track magnetic tape.

One important problem with the replaceable data storage option is the frequency with which the data storage devices would be collected from the instrumented dwellings. It is only by analysing the data from a particular dwelling that a check can be made that the solar collector system, the instrumentation and the data acquisition systems are correctly functioning. If this check can only be performed once each month then there is a distinct possibility that there will be breaks in the recorded data due to the inability to identify errors immediately. However, if the data storage devices are collected each week then their capacity to store data will be seriously under utilised and the personnel cost involved in collecting the devices increased by a factor of four.

Removeable memory module devices may in the future prove to be a useful solution to this problem because:

i) They do not require expensive mechanical writing devices such as those used in tape decks and disk drives

ii) their size can be varied to suit the application. However, at the moment these memory modules are more expensive than tape cartridges. This price disadvantage may be reduced in the future.

There have been a number of reviews of replaceable storage devices which have been considered in the proposed system concept. These are summarised below:

TABLE 12 Review of Replaceable Storage Devices

| Devices | Advantages | Disadvantages |
|---|---|---|
| Paper tape | Cheap | Too bulky, easily torn |
| Digital Tape Cassette | Relatively cheap | Unsatisfactory error rate |
| Floppy disk | Good error rates and storage capacity | The random access facility provided is not required in this application; also expensive |
| Tape cartridge | Good error rates and storage capacity | Confusion over the interpretation of international standards, also expensive |
| Disk cartridge) 9 and 9 track ) $\frac{1}{2}$'' tape ) | | Capacity is too large and systems are too expensive for this application |
| Removable Memory Module | | Too expensive |

Modem/Telephone Data Collection Option for an Individual Dwelling

In this option a number of hours' data is stored within the data acquisition system. A data communication system is used so that data can be retrieved by a central or regional data collection point, without the transport of replaceable data storage devices. This scheme uses a modem in each data acquisition system. With such devices data can be transmitted down normal telephone lines (or specially rented private lines) to a regional computer which would act as the data collection point.

The main frame computer need not be dedicated to the task of data collection. It is proposed that data transfer occurs at night between say, midnight and 6.00 a.m. Therefore, the computer system should have the following facilities:

(i) Can be linked to auto-calling modems

(ii) Can be used between midnight and 6.00 a.m.

(iii) Has tape drives capable of producing the output in the specified format

(iv) Can be used to run a data gathering program.

This can be achieved by using an automatic answering modem and a timing program in the data acquisition system to switch on the automatic answering modem only between midnight and 6.00 a.m. Thus, when the regional data collection centre calls, the modem is in automatic answering mode and the occupier is not disturbed by the telephone call:

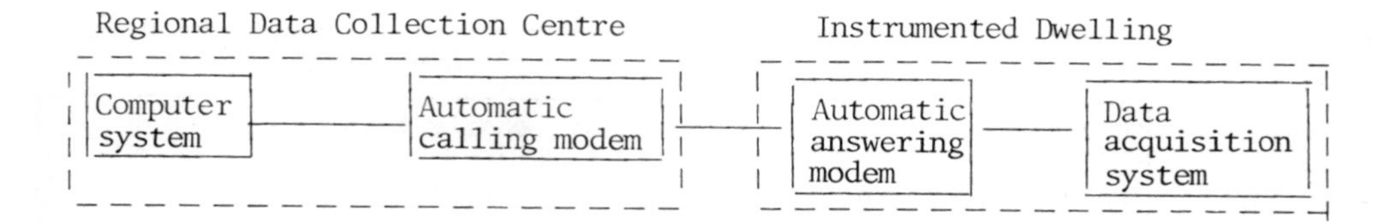


Modem/telephone data collection: option for a group of dwellings on the same site. The variation of the modem/telephone option occurs when there is a number of dwellings being monitored on the same site. Under these conditions it may be cost effective not to have a modem and telephone in each dwelling but to link all the data acquisition systems by dedicated serial links to one control system. This single control system is then linked to the regional data collection centre by a modem/telephone system.

Evaluation of modem/telephone data collection in the CEC. The use of modem and domestic telephone systems as the basis for gathering data from instrumented dwellings has been evaluated for the CEC countries. Such a system would use a main frame computer as the regional data collection centre with an auto calling modem equivalent to the British Post Office DCE 1 device.

The modems are marketed under the DATEL name. Datel services at both 200 and 600 band rates are available in all CEC countries.

The time for double transmission for an instrumented dwelling recording 61 variables is approximately 13 minutes. On the same basis transmission time for collection of daily data from a data acquisition system recording 20 variables will be in the order of 4 - 5 minutes per dwelling.

For the two preferred basic systems replaceable data storage and modem/telephone data collection, certain cost parameters must be considered.

1. Replaceable data storage. The cost of:

   - data recording device in each data acquisition system (e.g. tape cartridge transport).
   - replaceable data storage devices (at least twice as many as there are dwellings being monitored).
   - collection of data storage devices (this could be part of other normal inspection activities carried out by the monitoring organisation). This cost increases if dwellings are geographically dispersed.
   - supervising transfer of data from data storage devices to long term storage (e.g. 9 track tape).
   - not identifying failure in a particular data acquisition system until that week's data is processed.

2. Modem/telephone data collection. The cost of:

   - modem system with each data acquisition system.
   - subsidising the use of the domestic telephone installed in each dwelling.
   - regional data gathering computer used to run data gathering program.
   - identifying if any particular data acquisition system has failed within the last 24 hours.

These cost considerations are summarised in Table 13, and the conclusions are:

1. If data is transmitted over a modem/telephone system.

   i) the data acquisition system must be capable of storing at least 7 days data in case there is a failure in the modem/telephone system

   ii) the Regional Data Collection Centre Computer should interrogate the Data Acquisition System in each dwelling at not less than a daily frequency

   iii) data is transmitted twice to ensure integrity of operation.

2. If data is collected on replaceable data storage device (e.g. cartridge).

   i) the data storage device should have the capacity for two weeks' data

   ii) the data storage device should be exchanged at not less than a weekly frequency

   iii) data is recorded twice to ensure integrity of operation.

TABLE 13 Comparative Cost of Data Collection Using Either Tape Cartridge or Modem/Telephone Systems

| SCHEME | No. of Sites | Acquisition systems/site | Total acquisition systems | CARTRIDGE/CASSETTE SYSTEM | | | | | MODEM/TELEPHONE SYSTEM | | | | |
|---|---|---|---|---|---|---|---|---|---|---|---|---|---|
| | | | | Total Cartridge systems | Cartridge cost/house | Man-days/house per week | Labour for 100 weeks at £100/day | Cost/house for 2 years | Modems No. | Acquisition systems No. | Modem hire/house | Telephone at 10 mins/day | Total cost/house for 2 years |
| | | | | | £ | | £ | £ | | | £ | £ | £ |
| 25 houses on isolated sites each with a data acquisition system | 25 | 1 | 25 | 25 | 1000 | 0.2 | 2000 | 3000 | 25 | 1 | 800* | 250* | 1050 |
| 5 sites, 5 dwellings on each site, one data acquisition system per site | 5 | 1 | 5 | 5 | 200 | 0.08 | 800 | 1000 | 5 | 1 | 160 | 250 | 410 |
| 5 sites, 5 dwellings on each site, 3 data acquisition systems per site | 5 | 3 | 15 | 15 | 600 | 0.12 | 1200 | 1800 | 5 | 3 | 160 | 250 | 410 |
| 5 sites, 5 dwellings per site, one data acquisition system per dwelling | 5 | 5 | 25 | 25 | 1000 | 0.16 | 1600 | 2600 | 5 | 5 | 160 | 250 | 410 |
| 25 dwellings on one site with one data acquisition system for each group of five dwellings | 1 | 5 | 5 | 5 | 200 | 0.04 | 400 | 800 | 1 | 5 | 160 | 250 | 410 |
| 25 dwellings or one site with a separate data acquisition system per dwelling | 1 | 25 | 25 | 25 | 1000 | 0.16 | 1600 | 2600 | 1 | 25 | 32 | 250 | 282 |

*Cost for Telecommunication services provided by PTTs in CEC countries, from the Euro Data Foundation Year Book 1979. (Worst Case).

++ Assuming ½ hour per site plus 1 hour per data acquisition system

## DEVELOPMENT CONSIDERATIONS FOR DATA ACQUISITION SYSTEMS

The system concept has been formulated to reduce the overall cost of monitoring instrumented dwellings. However, even if all the component parts (hardware) of the data acquisition systems could be purchased there is still an important development task involved in interfacing all the components together, including the writing of the necessary software.

An outline for such development budget gives (1979 values):

Micro-Processor Based Data Acquisition Systems

| | |
|---|---|
| Microprocessor based development system | £20,000 |
| Instrumentation and testing | 5,000 |
| Testing on site | 10,000 |
| Documentation in all CEC working languages | 5,000 |
| Staff cost (2 man years) | 40,000 |
| | £80,000 |

Regional Data Collection Centre Software Package

| | |
|---|---|
| Hire of computer time | £20,000 |
| Lease of modem equipment | 2,000 |
| Documentation in all CEC working languages | 5,000 |
| Staff cost (1 man year) | 20,000 |
| | £47,000 |
| Overall total | £127,000 |

The development work for the design and production of microprocessor-based data acquisition systems should include for hardware and software design.

Hardware design will include:

- interface to transducers,
- interface to either a data storage device or a modem/telephone system,
- interface to a teletype for use during commissioning.

Software development should include a full commented software documentation. The software will need to be written in a high level language and will include programs for the following:

- selecting transducers for reading,
- performing analogue to digital conversion,
- scaling voltages,
- performing statistical operations,

- writing data to storage devices,

- communicating to external devices via a modem to a data collection computer and via a teletype device during commissioning,

- checking programs to exercise and test all functions of the data acquisition system to be used from a teletype device during commissioning.

For data collection centres this work should include:

- software development in a machine independent high level language;

- testing and validation with an example of the microprocessor based data acquisition system. This should be done in conjunction with the supplier of the data acquisition systems;

- documentation of the software package in all CEC working languages.

The standard microprocessor system and software package should be used on each project. It is the responsibility of the project group to install the data acquisition system in each dwelling and to connect it to the transducers. It is the responsibility of the project group to hire time on an appropriate time-sharing computing system and mount and test the regional data collection software package on that system. If a modem/telephone based data collection system is used then the project group must liaise with the relevant PTT to ensure that all data acquisition systems and the computer used as the regional data collection centre are correctly interfaced to the appropriate data communication equipment. It is the responsibility of the project group to demonstrate that the performance of the instrumented dwellings are accurately recorded in the final data format on a nine track magnetic tape.

Developments in the microprocessor field may provide a number of standard systems from various suppliers which satisfy the performance criteria for the proposed data acquisition systems. This would permit the cost of the 'Microprocessor based development system' to be applied in part to the selection of such standard systems.

# GLOSSARY

Absorber. The part of a solar collector which converts the incident solar radiation into heat and from which the heat is removed by the transfer fluid. If an absorbing liquid is used then this may be both the absorber and the heat transfer fluid.

Aperture area of collector. The opening or projected area of a collector through which the unconcentrated solar energy is admitted.

Aperture cover. The transparent part of a solar collector, normally positioned at the aperture, which is used to reduce the heat loss from the absorber, and to provide some protection from the weather.

Auxiliary. The conventional (i.e. non-solar) contribution to the total load (e.g. gas boiler, immersion heater etc.).

Base temperature. See degree days

Collection efficiency. The percentage of solar energy incident on the collector transferred into the storage and contributing to its rise in temperature.

Degree days. The product of the number of degrees below a given base temperature and the number of days when that difference occurs. The base temperature is usually defined between 15.5 to 21°C.

Diffuse solar radiation. The solar radiation as received on a surface from a solid angle of $2\pi$ with the exception of the solid angle subtended by the sun's disk.

Direct solar gains. The result of direct sunshine passing through glass areas (mainly south facing). A proportion of this energy contributes to space heating (kWh).

Direct solar radiation. The solar radiation coming from the solid angle of the sun's disk.

Direct system. One where fluid in the storage tank circulates through the collectors without a heat exchanger in the primary circuit.

Drain down collector or system. A system where the circulating fluid is evacuated (i.e. replaced by air) to prevent freezing in very cold nights or to avoid boiling under conditions of high insolation and storage temperature.

Evacuated tube collector. A collector where the complete absorber is inside an evacuated glass tube. This virtually eliminates heat losses by convection.

Flat plate collector. A collector, essentially planar in which the aperture area is practically identical to the absorber area.

Free heat gains. The sum of direct solar gains and internal gains (kWh).

Gross area of collector. The overall projected area of the collector with its containing box, if present.

Gross heat load. The energy including free heat gains needed for domestic hot water and for space heating (kWh).

Heat emitter. The part of the heating circuit designed to transfer heat from the system into the heated space. It can be a radiator, a warm air unit (a heat exchanger) or an underfloor circuit.

Heat pipe. A device for transferring heat by means of evaporation and condensation of a fluid in a sealed system. Heat pipes may be used as components of a solar collector.

Heat storage. The part of the heating system used to absorb excessive collected energy for periods when demand exceeds collection. It usually comprises a tank with water and heat exchanger, but may also be a solid wall or consist of rocks.

Heat transfer fluid. The medium by which the energy retained by a collector, as heat, is removed from the collector.

High efficiency collector. A collector where thermal losses from the absorber have been reduced to a minimum. Primary examples include evacuated tube and heat pipe collectors.

Internal gains. The energy dissipated inside the heated space by people (body heat) and appliances (lighting, cooker, etc.). A proportion of this energy contributes to the space heating requirements (kWh).

Irradiation (of a surface). The time integral of the irradiance at that surface. Irradiation is often termed radiant exposure.

Liquid heating collector. A liquid heating collector is a solar collector which employs a liquid as the heat transfer fluid.

Overload heat exchanger. A heat exchanger used to dissipate energy incident on the collectors when the storage temperature is reaching boiling point.

Phase change material (p.c.m.). A heat storage medium which relies on the latent heat of fusion and solidification for absorbing and releasing heat in the heating system.

Primary circuit. This includes the collector array, the interconnecting pipework and (if existing) the heat exchanger in the storage tank.

Pyranometer. An instrument for measuring the solar irradiance on a plane surface from a solid angle of $2\pi$. When the solar radiation coming from the solid angle of the sun's disk is obscured from the instrument, a pyranometer can be used to determine the irradiance on a plane surface of diffuse solar radiation.

Selective surface (absorber). A surface absorbing essentially all incident solar radiation (short wave - high temperature source), while emitting a small fraction of radiation (long-wave).

Solar collector. A device which absorbs solar radiation, converts it into heat and passes this heat onto a heat transfer fluid.

Solar energy used. The amount of solar energy contributing to the total heat load. Unless otherwise stated, it includes useful losses. It is expressed in absolute figures (kWh) or per unit collector area (kWh/m$^2$).

Solar fraction (or percentage solar). The percentage of the total heat load supplied by the solar heating system and includes useful losses from the storage tank.

Solar radiation. The radiation emitted by the sun. (Approximately all of the incident solar energy is at wavelengths less than 4.0 µm and is often termed shortwave radiation).

Solar system output. The annual or monthly energy provided by the solar heating system to satisfy the heat load. It is expressed in absolute figures (kWh) or per collector area (kWh/$m^2$) and used for comparing projects.

Solarimeter. A specific type of pyranometer based upon the Moll-Gorczynski thermopile design.

Stagnation temperature. The temperature of the absorber under no-flow conditions and high solar irradiation.

Storage efficiency. The percentage of solar energy input to the heat storage, subsequently used in the heat distribution system (i.e. excludes un-wanted heat losses)(%).

Stratification (temperature). The temperature differential which can be maintained in an undisturbed water tank, between the hotter and lighter fluid which rises to the top and the heavier and colder fluid which remains at the bottom.

System efficiency. The percentage of solar radiation incident on the collector which is used for space heating and/or domestic hot water heating (%).

Thermal mass. The mass of the building within the insulation, expressed per volume of heated space (kg/$m^3$).

Total heat load. The energy, excluding free heat gains, required for space heating and hot water supply coming from solar or auxiliary energy.

Transmission (fabric) losses. The conduction losses from the heated space through the building envelope (kWh).

Trickle collector. A collector where the circulating fluid is sprayed at the top edge of the absorber, trickling down to a gutter where the fluid is collected and re-circulated through the storage tank.

Useful heat losses. The conduction and radiation heat losses from the primary circuit and storage tank into the heated space occurring at times when there is a space heating load (kWh).

Ventilation losses. The heat losses associated with the continuous replacement of warm, stale air by fresh cold air.

Volumetric heat loss coefficient (G). The total heat loss of a dwelling (through the fabric and ventilation), divided by the heated volume and the temperature differential at which the loss occurs (in Watts/$m^{2o}C$).

Water content (of collector/primary circuit). The volume of circulating fluid in the collector array (often including the complete primary circuit).

Wind speed. The speed of the air measured in accordance with the recommendations of the World Meteorological Organisation, normally measured ten metres above ground level.

Ref: CEC Recommendations for European Solar Collector Test Methods, drafted and edited by A Derrick, W B Gillet, January 1980

## CHAPTER 7

# REPORTS ON 31 PROJECTS

The need for a standard reporting format became apparent after an earlier study which Stephen George and Partners undertook for the Commission of the European Communities in 1978. That study, "European Solar Houses", showed that the 24 houses which were then looked at, did not provide data which could be compared.

Working Group A of Directorate-General XII of the Commission, is concerned with Performance Monitoring. The present contributors, who are listed at the front of this volume, participate in this Working Group. They agreed after that first study that a new reporting format should be devised. This format was evolved by David Turrent from the CCMS (Conference on the Challenges of Modern Society) format and from that of the IEA (International Energy Agency). It has now been completed by all the thirty-one projects herein with the guidance of the present contributors who acted as sub-contractors to Stephen George and Partners. The earlier chapters of this volume begin the process of analysis of these formats in a way which was not possible with the submissions based upon earlier formats. These were either too long for many to complete, or else were too open, so that individual entries were not comparable.

The following pages are abbreviations of the original thirty-one formats. Blank space and information not repeated in enough projects to allow for analysis has been eliminated. New data since September 1979, where available, has been substituted, and corrections to tables and figures, where necessary, have been made.

Researchers who wish to consult the original full submissions by sub-contractors are referred to Chris Whittaker, 5 Dryden Street, London WC2.

The main headings of each of the formats which follow are:

PROJECT DESCRIPTION
Main participants
Project Description
Project objectives

SITE LOCATION
Map
Nearby obstructions
Distance from nearest city
Photograph

SUMMARY SHEET
A1 Climate
B1 Building
C1 System
Performance monitoring

A2 LOCAL CLIMATE

B2 BUILDING DESCRIPTION
Design concepts
Construction
Energy conservation measures
Section

C2 SYSTEM DESCRIPTION
Collector/storage
Space heating/domestic hot water
Other points/heat pump
Control strategy/operating modes
Diagram of system

TECHNICAL APPRAISAL/PRACTICAL EXPERIENCE
System/design
System/installation
Component performance
System operation/controls, electronics

PERFORMANCE EVALUATION/CONCLUSIONS
Comparison of measured with predicted performance
Modifications to system
Occupants/response
Conclusions
Future work

## LIST OF PROJECTS

| PROJECT DESCRIPTION | BLOVSTROD | REF. | DK | BL | 78 | 79 | 1 |
|---|---|---|---|---|---|---|---|

Main Participants:

| | |
|---|---|
| OWNER | : Private family. |
| DESIGN (solar system) | : Technological Institute, Copenhagen, and Thermal Insulation Laboratory, Technical University of Denmark |
| CONSTRUCTION | : Private firm. |
| FINANCIAL SUPPORT | : Ministry of Commerce (100% for installation and monitoring programme). |
| MONITORING | : Technological Institute, Copenhagen<br>phone (02) 996611<br>contact: P. Steensen |

Project Description:

This is a retrofit of a solar water heating system only on a single family house built in 1968, with a structural addition in 1976. The solar system was constructed at the end of 1977.
The collector of 10 $m^2$ is situated on the roof, and the storage tank of 500 l is situated in the scullery.

Project Objectives:

The project objective is to demonstrate the retrofit of a solar water heating system on a single-family house. This includes a monitoring programme with survey measurements where the main heat flows and temperatures are monitored.

The monitoring period is 2 years, starting March 1978.

| SITE LOCATION MAP | REF. | DK | BL | 78 | 79 | | 1 |
|---|---|---|---|---|---|---|---|

Distance to main city 24 km north west of Copenhagen city.
Nearest obstructions: In the neighbourhood there are no obstructions higher than the house itself.

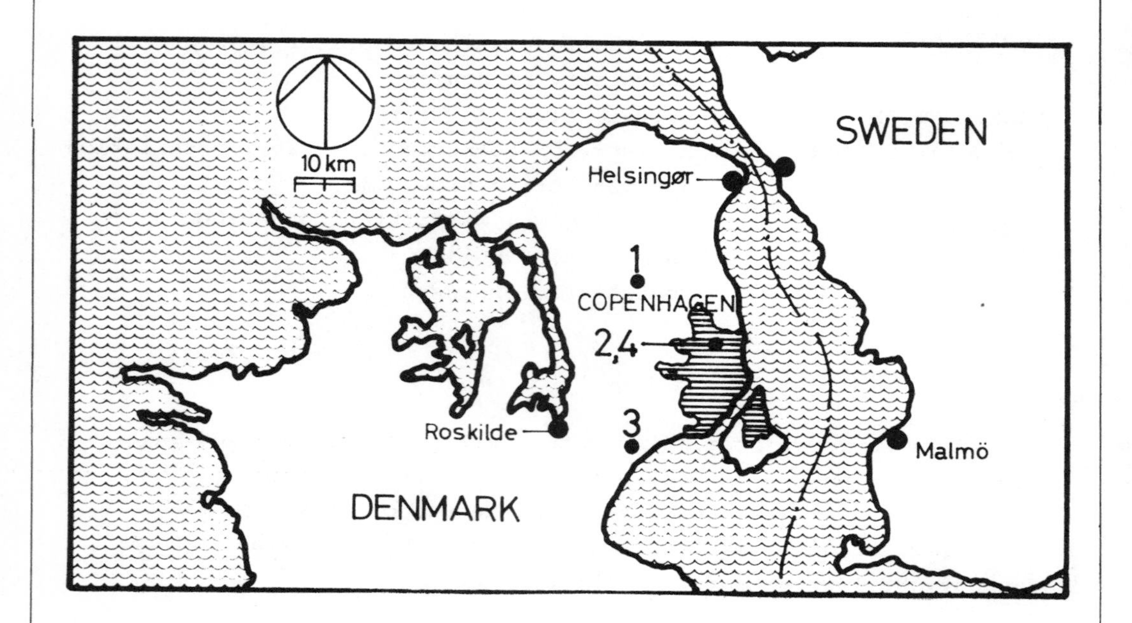

Photograph of Project

| SUMMARY SHEET | BLOVSTROD | REF. | DK | BL | 78 | 79 | 1 |
|---|---|---|---|---|---|---|---|

**Design Data**

**A1 CLIMATE**

1) Source of Data: Danish Test Reference Year
2) Latitude: 55° 53′ N Longitude: 12° 24′ E Altitude: ~ 50 metres
3) Global Irradiation (horiz plane): 1024 kWh/m²·year % Diffuse: 47
4) Degree Days: 2829 Base Temp: 17 °C
5) Sunshine Hours: July: 226 January: 70 Annual: 1580

**B1 BUILDING**

1) Building Type: single-family house No Occupants: 3
2) Floor area: 160 m² Heated Volume: m³
3) Design Temperature: External: °C Internal: HOT WATER SYSTEM ONLY °C
4) Ventilation Rate: HOT WATER SYSTEM ONLY a.c.h. Vol. Heat Loss: W/m³·K
5) Space Heat Load: kWh Hot Water Load: 4600 or 2500 kWh *)

*) 250 l DHW/day (4 persons) or 125 l DHW/day (2 persons)

**C1 SYSTEM**

1) Absorber Type: iron, single sheet and tube, black enamel
2) Collector Area 10 m² (Aperture) Coolant: ethylene glycol
3) Orientation: 190° Tilt: 20° Glazing: single
4) Storage Volume: 0.5 m³ Heat Emitters:
5) Auxiliary System: oil fired boiler (unit) Heat Pump: none

**PERFORMANCE MONITORING**

1) Is there a Computer Model? Yes 2) Start Date for Monitoring Programme: March 1978
3) Period for which results available March 1978 - February 1979
4) No. of Measuring Points per house 13
5) Data Acquisition System: 12 channel chart recorder + separate heat flow-meter. Solar incident and heat flows (3) are integrated.

*) Auxiliary and solar to DHW is output from DHW-tanks.
**) See note on page following sheet P1.

| | Space Heating | Hot Water *) | Total |
|---|---|---|---|
| 6) Predicted | % of kWh | 49 % of 1938 kWh**) | % of kWh |
| 7) Measured | % of kWh | 47 % of 1938 kWh*) | % of kWh |

8) Solar Energy Used: Including useful losses 92 kWh/m²·year*)
Excluding losses: 92 kWh/m²·year*)

9) System Efficiency: Solar Energy Used / Global Irradiation on collector × 100 = 8 % *)

| A2 LOCAL CLIMATE | REF. | DK | BL | 78 | 79 | 1 |
|---|---|---|---|---|---|---|

1 – Average Cloud cover: 5 Octas

2 – Average daily max temperatures (July): 21.1 °C

Average daily min temperatures (Jan): -6.5 °C

3 Source of Weather Data: Danish Test Reference Year ("Meteorological Data for Design of Building and Installation: A Reference Year", by H. Lund et al, TIL 1974).
Basis for the Test Reference Year is weather observations from a site 25 km from Blovstrod.

4 – Micro Climate/Site Description:

The house is situated within a flat urban area. To the south of the house there is a football ground (no obstructions). Apart from this the house is surrounded by single-family houses. The distance to the Sound ("Øresund") is 15 km.

*) "Degree Days" are based on the "Danish Normal Year" 1901-1940.

| MONTH | Test Ref. YEAR | Irradiation on horizontal plane | | *) Degree Days | Precipitation | Average Ambient Temperature | Average Relative Humidity | Average Wind Speed | Prevailing Wind Direction |
|---|---|---|---|---|---|---|---|---|---|
| | | Global | Diffuse | Base. 17 °C | | | | | |
| JAN | | 19 | 8 | 509 | 23 | 0.2 | 89 | 5.1 | |
| FEB | | 36 | 16 | 463 | 19 | -0.4 | 85 | 5.8 | |
| MARCH | | 83 | 31 | 433 | 19 | 2.0 | 86 | 5.9 | |
| APRIL | | 122 | 49 | 282 | 27 | 5.7 | 78 | 6.7 | |
| MAY | | 149 | 66 | 45 | 42 | 11.4 | 78 | 6.0 | |
| JUNE | | 170 | 78 | – | 38 | 16.0 | 77 | 5.0 | |
| JULY | | 161 | 79 | – | 79 | 16.4 | 74 | 3.3 | |
| AUGUST | | 123 | 66 | – | 91 | 16.1 | 84 | 4.4 | |
| SEPT | | 83 | 44 | 29 | 59 | 13.7 | 85 | 5.2 | |
| OCT | | 44 | 24 | 236 | 19 | 9.2 | 87 | 3.5 | |
| NOV | | 19 | 12 | 366 | 35 | 5.0 | 91 | 5.0 | |
| DEC | | 15 | 8 | 466 | 40 | -0.4 | 89 | 5.1 | |
| Totals | | 1024 | 481 | 2829 | 491 | | | | |
| Averages | | | | | | 8.0 | 84 | 5.1 | |
| Units | – | kWh/m²·month | | °C days | mm | °C | % | m/s | – |

B2 BUILDING DESCRIPTION REF. DK BL 78 79 1

## Design Concepts

The house is a typical Danish single-family house from the sixties. The solar collector is assembled on the outside of the roofing (asbestos cement tiles ).

The storage tank is situated in the scullery next to the boiler.

## Construction

1 · Glazing: ............ % Area: ............ $m^2$ 'U' value: ............ $W/m^2 \cdot K$

2 · 'U' values: Walls: HOT WATER SYSTEM ONLY $W/m^2 \cdot K$ Floor: ............ $W/m^2 \cdot K$ Roof/loft: ............ $W/m^2 \cdot K$

3 · Thermal mass: ............ $kg/m^3$ Annual Energy Demand: ............ kWh *)

*) Annual Energy Demand: A family of four is expected to use 250 l DHW/day corresponding to 4600 kWh/year. A family of two is expected to use 125 l DHW/day, 2500 kWh/year. Both figures on energy consumption include heat loss from the hot-water tank (included in the boiler-unit).

## Section Through House Showing main solar components

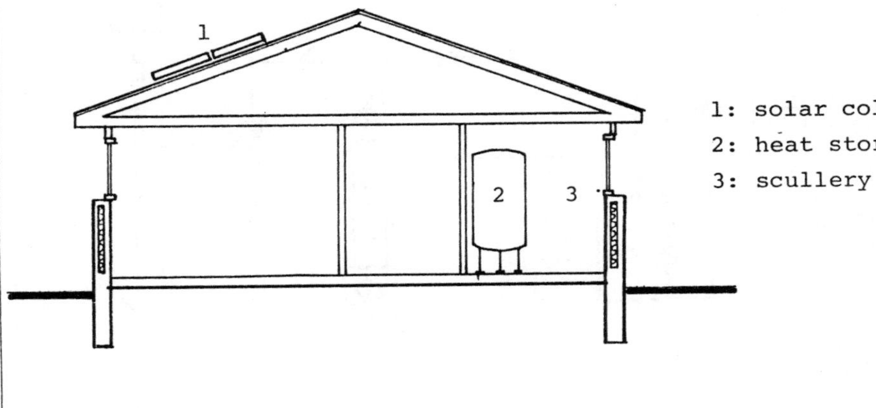


1: solar collector 10 $m^2$

2: heat storage .5 $m^3$

3: scullery

C2 SYSTEM DESCRIPTION REF. | DK | BL | 78 | 79 | 1

**Collector/Storage**

1 – Primary circuit. Water content: .................. litres/m$^2$ collector area

2 – Insulation. Collector: 45 mm Storage: 100 mm Pipework: 30-40 mm

3 – Frost Protection: 30% ethylene glycol

4 – Overheating Protection: none

5 – Corrosion Protection: ethylene glycol

**Space Heating/Domestic Hot Water**

1 – Auxiliary (S.H.): .................. (DHW): oil fired boiler (unit)

2 – Heat Emitters: .................. Temp. Range: ..................

3 – DHW Insulation: .................. Set Temperature: 50°C on mixing valve
65°C on boiler thermostat

**Other Points/Heat Pump:** none

**Control Strategy/Operating Modes**

The pump in the primary circuit is controlled by the use of a conventional differential thermostat.
In summertime DHW goes directly from the storage tank to the taps.
In wintertime DHW is postheated in the DHW-tank which is included in the boiler-unit.
To avoid DHW-temperatures above 50°C, a thermostatic mixing valve is included.

**Diagram of System** Showing sensor locations, power ratings, storage volumes & flows.

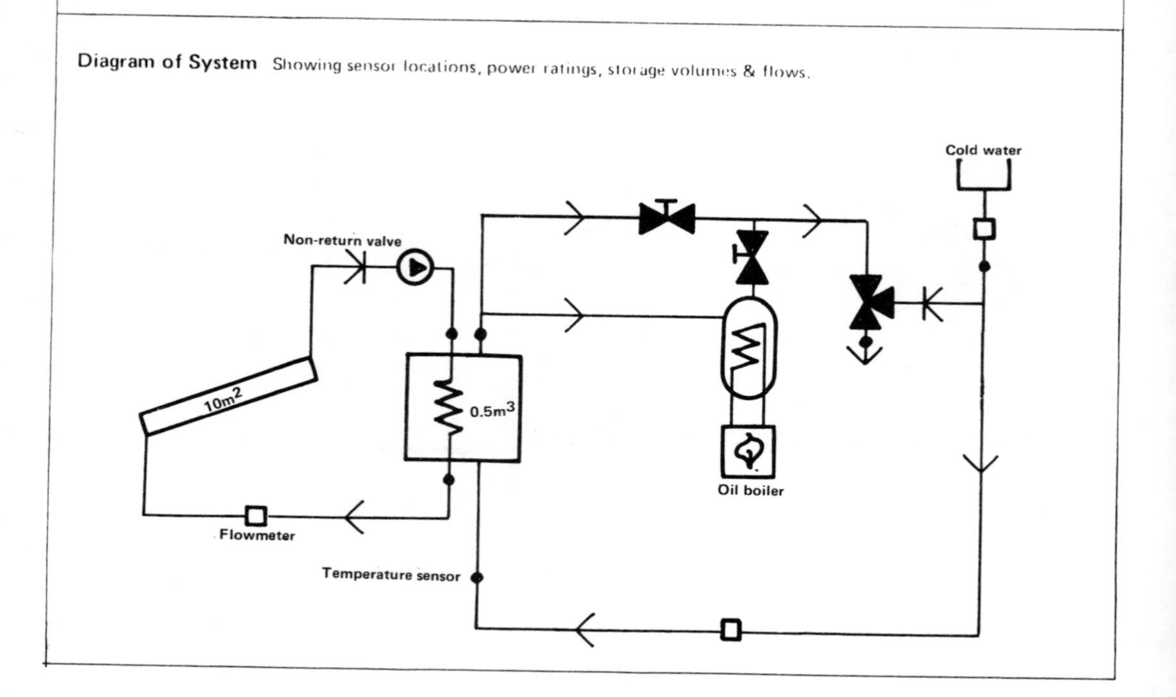

| TECHNICAL APPRAISAL/PRACTICAL EXPERIENCE | REF. | DK | BL | 78 | 79 | 1 |
|---|---|---|---|---|---|---|

System/Design:

The cold water supply pipe for the thermostatic mixing valve should include a spring loaded counterflow valve. Otherwise natural circulation from the top of the store through the mixing valve and back to the bottom of the store is possible.

Component Performance: (Collector, Heat Exchangers, Storage, Pipework, Valves, Fittings, Pumps, Auxiliary)

The counterflow valve in the primary circuit should be spring loaded, which was not the case initially.
There have been continuous problems with condensation in the collector.
The heat loss from the store is considerably larger than calculated from the geometry of the store and the insulation. Heat loss via pipes, phials etc. penetrating the insulation is not taken into account. Measured values are twice those calculated as described above. In addition the heat loss from the store has been increased in periods due to natural circulation in the primary circuit and through the thermostatic mixing valve.

System Operation/Controls, Electronics:

The position of the phials for the differential thermostat must be chosen with great care. The phial for the collector was initially placed in the loft on a pipe .3 m below the collector, which was unacceptable. The phial in the store should be positioned nearby the heat exchanger of the primary circuit.

PERFORMANCE EVALUATION/CONCLUSIONS REF. DK BL 78 79 1

**Comparison of Measured with Predicted Performance**
(State reasons for any discrepancy and summarise the main factors affecting performance)

The original predictions have been corrected regarding actual measured weather data, measured DHW-comsumption and measured DHW-temperature. When the boiler is on, the DHW set temperature is 65°C, because corrosion considerations demand at least 65°C in the boiler. Since the demanded temperature set by the mixing valve is only about 50°C, this will lower the output of the solar system.

When the predictions have been corrected as described above, there is a good agreement between "Percent Solar" predicted and measured, whereas the actual output from the solar system is significantly lower because of lower load. This means that the payment of interest is lower too.

The boiler was stopped in June and August. Reduced heat loss from the boiler-unit (in 70 days) is estimated to 1175 kWh.

**Modifications to System**

- Counterflow valve in primary circuit was replaced.
- Counterflow valve in the cold supply pipe for the mixing valve was installed.
- Position of solar collector phial for differential thermostat was improved.

**Occupants/Response**

The occupants are generally satisfied with the system.

**Conclusions**

Attention should be paid to:

Natural circulation.

Counterflow valves.

Condensation in the collector.

Position of phials for the differential thermostat.

**Future Work**

Nothing is planned yet except the completion of the two year monitoring programme.

| P1 COMMENTS ON PERFORMANCE FIGURES | REF. | DK | BL | 78 | 79 | 1 |
|---|---|---|---|---|---|---|

The performance figures must be taken as provisional, as they may be changed, for instance due to re-calibration of instruments.

Until and including May 1978 there has been natural circulation in the primary circuit. Until and including August 1978 there has been natural circulation through the thermostatic mixing valve. Both occurences are registered as an increased heat loss from the store.

Until and included September 1978 the heat meter HW 3 was not installed, and the contribution of solar to DHW has been calculated assuming that the heat loss from the store could be calculated with a reasonable accuracy. Since problems appeared on this point, the heat meter HW 3 was installed, so that heat loss from the store can be measured directly. Until and including September 1978 the heat loss from the store has been estimated on the basis of the heat loss coefficient measured in a period without input or output to the store.

Referring to sheet P1:

1) When there is a heat demand in the scullery, this heat demand is always covered by heat loss from the boiler (∿ 700 W).

2) Solar energy used for DHW is output from the store (the DHW-preheater).

3) Heat loss from store is negative in December (-16 kWh), January (-21 kWh) and February (-13 kWh). Since column nr. 7 is output from store, this effect increases the percentage solar in the months in question. The heat is taken from the scullery, which nearly always is too hot because of heat loss from the boiler.

Referring to "Summary Sheet":

**) "Predicted Percent Solar" is the original value from the computer calculation, but it is corrected regarding actual measured weather data, measured DHW-consumption and measured DHW-temperature by the use of the F-chart method.

## P1 MONTHLY PERFORMANCE

REF: DK BL 78 79 1

| Month | Year | No Days Data | Average Temp °C: Internal | Average Temp °C: External | SOLAR ENERGY kWh: Global Irradiation On Collector Area | Solar Energy Collected | Solar Energy Used: Space Heating Useful Losses 1) Incl. | Solar Energy Used: Space Heating Useful Losses 1) Excl. | Solar Energy Used: Domestic Hot Water 2) | Solar Energy Used: Input To Heat | Heat Pump Total Output kWh: Space Heating | Heat Pump Total Output kWh: Domestic Hot Water | AUXILIARY ENERGY kWh: Heat Pump | Space Heating | Domestic Hot Water | Pumps & Fans | Free Heat Gains kWh: Internal | Free Heat Gains kWh: Direct Solar | SUMMARY kWh: Total Load | SUMMARY %: Percentage Solar | SUMMARY %: Syst. Eff. |
|---|---|---|---|---|---|---|---|---|---|---|---|---|---|---|---|---|---|---|---|---|---|
| Column No | | | 1 | 2 | 3 | 4 | 5 | 6 | 7 | 8 | 9 | 10 | 11 | 12 | 13 | 14 | 15 | 16 | 17 | 18 | 19 |
| J | 79 | | | | 140 | 0 | | | 20 | 3) | | | | | 153 | | | | 173 | 12 | 14 |
| F | 79 | | | | 450 | 23 | | | 25 | 3) | | | | | 152 | | | | 177 | 14 | 6 |
| M | 78 | | | | 690 | 88 | | | (35) | | | | | | (128) | | | | 163 | (21) | (5) |
| A | 78 | | | | 1420 | 239 | | | (164) | | | | | | (31) | | | | 195 | (84) | (12) |
| M | 78 | | | | 1950 | 335 | | | (130) | | | | | | (0) | | | | 130 | (100) | (7) |
| J | 78 | | | | 1680 | 272 | | | 147 | | | | | | 0 | | BOILER OFF | | 147 | 100 | 9 |
| J | 78 | | | | 1450 | 225 | | | (81) | | | | | | (43) | | | | 124 | (65) | (6) |
| A | 78 | | | | 1470 | 249 | | | 155 | | | | | | 0 | | BOILER OFF | | 155 | 100 | 11 |
| S | 78 | | | | 740 | 115 | | | (57) | | | | | | (121) | | | | 178 | (32) | (8) |
| O | 78 | | | | 520 | 71 | | | 57 | | | | | | 93 | | | | 150 | 38 | 11 |
| N | 78 | | | | 200 | 24 | | | 19 | | | | | | 113 | | | | 132 | 14 | 10 |
| D | 78 | | | | 120 | 4 | | | 25 | 3) | | | | | 189 | | | | 214 | 12 | 21 |
| TOTAL | | | — | — | 10830 | 1645 | | | 915 | | | | | | 1023 | | | | 1938 | — | — |
| AVERAGE | | | | | | | | | | | | | | | | | | | | 47 | 8 |

TOTAL: kWh

1),2),3) : see next page

Notes:( ):calculated on the basis of an estimated heat loss from store. Further notes on next page.

| PROJECT DESCRIPTION | GENTOFTE | REF. | DK | GE | 78 | 79 | 2 |
|---|---|---|---|---|---|---|---|

Main Participants:

OWNER: Private family

DESIGN: (solar system) Institute of Technology and Thermal Insulation Laboratory, Technical University of Denmark

CONSTRUCTION: Private firm

FINANCIAL SUPPORT: Ministry of Commerce (100% for installation and the monitoring programme)

MONITORING: Thermal Insulation Laboratory,
Technical University of Denmark,
phone: (02) 88 35 11
contact: L.S. Joergensen or P. Kristensen

Project Description:

28 $m^2$ of solar collector was installed in 1978 on an existing single-family house from 1937. The 2000 l storage tank was made on site in the basement. The system is for space heating and DHW. The project paid all the installation costs.

Project Objectives:

One objective is to demonstrate the installation of a solar heating system on an existing single-family house. The performance of the different subsystems is monitored and appraised.

Another objective is to validate simulation programmes. This means detailed measurements and demands better accuracy of the monitoring system than the first objective.

The monitoring period is 2 years, starting August 1978.

| SITE LOCATION MAP | REF. | DK | GE | 78 | 79 | 2 |
|---|---|---|---|---|---|---|

Distance to main city: 10 km from Copenhagen City (northward)

Please indicate height of nearest obstructions

20 m south of the house there is another single-family house with ridge lower than the one of the solar house.

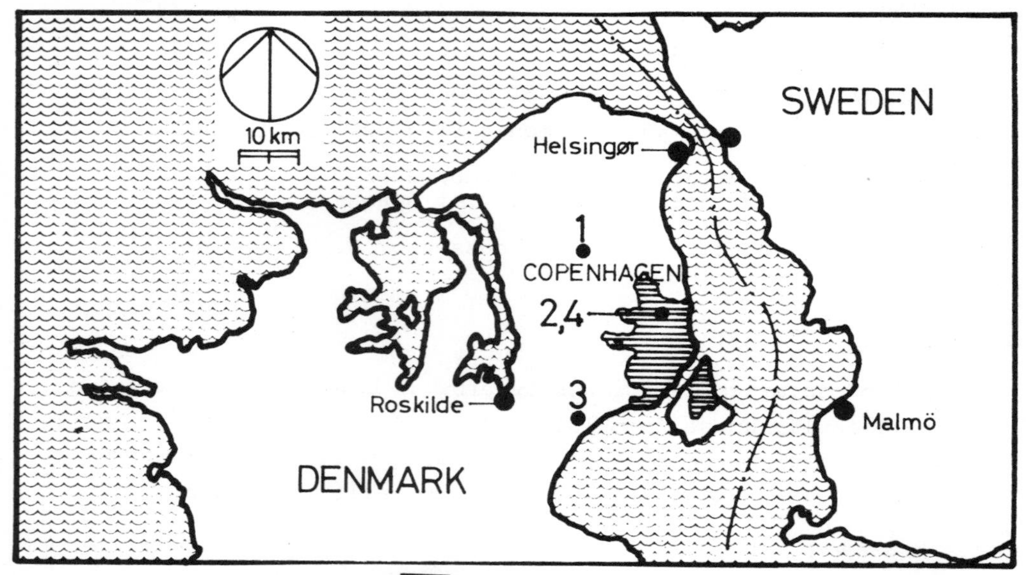

Photograph of Project

**SUMMARY SHEET** GENTOFTE REF. DK GE 78 79 **2**

**Design Data**

**A1 CLIMATE**

1) Source of Data: Danish Test Reference Year
2) Latitude: 55° 45′ N Longitude: 12° 33′ E Altitude: ~ sea level metres
3) Global Irradiation (horiz plane): 1024 kWh/m²·year % Diffuse: 47
4) Degree Days: 2829 Base Temp: 17 °C
5) Sunshine Hours: July: 226 January: 70 Annual: 1580

**B1 BUILDING**

1) Building Type: single-family No Occupants: 4
2) Floor area: 133 *) m² Heated Volume: 390 m³ *)
3) Design Temperature: External: -12 °C Internal: 21 °C
4) Ventilation Rate: a.c.h. Vol. Heat Loss: W/m³·K
5) Space Heat Load: 24,000 kWh **) Hot Water Load: 4,000 kWh **)

*) excluding basement **) preliminary estimate

**C1 SYSTEM**

1) Absorber Type: aluminium fins, steel tubes, black paint
2) Collector Area 28 m² (Aperture) Coolant: ethylene glycol
3) Orientation: 180 Tilt: 45° Glazing: double
4) Storage Volume: 2 m³ Heat Emitters: radiators
5) Auxiliary System: oil fired boiler Heat Pump:

**PERFORMANCE MONITORING**

1) Is there a Computer Model? under construction
2) Start Date for Monitoring Programme August 1978.
3) Period for which results available August 1978 on
4) No. of Measuring Points per house 40
5) Data Acquisition System: Computer-based data logging system (HP 3051 A), data transferred to tape puncher each hour.

| | Space Heating | Hot Water | Total |
|---|---|---|---|
| 6) Predicted | % of kWh | % of kWh | 30 % of 28000 kWh |
| 7)* Measured | % of kWh | % of kWh | 16 % of 22752 kWh (22534) corrected |

8)* Solar Energy Used: Including useful losses 127 kWh/m²·year; Excluding losses: 88 kWh/m²·year

9) *System Efficiency: $\frac{\text{Solar Energy Used}}{\text{Global Irradiation on collector}} \times 100 =$ 14 %

*provisional figures.

A2 LOCAL CLIMATE REF. DK GE 78 79 **2**

1 – Average Cloud cover: 5 Octas

2 – Average daily max temperatures (July): 21.1 °C

Average daily min temperatures (Jan): -6.5 °C

3 Source of Weather Data: Danish Test Reference Year ("Meteorological Data for Design of Building and Installation: A Reference Year", by H. Lund et al, TIL 1974)

4 – Micro Climate/Site Description:

The house is situated within an urban area 10 km north of Copenhagen city. The distance to the Sound (separates Zealand from Sweden) is about 5 kilometers.

The weather observations that form the basis for the Test Reference Year were taken at a site about 10 km north of this house.

There are no obstructions higher than the house itself nearby.

*) Degree days are based on the period 1901 - 1940.

| MONTH | Test Ref. YEAR | Irradiation on horizontal plane | | *) Degree Days | Precipitation | Average Ambient Temperature | Average Relative Humidity | Average Wind Speed | Prevailing Wind Direction |
|---|---|---|---|---|---|---|---|---|---|
| | | Global | Diffuse | Base. 17 °C | | | | | |
| JAN | | 19 | 8 | 509 | 23 | 0.2 | 89 | 5.1 | |
| FEB | | 36 | 16 | 463 | 19 | -0.4 | 85 | 5.8 | |
| MARCH | | 83 | 31 | 433 | 19 | 2.0 | 86 | 5.9 | |
| APRIL | | 122 | 49 | 282 | 27 | 5.7 | 78 | 6.7 | |
| MAY | | 149 | 66 | 45 | 42 | 11.4 | 78 | 6.0 | |
| JUNE | | 170 | 78 | 0 | 38 | 16.0 | 77 | 5.0 | |
| JULY | | 161 | 79 | 0 | 79 | 16.4 | 74 | 3.3 | |
| AUGUST | | 123 | 66 | 0 | 91 | 16.1 | 84 | 4.4 | |
| SEPT | | 83 | 44 | 29 | 59 | 13.7 | 85 | 5.2 | |
| OCT | | 44 | 24 | 236 | 19 | 9.2 | 87 | 3.5 | |
| NOV | | 19 | 12 | 366 | 35 | 5.0 | 91 | 5.0 | |
| DEC | | 15 | 8 | 466 | 40 | -0.4 | 89 | 5.1 | |
| Totals | | 1024 | 481 | 2829 | 491 | | | | |
| Averages | | | | | | 8.0 | 84 | 5.1 | |
| Units | – | $kWh/m^2 \cdot month$ | | °C days | mm | °C | % | m/s | – |

B2 BUILDING DESCRIPTION REF. DK GE 78 79 **2**

**Design Concepts**

Existing single-family house from 1937.
- 1 floor with occupied top storey and basement.
- Pitch roof, 45°, with tiles - south facing side with 28 $m^2$ of solar collector incorporated.

**Construction** (excluding basement)

1 – Glazing: 22 % Area: 29 $m^2$ 'U' value: 3.0 $W/m^2 \cdot K$

2 – 'U' values: Walls: 0.8 $W/m^2 \cdot K$ Floor: $W/m^2 \cdot K$ Roof/loft: 0.6 $W/m^2 \cdot K$

3 – Thermal mass: ~100 $kg/m^3$ Annual Energy Demand: ~28,000 kWh

Walls in lower floor consist of brickwork.
Exterior walls include an insulated cavity.
Walls in first floor are wooden with insulation except for the gables (brickwork).

**Energy Conservation Measures**

All radiators were fitted with thermostats when the solar heating system was installed.

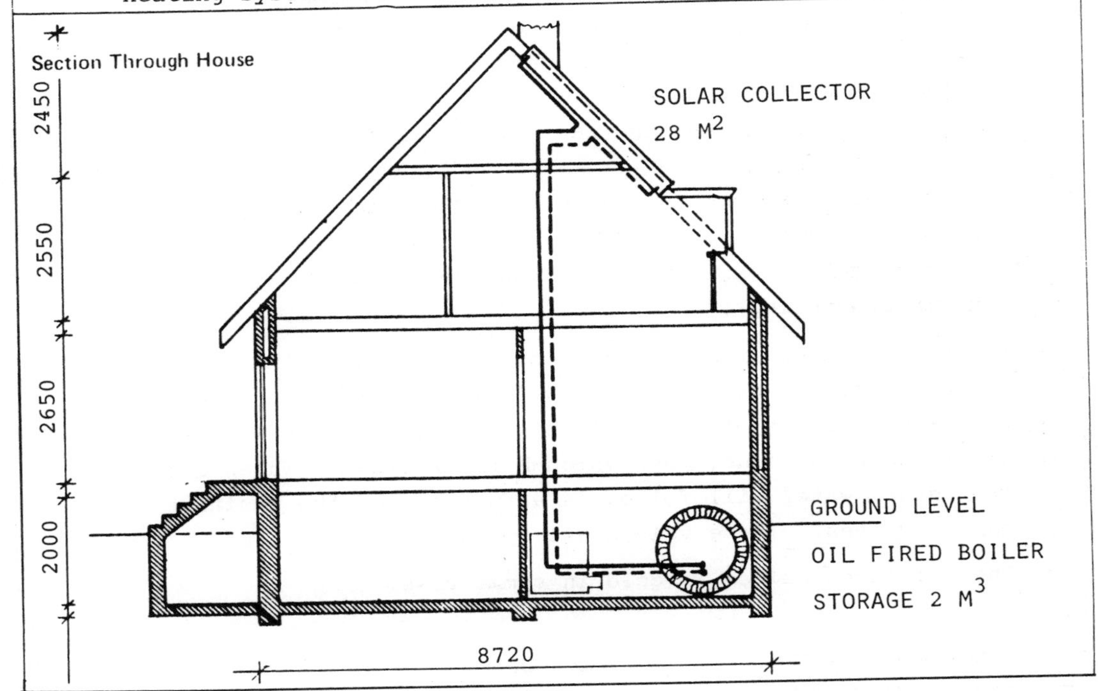

| C2 SYSTEM DESCRIPTION | REF. | DK | GE | 78 | 79 | 2 |
|---|---|---|---|---|---|---|

**Collector/Storage**

1 – Primary circuit. Water content: 5 litres/m$^2$ collector area

2 – Insulation. Collector: 75 mm Storage: 200 mm Pipework: 40 mm

3 – Frost Protection: 30% (vol) ethylene glycol

4 – Overheating Protection: Pump in prim. circuit is automatically operated if storage temperature exceeds 85°C

5 – Corrosion Protection: none

**Space Heating/Domestic Hot Water**

1 – Auxiliary (S.H.): oil fired boiler (DHW): oil fired boiler

2 – Heat Emitters: radiators Temp. Range: 20 - 80°C

3 – DHW Insulation: 80 mm Set Temperature: 55°C

**Other Points/Heat Pump:**

**Control Strategy/Operating Modes**

The pump in the primary circuit is controlled by the use of a differential thermostat including the overheating protection described above.

The pump between storage and DHW 1 ( as in diagram) is controlled by the use of a differential thermostat including the facility, that the pump is never operated if the temperature in DHW 1 exceeds 60°C.

The pump between boiler and DHW 2 is controlled by the use of an absolute thermostat (55°C).

In summertime DHW goes directly from DHW 1 to the taps. In wintertime DHW is postheated in DHW 2. The state is changed by the use of two manual valves. These valves are operated twice a year (autumn and spring).

The return water from the radiators is preheated in the storage. If necessary the preheated water is mixed with water from the boiler by the use of a manual three-way valve. In wintertime the return water does not pass through the storage. Which direction the return water will follow is a function of the position of two manual valves. These valves are operated twice a year.

The oil fired boiler is cold in summertime.

Diagram of System: see following page.

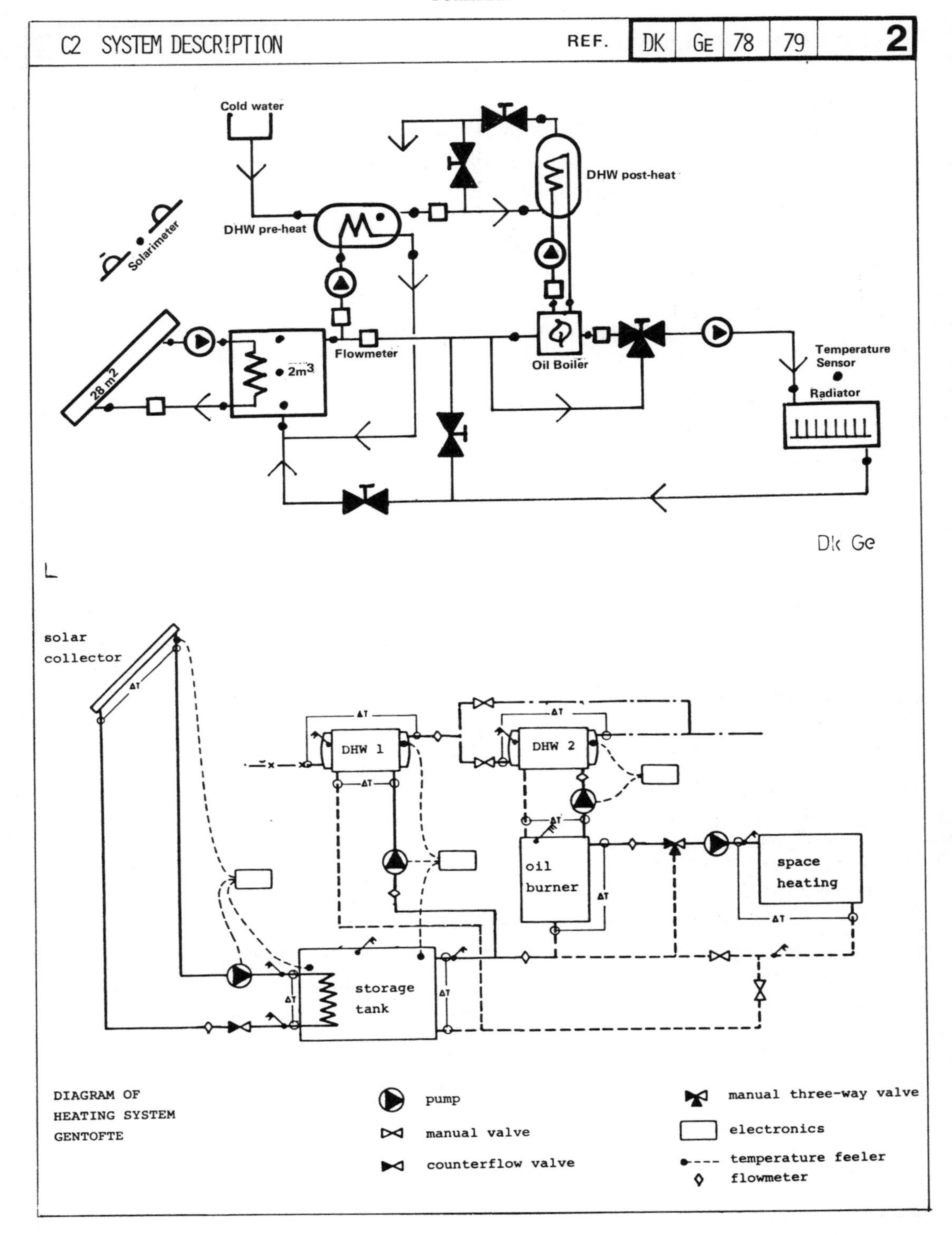
C2 SYSTEM DESCRIPTION
REF. DK GE 78 79 2
Cold water
DHW post-heat
DHW pre-heat
Solarimeter
28 m2
2m3
Flowmeter
Oil Boiler
Temperature Sensor
Radiator
Dk Ge
solar collector
DHW 1
DHW 2
oil burner
space heating
storage tank
DIAGRAM OF
HEATING SYSTEM
GENTOFTE
pump
manual valve
counterflow valve
manual three-way valve
electronics
temperature feeler
flowmeter

| TECHNICAL APPRAISAL/PRACTICAL EXPERIENCE | REF. | DK | GE | 78 | 79 | | 2 |
|---|---|---|---|---|---|---|---|

**System/Design:**

The temperature level in the heating system (radiators) is higher than expected. Supply temperatures up to 80°C have been necessary. This means high return temperatures too, and thereby less contribution of solar to space heating.

**System/Installation:**

Installation of the nine prefabricated solar collectors proved more time - consuming (several weeks) than expected (one week).
The 2000 l storage tank (steel) in the basement had to be made on site. This caused no problems.

**Component Performance:** (Collector, Heat Exchangers, Storage, Pipework, Valves, Fittings, Pumps, Auxiliary)

For several months there have been problems with condensation on the outer glass covering of some of the elements in the collector. The problem seems to be solved after one 8 mm hole has been drilled in each element.
The counter-flow valve in the primary circuit had to be replaced by another one suitable for ethylene-glycol . The original one had rubber bellows, the replacement has nylon chamber with a spring. Before this alteration, considerable heat was lost due to reverse circulation in the nighttime.

**System Operation/Controls, Electronics:**

The pump in the primary circuit had to be replaced by a larger one to give enough flow (1 ℓ/min x $m^2$).

The three-way valve had to be modified to allow the possibility of both 100% solar heating and 100% heating by the boiler. It had originally been incorrectly installed.
At high temperatures of the store, there is a tendency to natural circulation between the store and DHW 1. This means that DHW 1 may exceed 60°C. DHW 1 is situated at a level .5 m above the store.

Apart from this, system operation has been satisfactory.

PERFORMANCE EVALUATION/CONCLUSIONS REF. DK GE 78 79 2

Special remarks, Gentofte.

Heat loss from solar system:

a. The boiler room in the basement contains most of the solar installations. There is no wish to heat the boiler room itself. A heat balance it set up for this room.

b. Heat lost to the remaining rooms in the basement and to the rooms on the ground floor is useful if there is a heating demand (normally defined as "pump in heating system on").

c. Heat loss from the primary circuit goes directly to the house. In the period October - April inclusive, this heat loss is useful.

Heat loss from the boiler and the remaining pipework in the boiler room:

As well as heat loss from the solar system, only a part of this heat loss is useful. Most of the heat loss from the pipework is (for technical reasons with the monitoring system) included in auxiliary heating demand, 100% useful, whereas none of the heat loss from the boiler is included.

Occupants/Response

The family has a general interest in solar energy. Too much goodwill was tested in the construction phase, because construction lasted almost 5 months. Since then the family has been quite happy with the system. The operation of the manual valves has not caused problems.

Additional remarks on sheet P1, DK/GE/78/79/12-79 and DK/GR/78/79/12-79

1. DHW consumption (column no 7 and 13) is shown as output from DHW unit. The heat loss from the DHW preheater has been estimated, and thereby the input to the preheater. The influence of this on kWh/m$^2$.year is shown below.
2. Yearly output in kWh/m$^2$ is:

| | A | A·X | B | B·X |
|---|---|---|---|---|
| Greve | 82 | 93 | | |
| Gentofte | 88 | | 127 | 147 |

A:excluding useful losses (and solar DHW output DHW)
B:including useful losses ( " " " " )
C:solar DHW input DHW unit.

**P1 MONTHLY PERFORMANCE** — REF: DK GE 78 79 2

| Month | Year | No Days Data | Average Temp °C: Internal | Average Temp °C: External | SOLAR ENERGY kWh: Global Irradiation On Collector Area | Solar Energy Collected | Solar Energy Used: Space Heating Useful Losses Incl. | Space Heating Useful Losses Excl. | Domestic Hot Water I) | Input To Heat | Heat Pump Total Output kWh: Space Heating | Domestic Hot Water | AUXILIARY ENERGY kWh |  Space Heating II) | Domestic Hot Water I) | Pumps & Fans | Free Heat Gains kWh: Internal |  | SUMMARY kWh: Total Load | %: Percentage Solar | Syst. Eff. |
|---|---|---|---|---|---|---|---|---|---|---|---|---|---|---|---|---|---|---|---|---|---|
| Column No | | | 1 | 2 | 3 | 4 | 5 | 6 | 7 | 8 | 9 | 10 | 12 | 12 | 13 | 14 | 15 | 17 | 17 | 18 | 19 |
| J | 79 | 21 | 16.3 | -2.2 | 424 | 35 | 3 | 0 | 34 | | | | (3927) | 3950 | 182 | 2 | | (4143) | 4169 | 1 | 9 |
| F | 79 | 25 | 17.3 | -1.0 | 1250 | 254 | 67 | 0 | 99 | | | | (3400) | 3420 | 110 | 12 | not available | (3676) | 3696 | 4 | 13 |
| M | 79 | 31 | 18.0 | 3.2 | 1820 | 273 | 109 | 0 | 167 | | | | (2955) | 2980 | 107 | 11 | | (3338) | 3363 | 8 | 15 |
| A | 79 | 30 | 18.1 | 7.2 | 2830 | 610 | 345 | 140 | 172 | | | | (1370) | 1390 | 31 | 12 | | (1918) | 1938 | 27 | 18 |
| M | 79 | 23 | 19.9 | 12.7 | 3770 | 779 | 312 | 84 | 261 | | | | (399) | 409 | 0 | 18 | | (972) | 982 | 58 | 15 |
| J | 79 | 28 | 22.4 | 18.0 | 4090 | 591 | 38 | 13 | 250 | | | | | 0 | 0 | 15 | | | 288 | 100 | 7 |
| J | 79 | 31 | 19.9 | 16.3 | 3200 | 485 | 50 | 17 | 165 | | | | | 0 | 0 | 11 | | | 215 | 100 | 7 |
| A | 79 | 31 | 21.1 | 17.0 | 3530 | 638 | 43 | 17 | 269 | | | | | 0 | 0 | 16 | | | 312 | 100 | 9 |
| S | 79 | 29 | 19.4 | 14.3 | 2490 | 552 | 299 | 149 | 180 | | | | (128) | 137 | 0 (24) | 22 | | (631) | 640 | 75 | 19 |
| O | 79 | 30 | 18.3 | 9.7 | 2300 | 556 | 365 | 169 | 183 | | | | (814) | 828 | 24 (77) | 17 | | (1439) | 1453 | 38 | 24 |
| N | 78 | 30 | 18.4 | 8.8 | 615 | 116 | 52 | 0 | 79 | | | | (1628) | 1647 | 77 (166) | 7 | | (1925) | 1944 | 7 | 21 |
| D | 78 | 31 | 17.1 | 0.7 | 295 | 42 | 0 | 0 | 59 | | | | (3491) | 3516 | 166 (177) | 4 | | (3727) | 3752 | 2 | 20 |
| TOTAL | | | — | — | 26614 | 4931 | 1683 | 589 | 1918 | | | | (18112) | 18277 | 874 | 147 | | (22584) | 22752 | — | — |
| AVERAGE | | | 18.9 | 8.8 | | | | | | | | | | | | | | | | 16 | 14 |

TOTAL: 2,507 kWh

(Corrected figures in brackets)

Notes: I):output from DHW-unit; II):heat loss from a distribution system in basement inclusive.

| PROJECT DESCRIPTION | GREVE | REF. | DK | GR | 78 | 79 | 3 |
|---|---|---|---|---|---|---|---|

**Main Participants:**

OWNER: Private family.

DESIGN: (solar system) Institute of Technology and Thermal Insulation Laboratory, Technical University of Denmark

CONSTRUCTION: Private firm

FINANCIAL SUPPORT: Ministry of Commerce (100% for installation and the monitoring programme).

MONITORING: Thermal Insulation Laboratory, Technical University of Denmark, phone (02) 88 35 11 contact: Henrik Andersen or L.S. Joergensen.

**Project Description:**

This is a new single-family standard house fitted with 50 $m^2$ of flat-plate collector and a water storage tank of 5 $m^3$. The construction was finished early in 1978. The house has been unoccupied until April 1979.
The solar heating system is expected to cover 50% of the total demand for heating and DHW (about 18,000 kWh).

**Project Objectives:**

There are two main objectives with this project.

The first objective is to demonstrate the use of solar energy in a new standard house. This means to show how the system is installed in the house, to follow the behaviour of the system, and to obtain data on the output of the solar heating system.

The second objective is to validate simulation programmes. This objective means detailed measurements and demands better accuracy that the first objective.

The monitoring period is 2 years, starting July 1978.

| SITE LOCATION MAP | REF. | DK | GR | 78 | 79 | 3 |
|---|---|---|---|---|---|---|

Distance to main city: 25 km from Copenhagen City (south-westward)

Please indicate height of nearest obstructions

At a distance of about 20 m in front of the solar collector there is another single-family house with ridge at the same level.

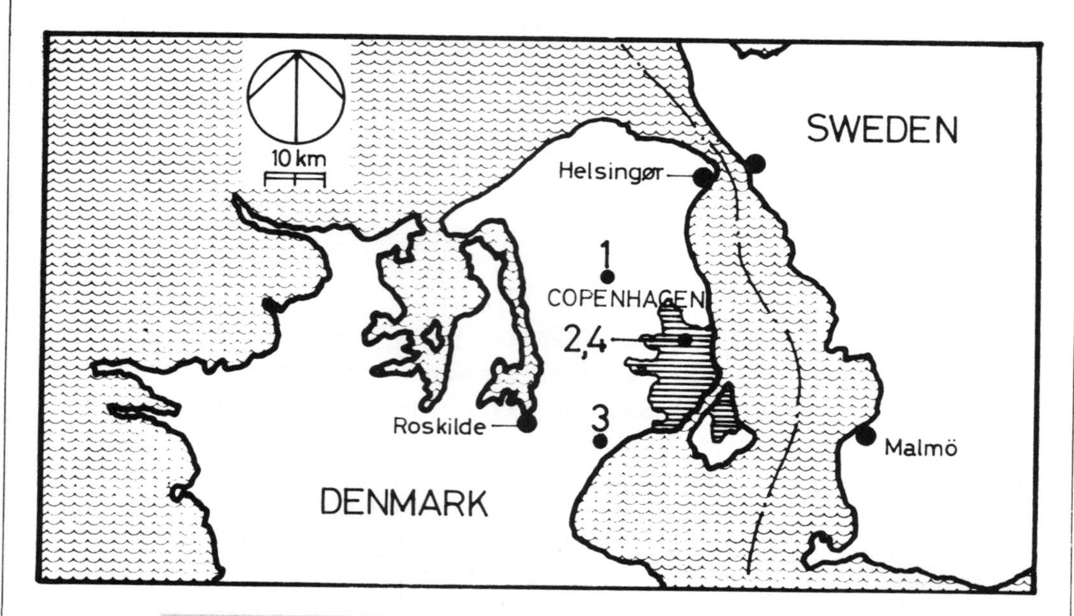

Photograph of Project

**SUMMARY SHEET** GREVE **REF.** DK | GR | 78 | 79 | **3**

**Design Data**

**A1 CLIMATE**

1) Source of Data: Danish Test Reference Year
2) Latitude: 55° 37′ Longitude: 12° 18′ Altitude: ~ sea level metres
3) Global Irradiation (horiz plane): 1024 kWh/m²·year % Diffuse: 47
4) Degree Days: 2829 Base Temp: 17 °C
5) Sunshine Hours: July: 226 January: 70 Annual: 1580

**B1 BUILDING**

1) Building Type: detached house No Occupants: 4 (2 adults + 2 child.)
2) Floor area: 110 m² Heated Volume: 264 m³
3) Design Temperature: External: -12 °C Internal: 21 °C
4) Ventilation Rate: ~ 1.0 a.c.h. Vol. Heat Loss: .78 W/m³·K
5) Space Heat Load: 14,000 kWh Hot Water Load: 4,300 kWh

**C1 SYSTEM**

1) Absorber Type: iron, two sheets with integrated channels, black
2) Collector Area 50 m² (Aperture) Coolant: Ethylene glucol
3) Orientation: 210° Tilt: 38° Glazing: double
4) Storage Volume: 5.0 m³ Heat Emitters: primary: radiators second.: underfloor
5) Auxiliary System: oil fired boiler Heat Pump: none

**PERFORMANCE MONITORING**

1) Is there a Computer Model? Yes. 2) Start Date for Monitoring Programme July 1978.

3) Period for which results available September 1978 on

4) No. of Measuring Points per House 40

5) Data Acquisition System: Computer-based data logging system HP 3051A data transferred to papertape puncher each hour.

| | Space Heating | Hot Water | Total |
|---|---|---|---|
| 6) Predicted | % of kWh | % of kWh | 50 % of 18000 kWh |
| 7) * Measured | 17 % of 15840 kWh | 51 % of 2587 kWh | 22 % of 18437 kWh |

8) * Solar Energy Used: Including useful losses 82 kWh/m²·year; Excluding losses: 82 kWh/m²·year

9) * System Efficiency: $\frac{\text{Solar Energy Used}}{\text{Global Irradiation on collector}} \times 100 =$ 8 %

*provisional data

| A2 LOCAL CLIMATE | REF. | DK | GR | 78 | 79 | 3 |
|---|---|---|---|---|---|---|

1 – Average Cloud cover: 5 Octas

2 – Average daily max temperatures (July) : 21.1 °C

Average daily min temperatures (Jan) : -6.5 °C

3 Source of Weather Data: Danish Test Reference Year ("Meteorological Data for Design of Building and Installation: A Reference Year", by H. Lund et al, TIL 1974).

Basis for the Test Reference Year is weather observations from a site 40 km from Greve.

4 – Micro Climate/Site Description:

The house is situated in a flat area with single-family houses.. There are no obstructions higher than a single-family house within 1 km. The distance to the Sound ("Øresund") is 5 km.

*) Degree days are based on the period 1901-1940.

| MONTH | Test Ref. YEAR | Irradiation on horizontal plane | | *) Degree Days | Precipitation | Average Ambient Temperature | Average Relative Humidity | Average Wind Speed | Prevailing Wind Direction |
|---|---|---|---|---|---|---|---|---|---|
| | | Global | Diffuse | Base. 17 °C | | | | | |
| JAN | | 19 | 8 | 509 | 23 | 0.2 | 89 | 5.1 | |
| FEB | | 36 | 16 | 463 | 19 | -0.4 | 85 | 5.8 | |
| MARCH | | 83 | 31 | 433 | 19 | 2.0 | 86 | 5.9 | |
| APRIL | | 122 | 49 | 282 | 27 | 5.7 | 78 | 6.7 | |
| MAY | | 149 | 66 | 45 | 42 | 11.4 | 78 | 6.0 | |
| JUNE | | 170 | 78 | 0 | 38 | 16.0 | 77 | 5.0 | |
| JULY | | 161 | 79 | 0 | 79 | 16.4 | 74 | 3.3 | |
| AUGUST | | 123 | 66 | 0 | 91 | 16.1 | 84 | 4.4 | |
| SEPT | | 83 | 44 | 29 | 59 | 13.7 | 85 | 5.2 | |
| OCT | | 44 | 24 | 236 | 19 | 9.2 | 87 | 3.5 | |
| NOV | | 19 | 12 | 366 | 35 | 5.0 | 91 | 5.0 | |
| DEC | | 15 | 8 | 466 | 40 | -0.4 | 89 | 5.1 | |
| Totals | | 1024 | 481 | 2829 | 491 | | | | |
| Averages | | | | | | 8.0 | 84 | 5.1 | |
| Units | – | kWh/m²·month | | °C days | mm | °C | % | m/s | – |

**B2 BUILDING DESCRIPTION** REF. DK GR 78 79 **3**

**Design Concepts**

A single-family standard house of typical size and construction. It meets the insulation demands in the new building code (these demands are mandatory from February 1979, and they fix a limit to the heat demand for a house of given geometry).
The loft is unoccupied and contains the storage tank.

**Construction**

1 – Glazing: 32 % Area: 35 $m^2$ 'U' value: 2.2 $W/m^2/°C$

2 – 'U' values: Walls: .27 $W/m^2/°C$ Floor: .40 $W/m^2/°C$ Roof/loft: .29 $W/m^2/°C$

3 – Thermal mass: 60 $kg/m^3$ Annual Energy Demand: 18,300 kWh

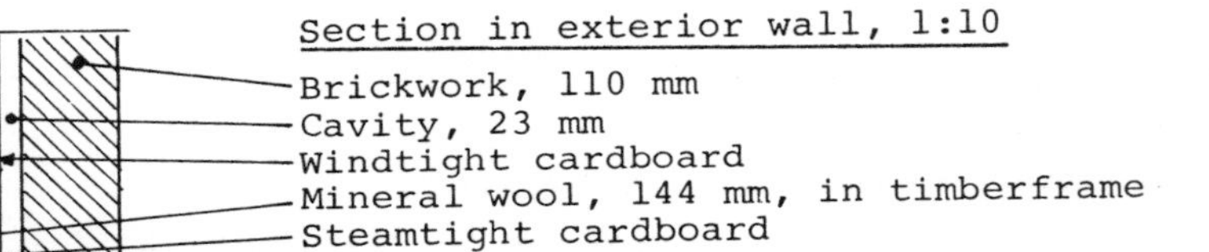


**Energy Conservation Measures** mechanical ventilation with heat recovery.

**Section Through House** Showing main solar components

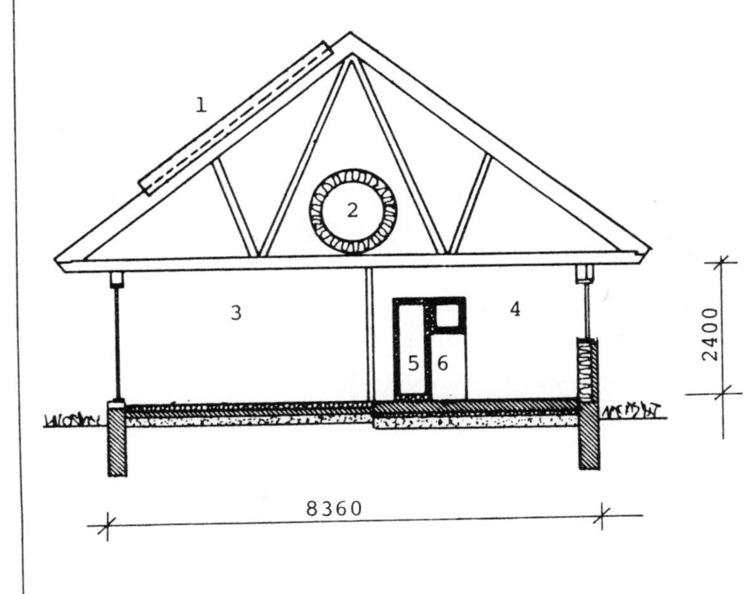


1: Solar collector 50 $m^2$
2: Storage tank 5 $m^3$
3: Sitting-room
4: Scullery
5: Solar preheater tank for DHW
6: Oil fired boiler including a post-heater tank for DHW

| C2 SYSTEM DESCRIPTION | REF. | DK | GR | 78 | 79 | 3 |
|---|---|---|---|---|---|---|

**Collector/Storage**

1 – Primary circuit. Water content: ........ litres/m$^2$ collector area

2 – Insulation. Collector: ........ Storage: 200 mm ........ Pipework: ........

3 – Frost Protection: 30 % ethyleme glycol

4 – Overheating Protection: Pump in prim. circuit is automatically operated if storage temperature exceeds 85 °C.

5 – Corrosion Protection: none

**Space Heating/Domestic Hot Water**

1 – Auxiliary (S.H.): Oil fired boiler (DHW): Oil fired boiler

2 – Heat Emitters: Radiators + floor (secondary) Temp. Range: 20 – 60 °C

3 – DHW Insulation: 50 mm (DHW 1) Set Temperature: 60 °C

**Other Points/Heat Pump:**

**Control Strategy/Operating Modes**

The pump in the primary circuit is controlled by the use of a differential thermostat including the overheating protection described above.

The pump between storage and DHW 1 (preheater - as in diagram) is controlled by the use of a differnetial thermostat including the facility that the pump is never operated if the temperature in DHW 1 exceeds 60 °C.

In summertime DHW goes directly from DHW 1 to the taps. In wintertime DHW is postheated in DHW 2. The state is changed by the use of two manual valves. These valves are operated twice a year (autumn and spring).

The return water from the space heating system is automatically preheated by the storage tank if possible (i.e. return temperature lower than storage temperature).

The supply temperature for the space heating system is governed automatically according to the ambient temperature with a maximum of 60 °C. During nighttime the supply temperature is lowered 20 °C. The temperature in each room is controlled separately by radiator thermostats.

Oil burner + boiler, DHW 2, pump and the valves for ambient temperature compensation is gathered in a unit. DWH 2 is heated by natural circulation.

The boiler and DHW 2 is cold in summertime.

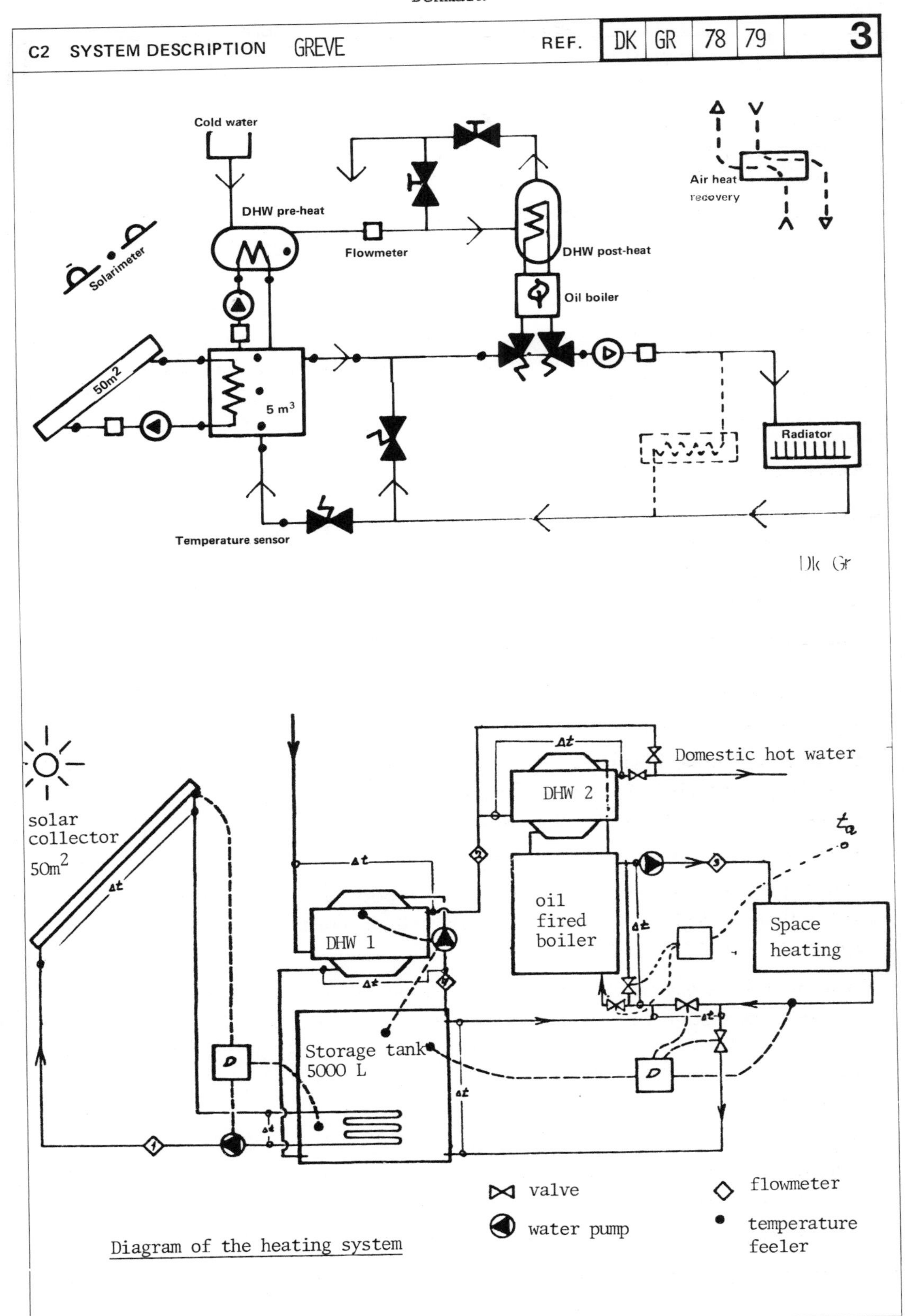


Diagram of the heating system

| TECHNICAL APPRAISAL/PRACTICAL EXPERIENCE | REF. | DK | GR | 78 | 79 | 3 |
|---|---|---|---|---|---|---|

**System/Design:**

The temperature level in the heating system is satisfactorily low (maximum 60 °C). Because the supply temperature is lowered during nighttime, storage temperatures down to 25 °C may be utilized for heating even in wintertime.

**System/Installation:**

The solar collector of 50 $m^2$ is divided into 30 prefabricated elements. This means a considerable piece of work to make the flashing around the elements and leaves many possibilities for leaks. A solution with an integrated collector built on site might have been a better solution.

**Component Performance:** (Collector, Heat Exchangers, Storage, Pipework, Valves, Fittings, Pumps, Auxiliary)

The inner glass covering in two elements of the collector broke, probably due to stagnation temperatures.

The pump in the primary circuit had to be replaced because it made an unacceptable noise.

The counterflow valve in the primary circuit has been found to be inactive. On the other hand, the pipe layout between collector and store forms a "natural counterflow valve" so that the problem of reverse circulation very seldom occurred.

**System Operation/Controls, Electronics:**

In general all the electronics had to be checked carefully before proper function was achieved. For instance, a considerable part of the wiring had to be rearranged.

The electronics governing the supply temperature for the heating system have operated without trouble and is considered to be very useful.

Apart from the initial problems described above, the system operation has been satisfactory.

The number of digits in the monitoring figures does not directly indicate the accuracy of the monitoring system.

PERFORMANCE EVALUATION/CONCLUSIONS REF. DK GR 78 79 3

**Comparison of Measured with Predicted Performance**
(State reasons for any discrepancy and summarise the main factors affecting performance)

Special remarks, Greve.
The proportion of heat that is lost from the solar system within the heated volume of the house is very small. The major part of the solar system, primary circuit and storage, is situated in the unheated roof space. Only the DHW preheater is situated within the heated volume. There is not accounted for any heat loss from the boiler, even though most of it is useful during the heating season.

**Occupants/Response**

The house was unoccupied untill April 1979.

Additional remarks on sheet Pl, DK/GE/78/79/12-79 and DK/GR/78/79/12-79.

1. DHW-consumption (column no 7 and 13) is shown as output from DHW-unit. The heat loss from the DHW-preheater has been estimated, and thereby the input to the preheater. The influence of this on kWh/m$^2$·year is shown below.
2. Yearly output in kWh/m$^2$ is:

| | A | A·X | B | B·X |
|---|---|---|---|---|
| Greve | 82 | 93 | | |
| Gentofte | 88 | | 127 | 147 |

A: excluding useful losses (and solar DHW ~ output DHW)
B: including useful losses ( -"- -"- )
X: solar DHW ~ input DHW-unit

**P1 MONTHLY PERFORMANCE** REF: DK GR 78 79 3

| Month | Year | No Days Data | Average Temp °C Internal | Average Temp °C External | SOLAR ENERGY kWh: Global Irradiation On Collector Area | Solar Energy Collected | Solar Energy Used: Space Heating Useful Losses Incl. | Space Heating Useful Losses Excl. | Domestic Hot Water | Input To Heat | Heat Pump Total Output kWh: Space Heating | Domestic Hot Water | AUXILIARY ENERGY kWh: (corrected) | Space Heating | Domestic Hot Water | Pumps & Fans | Free Heat Gains kWh: Internal | (corrected) | SUMMARY kWh: Total Load | %: Percentage Solar | Syst. Eff. |
|---|---|---|---|---|---|---|---|---|---|---|---|---|---|---|---|---|---|---|---|---|---|
| Column No | | | 1 | 2 | 3 | 4 | 5 | 6 | 7 | 8 | 9 | 10 | 12 | 12 | 13 | 14 | 15 | 17 | 17 | 18 | 19 |
| J | 79 | 29 | 16.4 | -3.2 | 875 | 16 | | -19 | 7 | | | | (2610) | 2480 | 322 | 1 | | (2920) | 2790 | 0 | - |
| F | 79 | 28 | 20.5 | -2.7 | 2640 | 531 | | 93 | 73 | | | | | 2650 | 283 | 16 | | (3099) | 3109 | 6 | 7 |
| M | 79 | 31 | 20.5 | 1.9 | 3030 | 398 | | 175 | 101 | | | | (2662) | 2290 | 268 | 14 | | (3206) | 2834 | 10 | 9 |
| A | 79 | 26 | 21.9 | 5.4 | 5190 | 1210 | | 806 | 140 | | | | (954) | 997 | 219 | 31 | | (2119) | 2162 | 44 | 18 |
| M | 79 | 24 | 22.8 | 14.0 | 7360 | 1390 | | 480 | 236 | | | | (133) | 162 | 83 | 38 | | (932) | 961 | 75 | 10 |
| J | 79 | 30. | 24.0 | 17.0 | 8130 | 1010 | | 177 | 215 | | | | (7) | 14 | 18 | 34 | | (417) | 424 | 92 | 5 |
| J | 79 | 30 | 22.1 | 15.2 | 6720 | 736 | | 155 | 133 | | | | (7) | 0 | 0 | 28 | | (295) | 288 | 100 | 4 |
| A | 79 | 31 | 22.5 | 16.4 | 6470 | 836 | | 67 | 232 | | | | | 0 | 0 | 26 | | (299) | 288 | 100 | 5 |
| S | 79 | 30 | 22.2 | 13.5 | 4700 | 853 | | 263 | 183 | | | | | 11 | 61 | 26 | | | 518 | 86 | 9 |
| O | 78 | 29 | 17.2 | 9.8 | 2760 | 670 | | 570 | 0 | | | | (30) | 40 | 0 | 19 | | (600) | 610 | 93 | 21 |
| N | 78 | 30 | 17.7 | 7.7 | 1150 | 172 | | 19 | 0 | | | | | 1290 | 0 | 8 | | | 1309 | 1 | 2 |
| D | 78 | 31 | 20.3 | -0.2 | 435 | 23 | | 0 | 0 | | | | | 3120 | 13 | 1 | | | 3133 | 0 | 0 |
| TOTAL | | | — | — | 49460 | 7845 | | 2786 | 1320 | | | | (13474) | 13054 | 1267 | 242 | | (18847) | 18437 | — | — |
| AVERAGE | | | | | | | | | | | | | | | | | | | | 22 | 8 |

TOTAL: 4106 kWh

Jan 9-April 1:simulated hot water consumption
Sep 27-Nov 9 (1978):boiler defective (note internal temps)

(Corrected figure in brackets)

House unoccupied until April 1st.
1):all data are corrected for missing days of data; 7,13):output from DHW-unit; 14):electricity for pump in prim. circuit.

| PROJECT DESCRIPTION | LYNGBY | REF. | DK | LY | 76 | 77 | 4 |
|---|---|---|---|---|---|---|---|

Main Participants:

Thermal Insulation Laboratory

Vagn Korsgaard, Professor, M.Sc.
M.R. Byberg, Senior Lecturer, M.Sc.
Torben V. Esbensen, M.Sc.

Institute of Building Design

Knud Peter Harboe, Professor, Architect
Søren Koch, Senior Lecturer, Architect
Klavs Helweg-Larsen, Senior Lecturer, Architect
Inger Nygaard, Civil Engineer, M.Sc.

Laboratory of Heating and Air Conditioning

Peder Kjerulf-Jensen, Senior Lecturer, M.Sc.

Project Description:

The Zero-Energy House was built at the Technical University of Denmark in the spring of 1975 as an experimental house.

From October 1976, to October 1977, the house was inhabited by a so-called standard family cinsisting of two adults and two school-age children.
Measurements were taken in the period 1976-78, after which the house was used as a guest residence for the Technical University.

Project Objectives:

The objectives of the project were to construct a single-family house, which can, for the most pert, be heated by the free heat from the residents, electric utensils etc., and to demonstrate the technical feasibility of utilizing a solar heating system with seasonal heat storage, to supply the necessary additional heat, including heat for domestic hot water, under Danish climatic conditions.

| SITE LOCATION MAP | REF. | DK | LY | 76 | 77 | | 4 |
|---|---|---|---|---|---|---|---|

Distance to main city: 10 km from Copenhagen Downtown

Please indicate height of nearest obstructions: The area is flat, and there are no solar obstructions.

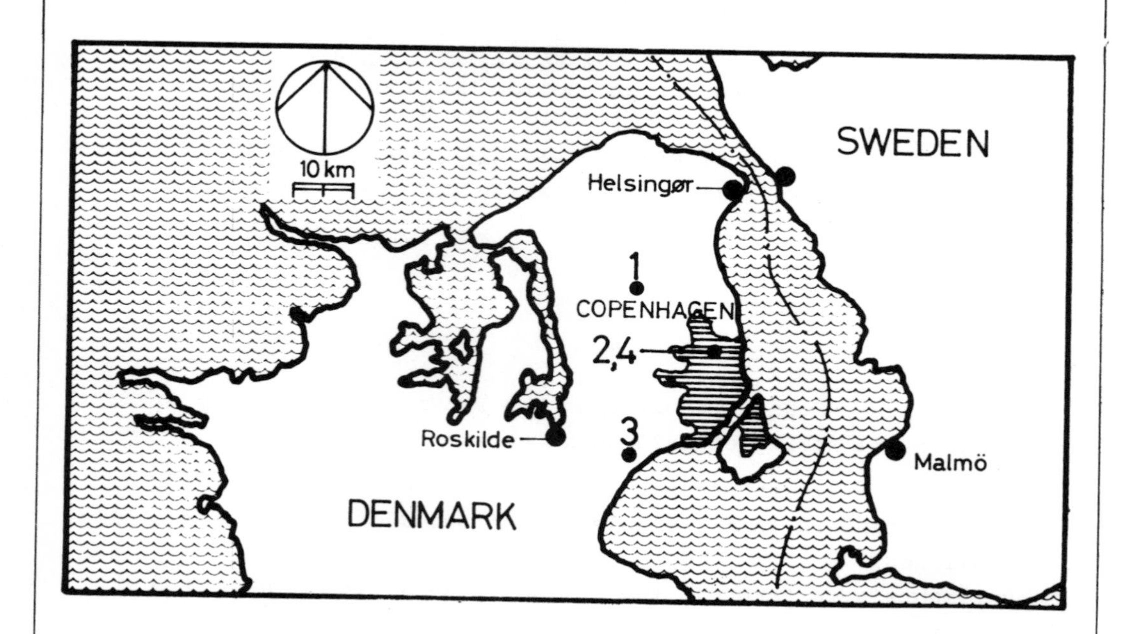

Photograph of Project

**SUMMARY SHEET** LYNGBY REF. DK LY 76 77 **4**

**Design Data**

**A1 CLIMATE**

1) Source of Data: Reference Year
2) Latitude: 55° 47′ N Longitude: 12° 31′ E Altitude: sea level metres
3) Global Irradiation (horiz plane): 1024 kWh/m²·year % Diffuse: 47
4) Degree Days: 2829 Base Temp: 17 °C
5) Sunshine Hours: July: 226 January: 70 Annual: 1580

**B1 BUILDING**

1) Building Type: one family dwelling No Occupants: 4
2) Floor area: 120 m² Heated Volume: 288 m³
3) Design Temperature: External: -12 °C Internal: 20 °C
4) Ventilation Rate: 0.7 a.c.h. Vol. Heat Loss: 0.41 W/m³·K
5) Space Heat Load: 2300 kWh Hot Water Load: 4600 kWh

**C1 SYSTEM**

1) Absorber Type: roll-band steel absorber
2) Collector Area 42 m² (Aperture) Coolant: water
3) Orientation: 180° Tilt: 90° Glazing: double thermopane
4) Storage Volume: 30 m³ Heat Emitters: fan-coil units
5) Auxiliary System: Electric water heating Heat Pump: none

**PERFORMANCE MONITORING**

1) Is there a Computer Model? Yes 2) Start data for Monitoring Programme: January 1976
3) Period for which results available: 1 October 1976 to 30 September 1977.
4) No. of measuring points per house: 90
5) Data Acquisition System: Solatron data logger with 70 channels. The channels registration the operation of the solar heating system and the meteorological data are scanned each 10 mins., the other channels are scanned each hour.

| | Space Heating | Hot Water | Total |
|---|---|---|---|
| 6) Predicted | 100 % of 2300 kWh | 100 % of 2300 kWh | 100 % of 4600 kWh |
| 7) Measured | 41 % of 6060 kWh | 23 % of 3560 kWh | 35 % of 9620 kWh |

8) Solar Energy Used: Including useful losses 79 kWh/m²·year; Excluding losses: 79 kWh/m²·year

9) System Efficiency: $\frac{\text{Solar Energy Used}}{\text{Global Irradiation on collector}} \times 100 =$ 11 %

## A2 LOCAL CLIMATE

REF. DK | LY | 76 | 77 | **4**

1 – Average Cloud cover: 5 Octas

2 – Average daily max temperatures (July) : 21.1 °C

Average daily min temperatures (Jan) : -6.5 °C

3 – Source of Weather Data:

The Reference Year is used as weather data in this description. It is based on Danish meteorological observations through eleven years, and it is built up by typical months from the eleven years period. Each month contains all observations made at one location.

4 – Micro Climate/Site Description:

The house is located on the experimental field of the Technical University in Lyngby, 10 km north of Copenhagen. The area is flat, there are no solar obstructions, and the air is clear of industrial pollution.

| MONTH | Reference YEAR | Irradiation on horizontal plane Global | Irradiation on horizontal plane Diffuse | Degree Days Base: 17°C | Precipitation | Average Ambient Temperature | Average Relative Humidity | Average Wind Speed | Prevailing Wind Direction |
|---|---|---|---|---|---|---|---|---|---|
| JAN | 1961 | 19 | 8 | 509 | 23 | 0.2 | 89 | 5.1 | |
| FEB | 1964 | 36 | 16 | 463 | 19 | -0.4 | 85 | 5.8 | |
| MARCH | 1960 | 83 | 31 | 433 | 19 | 2.0 | 86 | 5.9 | |
| APRIL | 1960 | 122 | 49 | 282 | 27 | 5.7 | 78 | 6.7 | |
| MAY | 1967 | 149 | 66 | 45 | 42 | 11.4 | 78 | 6.0 | |
| JUNE | 1961 | 170 | 78 | 0 | 38 | 16.0 | 77 | 5.0 | |
| JULY | 1963 | 161 | 79 | 0 | 79 | 16.4 | 74 | 3.3 | |
| AUGUST | 1960 | 123 | 66 | 0 | 91 | 16.1 | 84 | 4.4 | |
| SEPT | 1965 | 83 | 44 | 29 | 59 | 13.7 | 85 | 5.2 | |
| OCT | 1962 | 44 | 24 | 236 | 19 | 9.2 | 87 | 3.5 | |
| NOV | 1964 | 19 | 12 | 366 | 35 | 5.0 | 91 | 5.0 | |
| DEC | 1961 | 15 | 8 | 466 | 40 | -0.4 | 89 | 5.1 | |
| Totals | | 1024 | 481 | 2829 | 491 | | | | |
| Averages | | 85 | 40 | 236 | 41 | 8.0 | 84 | 5.1 | |
| Units | | kWh/m²·month | | °C days | mm | °C | % | m/s | – |

B2 BUILDING DESCRIPTION REF. DK LY 76 77 4

### Design Concepts

The house is designed as two dwelling units of 60 $m^2$ each, separated by a glass-roofed atrium of 70 $m^2$. The south facing upper vertical part of the atrium contains a flat-plate solar collector of 42 $m^2$. The two dwelling units are constructed of prefabricated units in the walls, floor and roof.

### Construction

1 – Glazing: 13 % Area: 16 $m^2$ 'U' value: 3.1 $W/m^2 \cdot K$

2 – 'U' values: Walls: 0.14 $W/m^2 \cdot K$ Floor: 0.11 $W/m^2 \cdot K$ Roof/loft: 0.11 $W/m^2 \cdot K$

3 – Thermal mass: 100 $kg/m^3$ Annual Energy Demand: 2300 kWh

### Energy Conservation Measures

The windows are provided with insulated shutters to increase the insulation value during the night. The house is very air-tight, infiltration is measured to 0.15 air changes per hour. The fresh air ventilation system is equipped with a heat recovery device on the opposing current principle.

Free Heat Gains: Internal: 8300 kWh/year Direct Solar: 3700 kWh/year

### Section Through House Showing main solar components

1: 42$m^2$ Solar collector
2: Circulation pump
3: Heat storage tank, 30 $m^3$
4: Heating unit in each room
5: Heat recovery unit

**C2 SYSTEM DESCRIPTION** REF. DK LY 76 77 **4**

**Collector/Storage**

1 – Primary circuit. Water content: 2.7 litres/m² collector area

2 – Insulation. Collector: 0.25 Storage: 0.60 Pipework: 0.05

3 – Frost Protection: automatic collector draining

4 – Overheating Protection: automatic collector draining

5 – Corrosion Protection: inhibitor ELEMENTIN

**Space Heating/Domestic Hot Water**

1 – Auxiliary (S.H.): electric water heater (DHW): electric water heater

2 – Heat Emitters: fan-coil units Temp. Range: 40° (SH) 45° (DWH)

3 – DHW Insulation: Set Temperature: 45°C

**Control Strategy/Operating Modes**

When the empty collector under exposure to radiation reaches a temperature, $t_1$, equal to the temperature in the storage tank, $t_4$, the control system starts. It runs through a cycle which consists of the following intervals:

1. Filling of the collector for 10 min., fixed value (the lifting pump $P_1$ is at work).
2. Stabilization of the temperatures for 10 min., fixed value (the circulation pump $P_2$ is at work).
3. Working period. Lasts as long as the sun can supply energy to the system. ($t_2 > t_3$).
4. Delayed emptying of the collector for 10 min., fixed value.
5. Emptying and awaiting the next start. (Both pumps are stopped, and the magnetic valve MV is open, so that the collector can be emtied by supply of air from the top of the tank to the top of the collector).

**Diagram of System** Showing sensor locations, power ratings, storage volumes & flows.

TECHNICAL APPRAISAL/PRACTICAL EXPERIENCE REF. DK LY 76 77 **4**

**System/Design:**

The solar energy system was designed to cover the total heat requirement for space heating and hot water supply during the whole year. It is our appraisal that it is very difficult and very expensive to perform a 100% solar house. It is much less expensive to design a solar system which covers 80-90% of the total heat requirement.

**System/Installation:** During the installation of the solar collector, when no water was circulating through the absorber, the temperatures in the collector got very high. The temperature difference between the inner and outer glass of the collector was measured to 60°C with the result that the inner glass was broken in almost 25% of the collector area. Therefore it is very important that the cover glass is mounted in such a way that it is possible for the glass to move under the varying temperatures.

**Component Performance:** (Collector, Heat Exchangers, Storage, Pipework, Valves, Fittings, Pumps, Auxiliary)

The accumulator is designed as a cylindrical steel tank, 2.5 m in diameter and 6.5 m long with a volume of 30 $m^3$. The tank is insulated with 60 cm of mineral wool. The ground water level is far below the tank bottom. To prevent rain water from penetrating the insulation, an earth-covered roof is built on the top of the insulation separated by a mechanically ventilated air space.

In the first year of operation the heat loss from the accumulator has been about twice as big as calculated. The cover of the 400 litres DHW-storage tank, which is placed directly in the accumulator, has been leaking due to temperature changes, and the surrounding insulation got wet. For this reason we must warn against bringing a similar system under pressure before the highest temperatures have been reached, and the bolts of the cover have been tightened once more.

**System Operation/Controls, Electronics:**

The control system ("Solar Energy Control") developed by Danfoss Ltd has been operating for the last four years without problems at all.

| PERFORMANCE EVALUATION/CONCLUSIONS | REF. | DK | LY | 76 | 77 | 4 |
|---|---|---|---|---|---|---|

**Comparison of Measured with Predicted Performance**
(State reasons for any discrepancy and summarise the main factors affecting performance)

With the energy conservation arrangements in the house the heat requirement for space heating was calculated to 2300 kWh per year. During the period October 1976 - October 1977, when the house was accupied by a typical family, the heat requirement, however, was measured to 6000 kWh. The main reason for this difference is an unintentional infiltration, a higher rate of mechanical ventilation and a bigger heat transmission loss than originally calculated, especially through the insulating window shutters.
The solar heating system was dimensioned to cover the heat requirement for space heating and hot water supply during the whole year. During the 12 months occupied period the solar radiation has only been about 70% of the solar heat gain during the corresponding period in the Reference Year. Due to this reduction in solar radiation and due to the bigger heat requirement, the solar heating system has covered only 41% (2500 kWh) of the heat requirement for space heating (6000 kWh) and 28% (1000 kWh) of the hot water supply (3500 kWh).

**Modifications to System**

There has been no modification to the system since the operation started in 1975. It might be an idea to install a heat pump in the system to use the temperature range below 40°C in the storage tank.

**Occupants/Response**

It was very interesting to see if the family would feel comfortable in this high-insulated, very tight house during the winter period.
However the family has been very satisfied with the indoor climate. The ventilation system has supplied the house with a sufficient amount of fresh air, and it has not at any moment been necessary opening the windows to ventilate the room. The only thing which is different by living in the Zero Energy House is to remember closing and opening the insulating shutters.

**Conclusions**

The measured performance shows that it has not been possible to design a 100% solar house (a zero energy house). However, the total heat demand from auxiliary has been reduced from 20,000 kWh per year for a normal insulated house to 6,000 kWh for this Zero Energy House. It might have been a better solution to reduce the storage volumen to 5-10 $m^3$ and install a heat pump in the solar heating system.

**Future Work**

Since this Zero Energy House was constructed in 1974 as the first solar energy house in Northern Europe, several solar assisted low-energy houses have been constructed in Denmark. In 1977 nine low-energy houses were constructed at one site in Skive, Jutland, and in 1978 six low-energy houses were constructed in Lyngby near the Technical University. In this project a low energy house is defined as a 120 $m^2$ one-family dwelling with a maximum energy supply of 5000 kWh/year covering space heating and hot water supply for domestic purposes.

## P1 MONTHLY PERFORMANCE

REF: DK LY 76 77 4

| Month | Year | No Days Data | Average Temp °C: Internal | Average Temp °C: External | SOLAR ENERGY kWh: Global Irradiation On Collector Area | Solar Energy Collected | Solar Energy Used: Space Heating Useful Losses Incl. | Space Heating Useful Losses Excl. | Domestic Hot Water | Input To Heat | Heat Pump Total Output kWh: Space Heating | Domestic Hot Water | AUXILIARY ENERGY kWh: Heat Pump | Space Heating | Domestic Hot Water | Pumps & Fans | Free Heat Gains kWh: Internal | Direct Solar | SUMMARY kWh: Total Load | %: Percent-age Solar | Syst. Eff. |
|---|---|---|---|---|---|---|---|---|---|---|---|---|---|---|---|---|---|---|---|---|---|
| Column No | | | 1 | 2 | 3 | 4 | 5 | 6 | 7 | 8 | 9 | 10 | 11 | 12 | 13 | 14 | 15 | 16 | 17 | 18 | 19 |
| J | 77 | 31 | 19.0 | -1.3 | 600 | 150 | 0 | 0 | 0 | | | | | 1130 | 390 | 2 | 782 | 75 | 1520 | 0 | 0 |
| F | 77 | 28 | 19.6 | -0.9 | 1360 | 400 | 0 | 0 | 0 | | | | | 980 | 320 | 5 | 682 | 170 | 1300 | 0 | 0 |
| M | 77 | 31 | 19.9 | 2.3 | 2050 | 500 | 450 | 450 | 0 | | | | | 350 | 300 | 8 | 742 | 270 | 1100 | 41 | 22 |
| A | 77 | 30 | 20.6 | 3.7 | 2650 | 650 | 500 | 500 | 0 | | | | | 200 | 310 | 12 | 722 | 350 | 1010 | 50 | 19 |
| M | 77 | 31 | 24.3 | 11.4 | 4350 | 1400 | 50 | 50 | 80 | | | | | 0 | 260 | 21 | 732 | 560 | 390 | 33 | 3 |
| J | 77 | 30 | 26.2 | 16.3 | 3450 | 750 | 0 | 0 | 150 | | | | | 0 | 140 | 20 | 575 | 450 | 290 | 52 | 4 |
| J | 77 | 31 | 24.2 | 16.1 | 3450 | 650 | 0 | 0 | 40 | | | | | 0 | 0 | 14 | 502 | 450 | 40 | 100 | 1 |
| A | 77 | 31 | 24.9 | 15.5 | 4050 | 1000 | 0 | 0 | 280 | | | | | 0 | 0 | 18 | 635 | 520 | 280 | 100 | 7 |
| S | 77 | 30 | 23.1 | 11.3 | 4150 | 850 | 0 | 0 | 280 | | | | | 200 | 0 | 15 | 652 | 540 | 480 | 40 | 7 |
| O | 76 | 31 | 21.2 | 7.1 | 1650 | 200 | 400 | 400 | 0 | | | | | 0 | 290 | 1 | 702 | 80 | 690 | 58 | 3 |
| N | 76 | 30 | 20.3 | 3.9 | 1250 | 300 | 700 | 700 | 0 | | | | | 0 | 360 | 5 | 802 | 160 | 1060 | 66 | 56 |
| D | 76 | 31 | 19.0 | -1.6 | 600 | 100 | 400 | 400 | 0 | | | | | 700 | 360 | 2 | 772 | 75 | 1460 | 27 | 67 |
| TOTAL | | | — | — | 29610 | 6950 | 2500 | 2500 | 830 | | | | | 3560 | 2730 | 123 | 8300 | 3700 | 9620 | — | — |
| AVERAGE | | | 21.9 | 7.0 | | | | | | | | | | | | | | | | 35 | 11 |

TOTAL: 3,330 kWh

| PROJECT DESCRIPTION | ARAMON (GARD-FRANCE) | REF. | F | Ar | 76 | 77 | 5 |
|---|---|---|---|---|---|---|---|

**Main Participants:**

RESEARCH/ENGINEERING: "ADE', Research Department. Direction des Etudes et Recherches de Electricité de France (EDF). EDF Les Renardiéres. Route de Sens. Ecuelles. 77250 Moret-sur-Loing

BUILDING MANAGEMENT: Region d'Equipment Alpes-Marseille (EDF). Ministere de l'Environnement, 140 Av Viton, 13000 Marseille.

ARCHITECT: M Chouleur. 6 rue Fresque, 30000. Nimes.

FUNDING AUTHORITIES: Agence Nationale de Valorisation de la Recherche (ANVAR), 13 rue Madeleine - Michelis. 92522. Neuilly-Seine, Societé pour l'Edification de logements Economique (SELEC) 4 place Raoul-Dautry. 75015 Paris. Societé Centrale Immobiliese de la Caisse des dépots (SCIC). Direction regionale 18d Bonrepos 31000 Toulouse. Ministere de la qualité de la vie. Sofretes. Zone Industrielle. d'Amilly 45203. Montargis.

**Project Description:**

The Aramon estate was built in 1974/75 and consists of twenty private houses for EDF staff. There are five solar houses* and fifteen others built to the same insulation standards, but electrically heated. They have all been occupied since 1975. The solar heating systems consist of water cooled flat plate collectors facing due south, water storage, and an underfloor heating system. Auxiliary heating is provided by electricity.

**Project Objectives:**

1. To study the possibility of using solar energy to provide part of the space and water heating requirements of houses insulated to normal standards.

2. To assess possible mass market applications of solar systems.

3. To obtain average measurements from the five solar houses and compare these with the fifteen non-solar reference houses.

4. To reach a correct estimate of the relative performance of various sub-systems in the solar energy system.

5. To estimate the accuracy of a simplified model used for calculations and of various computer programs prepared by ADE.

* Of the five solar houses, two have floor areas of 150m$^2$, G factor 1.3 W/m3K. Collector Area 38m$^2$ and three have floor areas of 130m$^2$, G factor 1.1 W/m$^3$K. Collector area 32m2. For this report, one of the 150m2 floor area houses has been chosen.

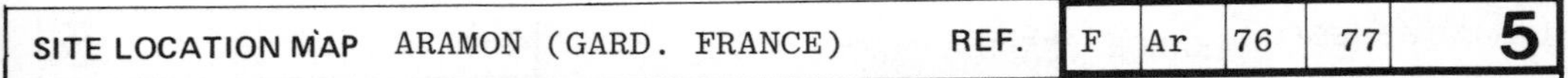

| SITE LOCATION MAP | ARAMON (GARD. FRANCE) | REF. | F | Ar | 76 | 77 | 5 |
|---|---|---|---|---|---|---|---|

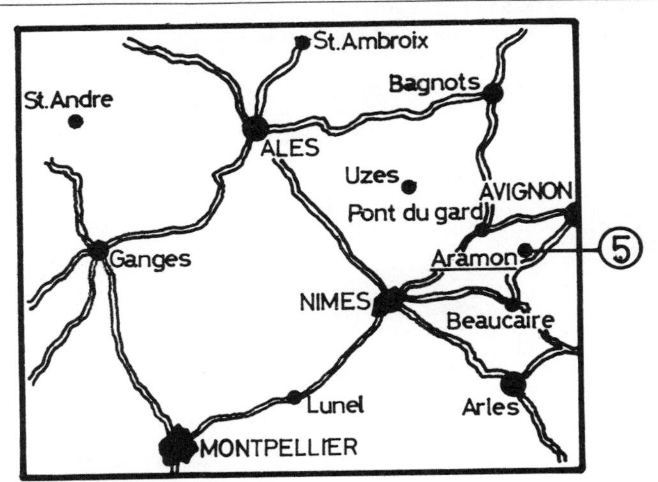


Please indicate height of nearest obstructions None

## A2 LOCAL CLIMATE

Average cloud cover:4.3 octas (1976/1977)
Average daily max temps (July):32.6°C ; average daily min temps (Jan): -1.05°C
Source of weather data: The meteorological data used for the original design refer to Montpellier station - 1961 (ref M*). This year was characterised by a mean temperature slightly below the average (from 1931 - 1960). The solar global radiation was computerised integrating meteorological data about the cloud cover ratio (ref M*).
Micro climate/site description: meteorological data from the Aramon station (ref A) has been available since 1976. The site has no particular microclimatic characteristic, except a mean wind velocity a little higher than the Montpellier station. There is close agreement for the mean outside temperature at both stations.

**SUMMARY SHEET** ARAMON (GARD-FRANCE) **REF.** F | Ar | 76 | 77 | **5**

**Design Data**

**A1 CLIMATE**

1) Source of Data: Met Station. Montpellier. Gard. 80km s.w of site

2) Latitude: 43 ° 37 ′ Longitude: 3 ° 54 ′ E Altitude: 35 metres

3) Global Irradiation (horiz plane): 1525 kWh/m$^2$·year % Diffuse: 21

4) Degree Days: 1788 Base Temp: 18 °C Ref Year 1961

5) Sunshine Hours: July: 324 January: 100 Annual: 2700

**B1 BUILDING**

1) Building Type: Detached House. 650 kg/m$^3$ No Occupants: 5

2) Floor area: 150 m$^2$ Heated Volume: 393 m$^3$

3) Design Temperature: External: −5 °C Internal: 20 °C

4) Ventilation Rate: 1 a.c.h. Vol. Heat Loss: 1.3 W/m$^3$·K

5) Space Heat Load: 18,200 kWh Hot Water Load: 2190 kWh

**C1 SYSTEM**

1) Absorber Type: Aluminium Roll Bond. Black Anodised

2) Collector Area 37.6 m$^2$ (total) Coolant: Water & Additives

3) Orientation: 180$^o$ Tilt: 90$^o$ Glazing: Double Plexiglass

4) Storage Volume: 4 m$^3$ Heat Emitters: Underfloor

5) Auxiliary System: Electric Fan Convectors 10kW. Electric for DHW Heat Pump:

**PERFORMANCE MONITORING**

1) Is there a Computer Model? Yes, two methods. 2) Start date for Monitoring Programme: July 1975

3) Period for which results available: 1976/77, 1977/78, 1978/79.

4) No. of Measuring Points per house: 24

5) Data Acquisition System: 12 sensors (11pt) connnected to Chart Recorder (Potentiometric Paper) and 12 (2pt) to a tape recording system. Also electricity consumption and power (SH & DHW) are taped.

| | Space Heating | Hot Water | Total |
|---|---|---|---|
| 6) Predicted | 48 % of 18182 kWh | 35 % of 2190 kWh | 46.6 % of 20372 kWh |
| 7) Measured 1976/77 | 38.7 % of 15645 kWh | 24.3 % of 1442 kWh | 37 % of 17087 kWh |

8) Solar Energy Used: Including useful losses kWh/m$^2$·year
Excluding losses: 170.4 kWh/m$^2$·year

9) System Efficiency: $\frac{\text{Solar Energy Used}}{\text{Global Irradiation on collector}} \times 100 =$ 17 %

| A2 LOCAL CLIMATE ARAMON (GARD.FRANCE) | REF. | F | Ar | 76 | 77 | 5 |
|---|---|---|---|---|---|---|

| | | Ref M* | | Ref A* | | Ref M* | Ref M | Ref A | |
|---|---|---|---|---|---|---|---|---|---|
| MONTH | YEAR | Irradiation on horizontal plane Global | Diffuse | Degree Days Base 18 °C | Precipitation | Average Ambient Temperature | Average Relative Humidity | Average Wind Speed | Prevailing Wind Direction |
| JAN | 77 | 47.2 | 13.4 | 358 | 73.2 | 5.3 | 77.9 | 3.8 | |
| FEB | 77 | 70.6 | 14.3 | 243 | 37.4 | 9.9 | 72.8 | 3.2 | N |
| MARCH | 77 | 141.5 | 20.3 | 216 | 86.6 | 10.7 | 63.0 | 3.6 | N.NE |
| APRIL | 77 | 141.1 | 31.1 | 155 | 23.2 | 14.5 | 72.3 | 4.1 | N.NW |
| MAY | 77 | 175.3 | 42.7 | 57 | 175.3 | 16.7 | 62.8 | 3.5 | NW |
| JUNE | 76 | 213.7 | 44.2 | 0 | 0.1 | 21.0 | 62.3 | 3.4 | S |
| JULY | 76 | 220.9 | 44.6 | 0 | 18.9 | 22.3 | 59.4 | 3.9 | S.SE |
| AUGUST | 76 | 192.5 | 38.4 | 0 | 56.9 | 21.7 | 59.9 | 3.2 | NE |
| SEPT | 76 | 136.8 | 28.5 | 0 | 141.0 | 21.2 | 72.8 | 3.7 | SE |
| OCT | 76 | 87.9 | 21.3 | 139 | 359.4 | 15.3 | 78.5 | 3.2 | |
| NOV | 76 | 52.8 | 14.9 | 273 | 62.1 | 9.5 | 75.4 | 5.1 | N |
| DEC | 76 | 44.6 | 12.4 | 347 | 73.8 | 7.7 | 73.1 | 3.8 | N |
| Totals | | 1524.8 | 326.1 | 1788 | 1107.9 | | | | |
| Averages | | | | | | 14.7 | 69.2 | 3.7 | N |
| Units | – | kWh/m²·month | | °C days | mm | °C | % | m/s | – |

## B2 BUILDING DESCRIPTION

### Design Concepts

A typical solar house in the Aramon Estate (called P7) is on two floors above the basement. It consists of seven main rooms - on the ground floor, kitchen and reception, with a large terrace, bathroom and three bedrooms; on the first floor, another bathroom and two bedrooms; in the basement, garage places and storage (solar) tank, heat exchangers etc.

### Construction

Glazing:10% (26m$^2$ area); 'U' values (W/m$^2$.K):3.33 (glazing), 0.4 (walls), 0.61 (floor), 0.25 (roof/loft) ; thermal mass:650 kg/m$^3$ ; annual energy demand:20,372 kWh.

Roof : ceilings insulated with 15cm fibreglass ; Walls : 27.5cm concrete blocks and 6cm insulation ; Floor : 17cm 'Stiplu' wooden mat and 14cm concrete.

### Energy Conservation Measures

The houses are well-insulated, to a similar standard required for houses with only electric energy. Windows are double glazed and include shutters.

| C2 SYSTEM DESCRIPTION ARAMON (GARD.FRANCE) REF. | F | Ar | 76 | 77 | 5 |
|---|---|---|---|---|---|

**Collector/Storage**

1 – Primary circuit. Water content: Absorber and pipework 5 litres/$m^2$ collector area

2 – Insulation. Collector: 3 cm Storage: 15cm Pipework: 3cm

3 – Frost Protection: Ethylene - Glycol

4 – Overheating Protection: Primary pump switched on. If necessary store depleted

5 – Corrosion Protection: 1.5% sodium Benzoate inhibitor

**Space Heating/Domestic Hot Water**

1 – Auxiliary (S.H.): Elec.fan convectors. 10kW (DHW): 250L. 2.5kW electric immersion

2 – Heat Emitters: Underfloor Temp. Range: 20°C - 50°C

3 – DHW Insulation: 10 cm Set Temperature: 60°C (design)

**Other Points/Heat Pump:** DWH is pre-heated by a 100 litre tank inside the main store. Collectors are connected in parallel with 2 sets of 6.

**Control Strategy/Operating Modes**

A differential thermostat operates the primary pump if $\Delta T > 5^{\circ}C$ and stops it if $\Delta T \simeq 0^{\circ}C$. A non-return valve prevents reverse thermocirculation at night. The heat exchanger inside the store is type CIAT Rx 3/1600T. 14kW.

A 2-level thermostat in the living room operates the secondary (underfloor system) pump at 21°C, and auxiliary fan convectors at 19°C. The living room temperature is adjustable so that solar energy, if available, is always given priority. Other room auxiliary convectors have adjustable thermostats.

**Diagram of System** Showing sensor locations, power ratings, storage volumes & flows.

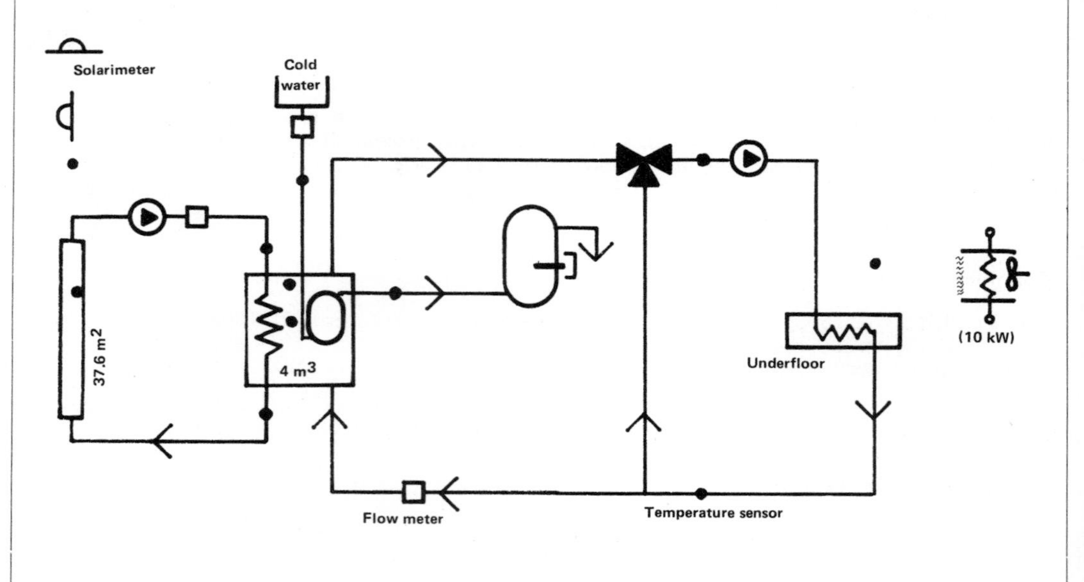

TECHNICAL APPRAISAL/PRACTICAL EXPERIENCE REF. | F | Ar | 76 | 77 | **5**

**System/Design:**

A simplified model, using abacuses for useful solar energy, was built up to estimate the global energy balance of solar house; and some 'solar sub-routines' were elaborated to be connected with the building thermal simulation programmes then available. The few balances checked during the first heating period over 1975-76 happened to be without significance because of some 'youthful errors'.

**System/Installation:**

On several occasions the primary water circuit had been leaking at the connection rubber pipes between solar panels and connecting water tubes. Clips had to be tightened up again and some primary fluid added - It is highly recommended to pay close attention (from the very beginning) to the regulation of all thermostats (operating the pumps) and also to the location of the Pt sensors (for diff. temp) because the whole subsequent energy consumption is strongly dependent on them.

**Component Performance:** (Collector, Heat Exchangers, Storage, Pipework, Valves, Fittings, Pumps, Auxiliary)

- No corrosion has been noticed, but additives ratios (for primary fluid) were carefully checked. Some plexiglass collector glazings were getting out of shape because of dilations (flexed up to 5cm in the middle; some glazings split lengthways) expansion seals were replaced. But optical characteristics remained unchanged.
- Energy losses throughout the storage tank and distribution pipes (however well insulated) were very much heavier than expected (about 34% instead of 20%). Maybe it would have been better having storage and main pipes inside the heated space...
- The DHW preheating heat exchanger (consisting of one 100l tank inside the storage tank) appeared to have much too low efficiency. It would be preferable to have a separate DWH tank in summer for storage (but with the same collectors) or rather to have an independent DHW solar system, with its own collectors but with a better tilt angle (eg, 45°)

**System Operation/Controls, Electronics:**

With some exceptions (*), no important troubles occurred in the whole monitoring equipment, for the sensors or for the recording and independent data analysis computerized systems.

Moreover, it is evident that a correct balance of the whole experiment cannot be reached without paying close attention to the good working order of all sub-systems, especially thermostats regulations, thermometers and pumps.

(*) Some failures happened in the tape recording system (in Summer 1976) but the chart recorder (potemtiometric paper) was found very reliable.

| PERFORMANCE EVALUATION/CONCLUSIONS | REF. | F | Ar | 76 | 77 | 5 |
|---|---|---|---|---|---|---|

**Comparison of Measured with Predicted Performance**
(State reasons for any discrepancy and summarise the main factors affecting performance)

- Vertical: Incident radiation results were found to be just 1% below the computed ones (yearly average), 3% below (heating period average) and 2% above (summer average); but for shorter averaging periods, eg, 1 month), the results were far different from -28% to + 15%).
- Original estimates about heating loads were 7% low, probably because the ventilation rates and the room temperature really required proved higher than the assumed ones; since 1976, the predicted calculations were revised (see the 'summary sheet'). The yearly average of solar energy stored (computed with the very short and accurate abacus method) was within 1% of results, but solar energy used in effect has been predicted 48% too high (1) underestimated storage and pipe losses, especially in summer, and very low solar DHW performance). In connection with it, the storage tank temp. was quite well computed except in mid-summer (high relative value of storage losses(*).

Modifications to System: **Anodised absorbers replaced by matt black**

Data from 1977-78 have led to quite similar yearly energy balances. New leaks in the primary circuit (over 1976-77 and 77-78) caused the addition of a topping-up automatic device (with the right additives ratios).

(*) Analysis of data from 1976-77 also led to some valuable comparisons between solar houses and 'all electric' houses, after results have been averaged over many similar houses (other variables recalculated to be equal). The first were 33% below for heating energy consumption, and 24% below for all energy uses. And the solar energy appeared to reach 38.5% of the global heating energy required (if the global energy required is recalculated to be equal).

Occupants/Response

Although employed in the neighbouring EDF power station, they were quite uninformed about solar energy. Since the very beginning of experiments, they have persisted in keeping mean room temperature above 20.75oC and in nearing 70oC for DHW regulation.

Conclusions

After this two year experiment, we may conclude that the energy balance (2) will be considered slightly disappointing ... If it had to be done again, many improvements would have to be achieved especially concerning storage losses as well as the sizes and locations of heat exchanges (to begin, with the DHW one)

Future Work

The complete experimenting and recording measurements were stopped in September 1978. Afterwards only electric auxiliary energy consumption (for SH and DHW) are to be checked yearly.

(1) But solar 22% too high if this prediction is compared with results averaged over the five similar solar houses; and the DHW average performance was 16% of the 35% expected.

(2) Solar contribution.

## P1 MONTHLY PERFORMANCE

REF: F | Ar | 76 | 77 | 5

| Month | Year | No Days Data | Average Temp °C: Internal | Average Temp °C: External | SOLAR ENERGY kWh: Global Irradiation On Collector Area | Solar Energy Collected | Solar Energy Used: Space Heating Useful Losses Incl. | Space Heating Useful Losses Excl. | Domestic Hot Water | Input To Heat | Heat Pump Total Output kWh: Space Heating | Domestic Hot Water | AUXILIARY ENERGY kWh: Heat Pump | Space Heating | Domestic Hot Water | Pumps & Fans | Free Heat Gains kWh: Internal | Direct Solar | SUMMARY kWh: Total Load | %: Percentage Solar | Syst. Eff. |
|---|---|---|---|---|---|---|---|---|---|---|---|---|---|---|---|---|---|---|---|---|---|
| Column No | | | 1 | 2 | 3 | 4 | 5 | 6 | 7 | 8 | 9 | 10 | 11 | 12 | 13 | 14 | 15 | 16 | 17 | 18 | 19 |
| J | 77 | 31 | 20.7 | 6.5 | 1866 | 681 | | 632 | 31 | - | | | - | 2265 | 134 | | | | 3062 | 22 | 35 |
| F | 77 | 28 | 20.7 | 9.3 | 2715 | 1285 | | 880 | 18 | - | | | - | 1541 | 205 | | | | 2644 | 34 | 33 |
| M | 77 | 25 | 20.7 | 11.0 | 3160 | 1258 | | 800 | 12 | - | | | - | 1081 | 130 | | | | 1960 | 41 | 26 |
| A | 77 | 30 | 20.7 | 12.8 | 4379 | 1305 | | 680 | 30 | - | | | - | 523 | 84 | | | | 1317 | 54 | 16 |
| M | 77 | 31 | 21 | 15.2 | 2542 | 852 | | 413 | 46 | - | | | - | 197 | 96 | | | | 752 | 61 | 18 |
| J | 76 | 15 | | 22.1 | 7160 | 859 | | 0 | 33 | - | | | - | 0 | 66 | | | | 99 | 33 | 0 |
| J | 76 | 15 | 24.8 | 24.4 | | | | 0 | | - | | | - | 0 | | | | | | | |
| A | 76 | 30 | | 22.1 | 3335 | 594 | | 0 | 52 | - | | | - | 0 | 0 | | | | 52 | 100 | 2 |
| S | 76 | 30 | | 17.5 | 3521 | 676 | | 0 | 46 | - | | | - | 0 | 49 | | | | 95 | 48 | 1 |
| O | 76 | 29 | 20.7 | 13.5 | 3114 | 576 | | 503 | 18 | - | | | - | 198 | 58 | | | | 877 | 59 | 17 |
| N | 76 | 25 | 29.7 | 8.9 | 3307 | 1438 | | 1270 | 29 | - | | | - | 1341 | 104 | | | | 2744 | 47 | 39 |
| D | 76 | 31 | 20.7 | 6.8 | 2294 | 972 | | 860 | 36 | - | | | - | 2406 | 166 | | | | 3468 | 26 | 33 |
| TOTAL | | | — | — | 37796 | 10497 | | 6057 | 351 | - | | | - | 9588 (1) | 1091 (2) | | 12000 (3) | | 17087 | — | — |
| AVERAGE | | | 21.2 | 13.5 | | | | | | | | | | | | | | | | 37 | 17 |

(For June–July 1976, the values 7160, 859, 33, 66, 99, 33 and 0 are printed once, covering both rows.)

**TOTAL:** 6,408 **kWh**

(1):average for five solar houses 10593 kWh (76-77)
10112 kWh (77-78)

(2):average for five solar houses 1906 kWh(76-77)

(3):of which 3900 kWh measured domestic electricity consumption.

*:excludes pumps;duty times measured as 1223 H (primary),2,370 (secondary); Storage:stratification max:28oC,min2oC; Average:7oC,max temp(middle):62oC (78oC for 1977-78).

| PROJECT DESCRIPTION | BLAGNAC | REF. | F | BL | 77 | 78 | 6 |
|---|---|---|---|---|---|---|---|

**Main Participants:**

Research, Measurements - "Institut National des Sciences Appliquees (I.N.S.A.) de Toulouse, Mr. Bernard Bourret, Dept. Genie Civil, Service Equipments Techniques, Toulouse". "Centre d'Etudes Techniques de l'Equipement (CETE) de Bordeaux, Division U.T.C., groupe Construction, Bordeaux".

Building : Management: "Societe Cooperative des H.L.M. 31400-Toulouse.

Architect: "Mr. Liebard, 7 Rue d'Argenteuil, 75001 - Paris".

Solar: "Enterprise Pereira, 26 Avue. de l'Urss, 31400 Toulouse".

Funding Authorities: Loan : "Societe H.L.M." (above).

Grant : "Plan Construction, Ministere de l'Equipement, 2 Avue. du Parc de Passy, 75016 - Paris.

Contribution: The future owners of the houses

**Project Description:**

Mm. Liebard and Pereira, as laureates of a competition proposed in 1975 by the "Plan Construction" called HOT (Habitation Originale par la Thermique), have built up, from March to December 1976 some private houses in Blagnac: 5 with both gas and solar energy for S.H. and D.H.W., 3 with gas for S.H. and gas and solar energy for D.H.W., 2 with gas only for S.H. and D.H.W..The solar system consists of flat water collectors, (trickle type), a water storage tank, heat distribution by air for S.H., and auxiliary energy supplied by a gas fired boiler for both S.H. and D.H.W.

**Project Objectives:**

- Studying the possibility of using solar energy to provide part of the space heating and domestic hot water in standard housing insulation conditions (just a little better than the ones currently enforced.
- The main economic objectives were:
  De-sophistication of the solar system in order to develop a mass produced equipment (i.e. ready to fit in every house);
  Lowering of the solar overcharge cost as against similar classical active systems;
  Solar system redemption time over ten or twelve years, as a reasonable aim.
- For average measurements, the ten private houses were built identically. Room temperatures and energy consumption were checked regularly in all (once a week) in order to have references for energy balances (from 1976 December 8th to 1978 May 31)
  The private house chosen here for description is one of the first type ones.

| SITE LOCATION MAP | REF. | F | BL | 77 | 78 | 6 |
|---|---|---|---|---|---|---|

– Distance to main city: about 5 km from Toulouse (to the North-west)

– Please indicate height of nearest obstructions 0 m (flat plain)

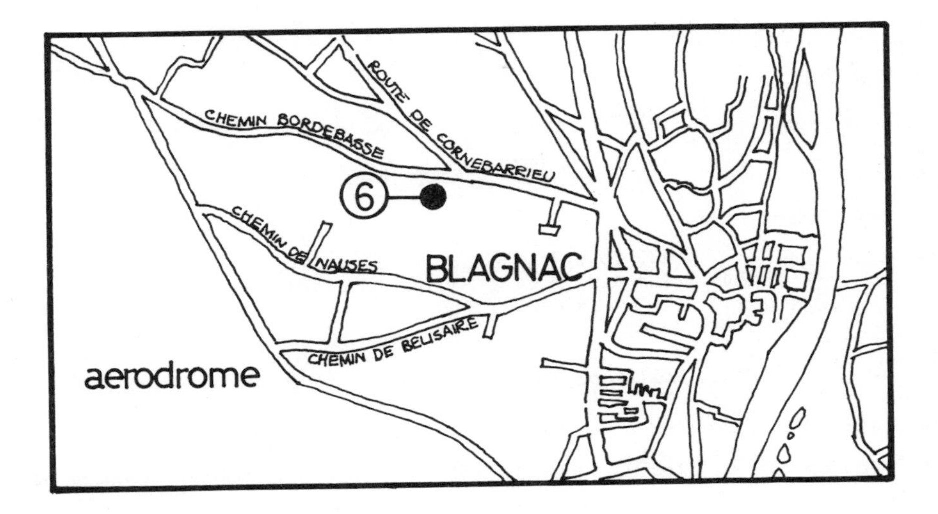

Photograph of Project

**SUMMARY SHEET** BLAGNAC **REF.** F BL 77 78 **6**

**Design Data**

**A1 CLIMATE**

1) Source of Data: Meteorological Station of Blagnac (Haute-Gazonne)

2) Latitude: 43° 38′ Longitude: 0° 45′ Altitude: 148 metres

3) Global Irradiation (horiz plane): 1285.3 kWh/m²·year % Diffuse: 27

4) Degree Days: 2,070 Base Temp: 18 °C

5) Sunshine Hours: July: 259.6 January: 77.3 Annual: 2,050

**B1 BUILDING**

1) Building Type: 2 storey detached No Occupants: from 4 to 5

2) Floor area: 103 m² Heated Volume: 277 m³

3) Design Temperature: External: -5 °C Internal: 20 °C

4) Ventilation Rate: 1.3 to 1.7 a.c.h. (54% fresh air) Vol. Heat Loss: 1.43 W/m³·K (initial design value 1.11)

5) Space Heat Load: 14,690 kWh (initial design: 12,495) Hot Water Load: 3,000 kWh

**C1 SYSTEM**

1) Absorber Type: Flat, Roll Bond Aluminium, matt black painted, Trickle

2) Collector Area 30 m² (Aperture) Coolant: Water

3) Orientation: 185° Tilt: 75° Glazing: Double, alveolar plastic

4) Storage Volume: 3 m³ Heat Emitters: Warm air (forced convection)

5) Auxiliary System: Gas fired boiler for SH & DHW Heat Pump: No

**PERFORMANCE MONITORING**

1) Is there a Computer Model? Yes (manual only)

2) Start Date for Monitoring Programme: December 1976

3) Period for which Results Available: SH from 1977 October 22nd to 1978 May 31st DHW only from 16.3.78 - 26.4.78 and at present time

4) No of Measuring Points per house: 9 (+ 3 for DHW balances over a six week period only)

5) Data Acquisition System Potentiometric recorder (12 ways) for the experimental solar house. Room temp. records in all houses, "Richard" systems (paper records), Gas consumption and internal ambient temp.checked weekly in all inhabited houses.

| | Space Heating | Hot Water | Total |
|---|---|---|---|
| 6) Predicted | 65 % of 12495 kWh | 69 % of 3000 kWh | 66 % of 15495 kWh |
| 7) Measured | 22.7 % of 14630 kWh | 29.6 % of 3500 kWh | 24.1 % of 18130 kWh |

8) Solar Energy Used: Including useful losses 145.4 kWh/m²·year

Excluding losses: ........ kWh/m²·year

9) System Efficiency: $\frac{\text{Solar Energy Used}}{\text{Global Irradiation on collector}} \times 100 =$ ____ %

## A2 LOCAL CLIMATE

REF. F | BL | 77 | 78 | 6

1 – Average Cloud cover: 5.0 Octas (reference : from October 1st 1977 to September 30th 1978, station of Blagnac)

2 – Average daily max temperatures (July) : 25.6 °C

3 – Source of Weather Data: For ext. ambient temp.: Blagnac meteorological station (reference B below) = averages from 1958 - 1977 period. As a reference, we are also giving here data from Toulouse station (reference T below) = year 1964, very close to averages over 1931 - 1960 period. (Toulouse station, Lat. 43°38', Long. 1°27'E, about 15 km from Blagnac). For solar data: Global and diffuse radiations were measured at a tilt angle 75° from horizontal only. Here, as a reference, we can also produce average data from Toulouse station (ref. T*)

4 – Micro Climate/Site Description:

The site was found not to have nay particular microclimatic characteristics.

It is to be noticed, too, in 1977 - 78 heating period, that the sunshine hours measured were below the average ones (over a 20 year period).

Sunshine hours per Day (average)

| Month | Oct | Nov | Dec | Jan | Feb | March | April | May | Average |
|---|---|---|---|---|---|---|---|---|---|
| Measd. | 4.8 | 3.5 | 3.3 | 2.5 | 2.2 | 4.5 | 5.9 | 4.7 | 3.94 |
| 20 year average | 5.2 | 3.0 | 1.9 | 2.5 | 4.0 | 5.6 | 6.2 | 7.6 | 4.51 |

Over the same period, average ambient temp was 9.2 (measured) against 9.5 (average over a 20 year period).

| MONTH | YEAR | Irradiation on horizontal plane Ref T* (Toulouse) Global | Diffuse | Degree Days Base: 18 °C | Blagnac 1977–78 Precipitation | Average Ambient Temperature Ref B | Ref T | Toulouse average T Average Relative Humidity | Blagnac 1977–78 Average Wind Speed | Prevailing Wind Direction |
|---|---|---|---|---|---|---|---|---|---|---|
| JAN | 1978 | 38.9 | 14.4 | | 91.8 | 5.4 | 2.7 | 65.8 | 3.9 | Prevailing Directions: N and NW 32%, S and SE 15% |
| FEB | " | 49.1 | 16.0 | | 75.5 | 6.8 | 6.8 | 78.1 | 4.3 | |
| MARCH | " | 83.6 | 24.6 | | 110.4 | 8.6 | 8.0 | 77.2 | 4.8 | |
| APRIL | " | 120.6 | 32.9 | | 81.6 | 11.3 | 11.3 | 77.7 | 3.9 | |
| MAY | " | 168.9 | 42.9 | | 96.9 | 14.9 | 16.7 | 73.2 | 3.6 | |
| JUNE | " | 176.2 | 48.1 | 0 | 116.7 | 18.4 | 18.9 | 69.7 | 3.2 | |
| JULY | " | 205.3 | 46.3 | 0 | 28.8 | 19.9 | 22.6 | 63.9 | 3.3 | |
| AUGUST | " | 169.7 | 40.5 | 0 | 12.8 | 20.5 | 21.1 | 66.0 | 2.9 | |
| SEPT | " | 121.9 | 29.9 | 0 | 43.9 | 18.2 | 19.7 | 74.8 | 3.5 | |
| OCT | 1977 | 71.0 | 22.8 | | 43.1 | 13.9 | 11.2 | 75.5 | 3.5 | |
| NOV | " | 41.1 | 16.3 | | 38.2 | 8.7 | 7.5 | 71.3 | 3.3 | |
| DEC | " | 38.9 | 13.0 | | 21.9 | 6.0 | 4.6 | 79.2 | 3.8 | |
| Totals | | 1285.3 | 347.4 | 2070 | 761.6 | | | | | |
| Averages | | See (1) | | See (2) | | 12.7 | 12.6 | 72.7 | 3.7 | |
| Units | – | $kWh/m^2 \cdot month$ | | °C days | mm | °C | | % | m/s | – |

(1) To be compared with the computed total over heating period (October 1st 1977 - May 20th 1978) at a tilt angle 75° from horizon ∴ 744 $kWh/m^2$, instead of 612 $kWh/m^2$ on horizon. plane.

(2) To be compared with the average value (over 1931-1960) for Toulouse station: 2353.

**B2 BUILDING DESCRIPTION** REF. | F | BL | 77 | 78 | **6**

**Design Concepts** A typical solar house in Blagnac estate is on two floors above the basement. It consists of five main rooms: on the ground floor, kitchen living room, garage and workroom, and a "technical room", arranged under the collectors, including preheating DHW and storage tanks, air fan, pipework etc. On the first floor, three bedrooms and a bathroom. The gas boiler is in the kitchen.

**Construction**

1 — Glazing: approx 19 % Area N:1.9 S:7.3 E:3.6 W:7.3 20.1 $m^2$ 'U' value: 2.9 $W/m^2 \cdot K$ including plastic shutters (double glazing)

2 — 'U' values: Walls: 1.0 $W/m^2 \cdot K$ (thickness: 33 cm) Floor: 2.56 $W/m^2 \cdot K$ (directly on the ground) Roof/loft: 0.29 $W/m^2 \cdot K$ (15cm glass-wool)

3 — Thermal mass: Heavy $kg/m^3$ (approx $150kg/m^3$) Annual Energy Demand: 14,700 kWh

- Middle class standard private house.
- Main walls in brickwork (bricks called "G12", $350kg/m^2$) giving a high thermal inertia.
- For the floor on the ground, surrounding insulation (2cm x 1.2 m polystyrene: k=1.25 W/m °C)
- One sided roof (about 30° from horizontal)
- Garage, workroom and porch as an outbuilding, northwards.
- Doors outwards: U = $4W/m^2 °C$

**Energy Conservation Measures**

Good housing insulation conditions, but about 15% lower than the minimum required by E.D.F. for "all electric" family houses. The G value is 1.42 $W/m^3 °C$ for a wind velocity 2.9 m/s, and 1.49 $W/m^3 °C$ for 5 m/s (a.c.h. measured respectively 1.3 and 1.7). For such private houses, the maximal G standard value is 1.6 $W/m^3 °C$.

Free Heat Gains: Total estimated (over 1977-78 heating period) approx. 5625 kWh
Usually expressed (in France) as a mean "minimal non-heating temperature here 15.9 °C

**Section Through House** Showing main solar components

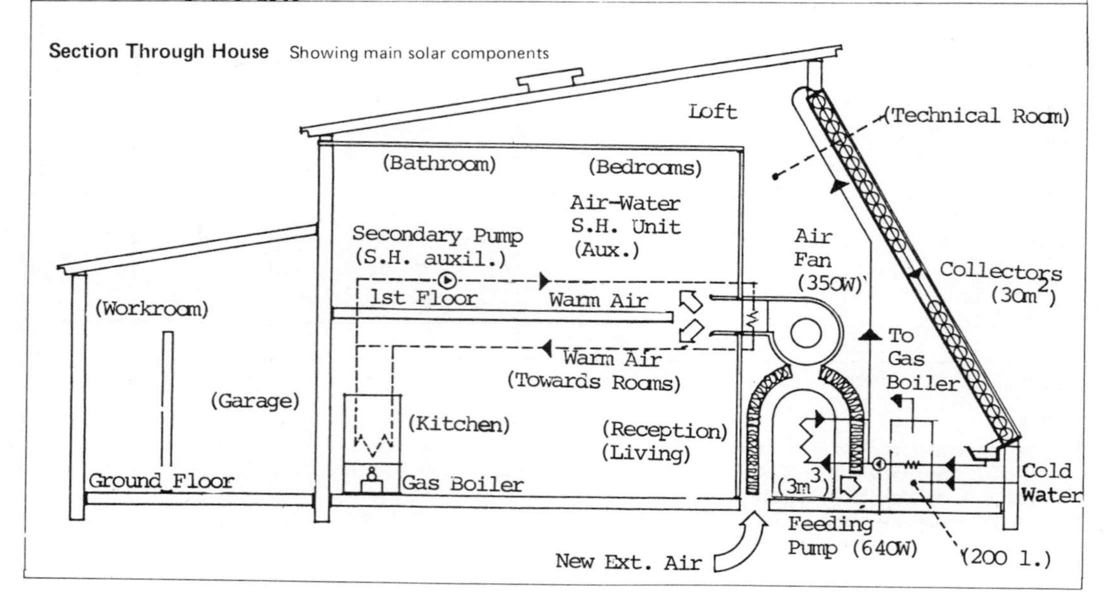

## C2 SYSTEM DESCRIPTION

REF. F | BL | 77 | 78 | 6

### Collector/Storage

1 – Primary circuit. Water content: 30 l (Balance tank content for the trickle collector) litres/m² collector area

2 – Insulation. Collector: 7.5 cm glasswool 10 cm polystyrene Storage: 10 cm polystyrene (none in the upper part) Pipework: none (inside)

3 – Frost Protection: None

4 – Overheating Protection: For storage: none; for S.H.: fan stopped by a thermostat if T air >22°C

5 – Corrosion Protection: Yes (classical type, probably sodium benzoate)

### Space Heating/Domestic Hot Water

1 – Auxiliary (S.H.) {air water unit, connected with the gas boiler: 14Kw for S.H. (DHW): Gas boiler (max. power 23.3 kW)

2 – Heat Emitters: Warm forced air; fan: 300-370W Temp. Range: adjustable general thermostat

3 – DHW Insulation: None Set Temperature: 50-60°C (initial: 70-80°C)

### Other Points

1. Water-water heat exchanger, in a 200l. tank for DHW pre-heat.
2. Water-water heat exchanger, in a 3m³ storage tank, for SH (water feeding air unit)
3. Air-water heat exchanger (with 32m² gills): forced air round the water storage tank.

### Control Strategy/Operating Modes

- A differential thermostat will stop the feeding (primary) pump if the water temp. in collectors gets higher than the outlet one. - Cold water (from mains) is pre-heated by solar water (through 1. above) and is topped up to DHW set temp. by the gas-fired boiler. - For SH solar energy is stored in the 3m³ storage tank (through 2. above); a three way valve by passes this loop in summer. Some new external air (approx. 225m³.hr.) plus some recaptured from the house (approx 190 m³.hr), filtered to 18-20°C, are preheated by forcing them round the water storage (through 3. above) and then blown into the rooms if air temperature issufficient. If not an electronic thermostat operates a three-way valve to adjust the water temperature in the air-water unit, connected with the gas-boiler. Another thermostat will stop the air fan if room temperature is above 22°C (case of solar overheating).

### Diagram of System

Showing sensor locations, power ratings, storage volumes & flows.

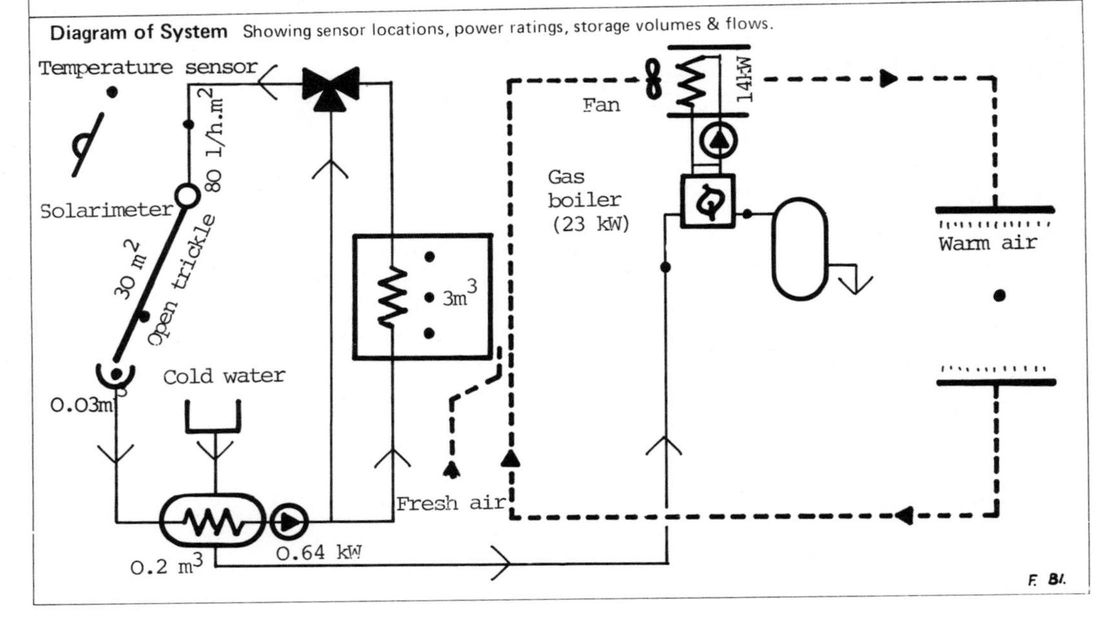

TECHNICAL APPRAISAL/PRACTICAL EXPERIENCE REF. | F | BL | 77 | 78 | 6

Component Performance: (Collector, Heat Exchangers, Storage, Pipework, Valves Fittings, Pumps, Auxiliary)

First comparisons between measurements and predictions (next page) led to measure the ventilation rate in winter, with North west winds of 1 m/s, and 5 m/s, using gas technique.
The low performance of the air-water heat exchanger (round the storage tank) conditioned to improve the heat exchange surface. This was found just as a penalty for global exchange power, at a fixed ventilation rate, because the flow dropped too much.
Room temperatures were finally decided to be recorded with chart recorders in all inhabited houses.
It was noticed during the first heating period:
Speedy steadying of corrosion and deposition phenomena in collectors. Some mishaps due to dilation: some connection plastic pipes broken between the bottom of solar panels and the balancing primary tank; some plastic collector covers got out of shape, at times, split locally. A few leaks occurred in collectors, possibly due to dilation.
The water-water heat exchanger (no. 2. in previous page) was found quite undersized, which was certainly affecting the collector efficiency (storage tank temperature remaining too high; see another clause below).
As the main failure, the operating noise of heating system appeared to be much too high (it was measured in different rooms): air fan, at max. flow from 36 to 46 dB, at min, flow from 34 to 39.5 dB; gas boiler: from 19 to 57 dB; Collector feeding pump: from 21 to 35 dB. Beside the great acoustic discomfort, nearly all of the owners were constrained to adjust the air fan at its minimal flow, or even at the intermittent work point. That was, in addition to the poor efficiency of the water-air heat exchanger, a heavy penalty for the whole solar system performance: storage-house heat exchangers dropping down, storage tank temperature rising too high and its unproductive losses enlarging too much, collector efficiency collapsing...etc.

System Operation/Controls, Electronics:

-The overheating protection thermostat (for warm air) was found to be well designed.

-The electronic air temperature controller appeared to be (and is still) very difficult to adjust in order to take maximum advantage of the solar energy stored.

-No important troubles occurred in the whole measureing equipment (strong enough) but, for a good reliability, the temperature sensor zero point and calibrations had been regularly checked.

-The potentiometric chart recording system proved to be very reliable for monitoring control and date recording, but far too long to work out for calculations; in fact, this kept an INSA technician in full time occupation.

-A lot of mishaps, failures and even breaks would have been avoided if one of the experimenters had been living on the site.

PERFORMANCE EVALUATION/CONCLUSIONS REF. | F | BL | 77 | 78 | 6

**Comparison of Measured with Predicted Performance**
(State reasons for any discrepancy and summarise the main factors affecting performance)

A 66% solar contribution to S.H. and DHW demand was predicted. The measured value was 22%. The difference was due to ; poor collector performance, under-sized heat exchangers, high ventilation rates. A re-calculation using measured ventilation rates and U-values gave a solar contribution of 24%. This is increased to 37% if an average climate year is used.

**Modifications to System**

- Repairing and exchanging pipework and collector parts.
- Measurement of ventilation rate and noise level and some components.
- Improvement of heat exchangers, and measurement of room temperatures.
- Gas consumption was measured in unoccupied solar and non-solar houses for a ten week period. An energy saving of 36% was found in the solar house.

**Occupants/Response :**

- Satisfaction with winter comfort conditions.
- Bedrooms under collectors overheated in summer.
- Complaints about noise of fan.
- Automatic controls and safety devices were considered essential.
- There was concern about maintenance costs.

**Conclusions**

The solar contribution appears disappointing, and such systems at the present time are uneconomic. There are still many technical improvements to be made. However, under ideal working conditions, over a 20 year period, each house would have saved 19,000 $m^3$ gas.

**Future Work**

If it had to be done again, improvements would have been made to heat exchangers, and more work would be done on finding the optimum thermal differential adjustment so that the pump is not working unnecessarily. The measurements stopped on May 31st 1978. Since then, room temperatures and gas consumption have been checked, leading to similar energy balances.

**P1 MONTHLY PERFORMANCE** REF: F BL 77 78 6

| Month | Year | No Days Data | Average Temp °C: Internal | Average Temp °C: External | SOLAR ENERGY kWh: Global Irradiation On Collector Area | collector efficiency % | Solar Energy Used – collected for SH only: Incl. | Solar Energy Used – collected for SH only: Excl. | Solar Energy Used: Domestic Hot Water | Solar Energy Used: Input To Heat | Heat Pump Total Output kWh: Space Heating | Heat Pump Total Output kWh: Domestic Hot Water | AUXILIARY ENERGY kWh: Heat Pump | AUXILIARY ENERGY kWh: Space Heating | AUXILIARY ENERGY kWh: Domestic Hot Water | AUXILIARY ENERGY kWh: Pumps & Fans | Free Heat Gains kWh: Internal | Free Heat Gains kWh: Direct Solar | SUMMARY kWh: Total Load | SUMMARY %: Percentage Solar S-heat | SUMMARY %: Syst. Eff. |
|---|---|---|---|---|---|---|---|---|---|---|---|---|---|---|---|---|---|---|---|---|---|
| Column No | | | 1 | 2 | 3 | 4 | 5 | 6 | 7 | 8 | 9 | 10 | 11 | 12 | 13 | 14 | 15 | 16 | 17 | 18 | 19 |
| J | 78 | 31 | 24.5 | 4.5 | 1665 | 23.4 | 389 | | | | | | | 3115.6 | | | | | | 11 | |
| F | 78 | 28 | 20.3 | 7.4 | 1626 | 18.6 | 303 | | | | | | | 2521.5 | | | | | | 11 | |
| M | 78 | 31 | 19.2 | 9.2 | 2697 | 24.1 | 650 | | (14 days) 32 | | | | | 1584.6 | 209.3 | | | | | 29 | |
| A | 78 | 30 | 20.7 | 9.8 | 2996 | 21.1 | 635 | | (26 days) 69 | | | | | 1213.1 | 202.5 | | | | | 34 | |
| M | 78 | 31 | 20.4 | 13.6 | 2812 | 21.3 | 599 | | | | | | | 717.8 | | | | | | 46 | |
| J | | | | | | | | | | | | | | | | | | | | | |
| J | | | | | | | | | | | | | | | | | | | | | |
| A | | | | | | | | | | | | | | | | | | | | | |
| S | | | | | | | | | | | | | | | | | | | | | |
| O | 77 | 10 | 20.2 | 15.8 | 955 | 25.4 | 243 | | | | | | | 73.5 | | | | | | 77 | |
| N | 77 | 25 | 14.8 | 9.4 | 1958 | 22.1 | 433 | | | | | | | 1415.7 | | | | | | 23 | |
| D | 77 | 8 | 17.2 | 6.4 | 314 | 23.2 | 73 | | | | | | | 660.3 | | | | | | 10 | |
| TOTAL | | | — | — | 15023 | - | 3325 | | (1036) | | | | | 11302.2 | (2464) | | 5625 | | | — | — |
| AVERAGE | | | 20.0 | 9.2 | ave:22.1 | | | estimated | | | | | | | | | | | | 24 | |

TOTAL: kWh

DHW: figures estimated from extrapolated measurements and EDF averages.

5,12:incl. 571 kWh pump electric pump consumption. 12:computed from gas consump. (SH only)checked, and boiler output and heating circuit regularly measured. 15,16: estimated:lighting and cooking 1085 kWh, direct solar + int gains:4540kWh.

PROJECT DESCRIPTION BOURGOIN, FRANCE. REF. | F | Bo | 78 | 79 | 7

**Main Participants:**

- Experiment : CNRS-CRTBT, BP 166, 38042 Grenoble-Cedex, France, Tel. (76) 96.98.37 MM. Kuhn et Pataud.
- Financing of solar system and experiment : Plan Construction, Ministere de l'Environnement et du Cadre de vie,Avenue du Parc de Passy, 75016 Paris-Cedex, Tel. (1) 524.52.34. CNRS PIRDES, 282 Bd St-Germain, 75007 Paris, Tel. (1) 705.77.15.
- Construction S.A. d'HLM de Voiron et des Terres Froides, 6 rue Genevoise, 38503 Voiron, Tel. (76) 05.40.66.
- Architects Audrain and Geneve, 6 rue Beyle Stendhal, 38000 Grenoble, Tel. (76) 54.09.71.
- Thermal system conception : CET (Mr Chavin) 47 chemin de la Taillat, 38240 Meylan. Tel. (76) 90.62.18.

**Project Description:**

The project consists of a three storey block, containing twelve flats in Bourgoin, Jallieu, Isere, France. The plan is symmetrical with six flats in each half of the building. The south face of the building is a curtain wall with forty bands of steel collectors, single glazed, with a total area of 305m$^2$. Each bandcan be operated independently, and the solar and anxiliary heating systems are also independent for each half. Heat storage is provided in six 10m$^3$ tanks, three in each half which can be used, or not, in series or in parallel.

**Project Objectives:**

1) To test theoretical computer models of thermal behaviour of solar systems for housing.

2) To study the effects of varying the following parameters:
   - Flow Velocity, Surface, and Inertia of the Collectors
   - Storage Capacity

The water content of the collector system can be changed by a factor of 4, by adding tanks in series, thus modifying the inertia. The collector fluid, and heat exchanger surface area between collectors and storage can also be changed.

The building is completely instrumented for weather data and energy balance measurements.

Results for the half building WEST, where parameters have been left un-changed, are available from September 1978.

| SITE LOCATION MAP | BOURGOIN, FRANCE | REF. | F | Bo | 78 | 79 | |
|---|---|---|---|---|---|---|---|

– Distance to main city: 45 km from LYON

– Please indicate height of nearest obstructions No obstruction by buildings. A small hill, 120m high, 500 metres to the East of the building creates some shadowing early in the morning. (Angle 13°).

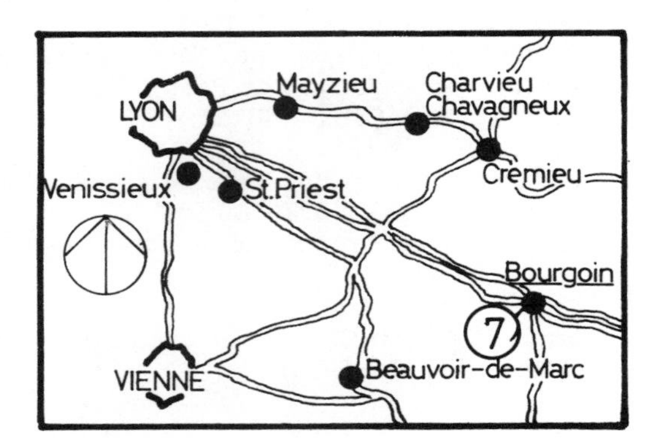

Photograph of Project

SUMMARY SHEET BOURGOIN, FRANCE REF. F | Bo | 78 | 79 | 7

**Design Data** * All Data for Half Building - West

**A1 CLIMATE**

1) Source of Data: Met. Station. Lyon-Bron

2) Latitude: 45 ° 30 ′ N Longitude: 5 ° 16 ′ Altitude: 230 metres

3) Global Irradiation (horiz plane): 1402 kWh/m²·year % Diffuse: 25

4) Degree Days: 2780 Base Temp: 18 °C

5) Sunshine Hours: July: January: Annual: 2050

**B1 BUILDING**

1) Building Type: 3 storey flats No Occupants: 23 in 6 flats

2) Floor area: 573 m² Heated Volume: 1375 m³

3) Design Temperature: External: -11 °C Internal: 20 °C

4) Ventilation Rate: 1 a.c.h. Vol. Heat Loss: 0.93 W/m³·K

5) Space Heat Load: 60,000 kWh Hot Water Load: 25,000 kWh

**C1 SYSTEM**

1) Absorber Type: Black painted steel

2) Collector Area 152 m² (Aperture) Coolant: Water & Glycol

3) Orientation: 180 Tilt: 90 Glazing: Single 6mm

4) Storage Volume: 30 m³ water ** Heat Emitters: Underfloor

5) Auxiliary System: Electric convectors Heat Pump: No

** Phase change material in other half of building

**PERFORMANCE MONITORING**

1) Is there a Computer Model? Yes (EDF) 2) Start date for Monitoring Programme: 1st Sept. 1978

3) Period for which results available: 1 Sept 1978 to 31 May 1979.

4) No. of Measuring Points per house 40

5) Data Acquisition System: 64 channel data logger with 6 integrated channels. Data is stored on Punched tape on Mag tape. Electricity consumption is recorded separately.

| | Space Heating | Hot Water | Total |
|---|---|---|---|
| 6) Predicted | 37 % of 54898 kWh | 45 % of 15389 kWh | 38 % of 70287 kWh |
| ** 7) Measured | 39 % of 50363 kWh | 20 % of 22164 kWh | 33 % of 72527 kWh |

8) Solar Energy Used:
* 1/09/78 - 31/05/79
**1/09/78 - /06/79

Including useful losses kWh/m²·year

Excluding losses: **159 kWh/m²·year

9) System Efficiency: Solar Energy Used / Global Irradiation on collector × 100 = 27 % **

| A2 LOCAL CLIMATE BOURGOIN, FRANCE. | REF. | F | Bo | 78 | 79 | 7 |
|---|---|---|---|---|---|---|

1 — Average Cloud cover: 5.3 Octas Lyon 1961-70

2 — Average daily max temperatures (July) : 26.4 °C

Average daily min temperatures (Jan) : -0.7 °C Lyon 1922-78

3 — Source of Weather Data:

Lyon Meteorological Station for average outside temperatures.
Solar data was computerised integrating Lyon met data about the cloud cover ratio.

4 — Micro Climate/Site Description:

Site measurements are close to the Lyon data for annual average outside air temperature (about 4% below) and 9% below for average temperatures over the heating season.

Computerised solar data is expected to be within 10-25% of site measurements.

| MONTH | YEAR | Irradiation on horizontal plane | | Degree Days | Lyon Average | Lyon 1966 Average Ambient Temperature | Lyon 1966 Average Relative Humidity | Lyon 51-75 Average Wind Speed | Prevailing Wind Direction |
|---|---|---|---|---|---|---|---|---|---|
| | | Global | Diffuse | Base: 18 °C | Precipitation | | | | |
| JAN | | 40.65 | 12.71 | | 53 | 1.3 | 82.0 | 2.97 | |
| FEB | | 51.75 | 14.98 | | 50 | 8.0 | 77.0 | 3.53 | |
| MARCH | | 122.84 | 21.01 | | 57 | 5.3 | 72.8 | 3.53 | |
| APRIL | | 134.70 | 31.40 | | 60 | 11.4 | 73.2 | 3.71 | |
| MAY | | 188.50 | 40.95 | | 75 | 14.1 | 71.5 | 3.39 | |
| JUNE | | 206.14 | 45.16 | O | 79 | 17.9 | 68.7 | 2.96 | |
| JULY | | 204.02 | 46.78 | O | 63 | 17.9 | 65.6 | 2.92 | |
| AUGUST | | 166.22 | 40.83 | O | 89 | 18.6 | 68.1 | 2.64 | |
| SEPT | | 140.08 | 27.46 | O | 86 | 17.3 | 73.2 | 2.62 | |
| OCT | | 80.05 | 21.11 | | 77 | 13.9 | 80.2 | 2.50 | |
| NOV | | 38.01 | 15.17 | | 80 | 4.3 | 86.3 | 3.01 | |
| DEC | | 29.09 | 12.75 | | 57 | 3.7 | 85.2 | 2.76 | |
| Totals | | 1402.04 | 330.30 | 2780 | 825 | | | | |
| Averages | | | | | | 11.6 | 75.3 | 3.04 | N.E. |
| Units | – | kWh/m²·month | | °C days | mm | °C | % | m/s | – |

| B2 BUILDING DESCRIPTION BOURGOIN, FRANCE. | REF. | F | Bo | 78 | 79 | 7 |
|---|---|---|---|---|---|---|

**Design Concepts**

For experimental purposes, the building is divided into two parts, the parameters can be varied so that two solar configurations can be compared.

**Construction**

1 — Glazing: 22 % Area: 307* $m^2$ 'U' value: 3.33 $W/m^2 \cdot K$

2 — 'U' values: Walls: 0.43 $W/m^2 \cdot K$ Floor: 0.56 $W/m^2 \cdot K$ Roof/loft: 0.42 $W/m^2 \cdot K$

3 — Thermal mass: Medium $kg/m^3$ Annual Energy Demand: ........ kWh

Windows are double glazed, and include shutters, giving a mean U-Value of 2.6 $W/m^2K$. (*) S: 184.6 E: 9.6 W: 13.6 N: 99 $m^2$

There is a double door at the entrance.

**Energy Conservation Measures**

No special measures, but the storage room is insulated.

Free Heat Gains: Internal: 9675 kWh/year Direct Solar: N.A. kWh/year

**Section Through House** Showing main solar components

| C2 SYSTEM DESCRIPTION BOURGOIN, FRANCE. | REF. | F | Bo | 78 | 79 | 7 |
|---|---|---|---|---|---|---|

**Collector/Storage**

1 — Primary circuit. Water content: Primary Circuit - 9001 2.5 litres/m² collector area

2 — Insulation. Collector: 10cm Storage: 10cm Pipework: 2-6cm

3 — Frost Protection: 30% Ethylene Glycol

4 — Overheating Protection: None

5 — Corrosion Protection: Additives

**Space Heating/Domestic Hot Water**

1 — Auxiliary (S.H.): Electric Convectors (DHW): 2 x 1500 litre tanks

2 — Heat Emitters: Underfloor Coils Temp. Range: 30oC - 22oC

3 — DHW Insulation: 10cm Set Temperature: 60oC

**Other Points/** One of the DHW tanks is pre-heated by a heat exchanger fed from the main store.

**Control Strategy/Operating Modes**

An outdoor photo-electric cell operates the primary pump during the day. The primary 3-way valve opens to the main heat exchanger, and the secondary pump operates when the temperature of the primary circuit is higher than store temperature. For space heating, the regulation takes into account, external temperature, as well as maximum and minimum water temperatures.

**Diagram of System** Showing sensor locations, power ratings, storage volumes & flows.

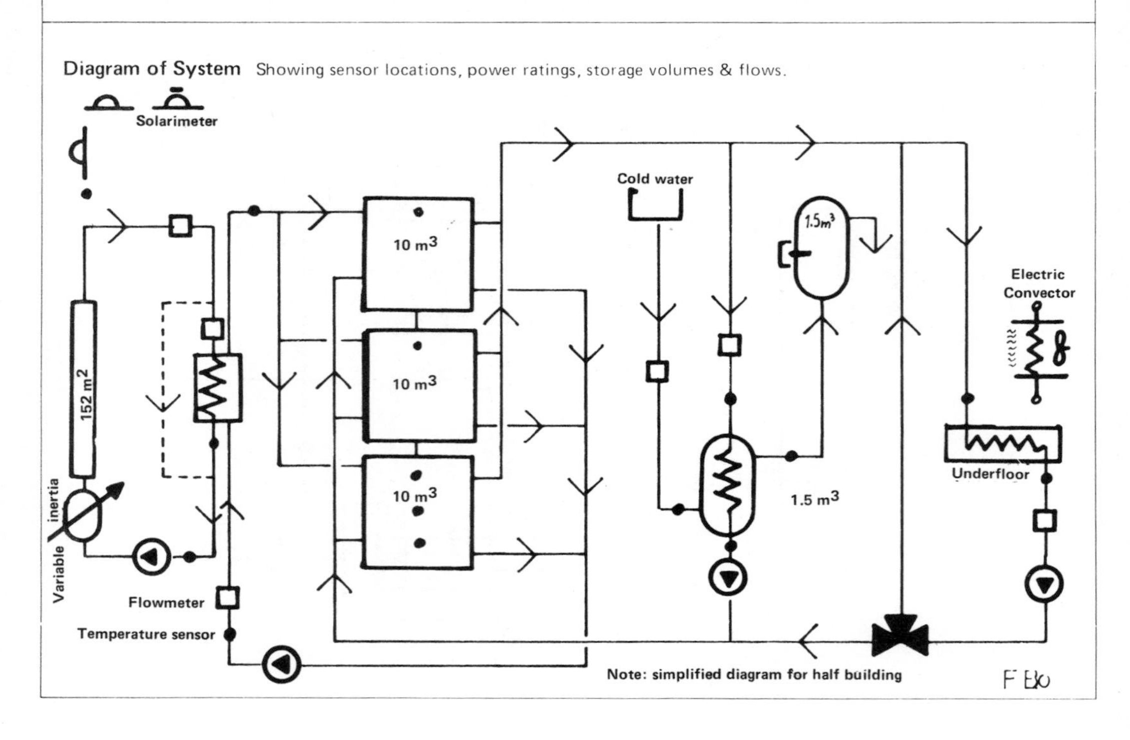

| TECHNICAL APPRAISAL/PRACTICAL EXPERIENCE | REF. | F | Bo | 78 | 79 | | 7 |
|---|---|---|---|---|---|---|---|

**System/Design:**

Some shading of the collectors by projecting balconies could have been avoided. Storage tanks are horizontal giving little scope for stratification. A better solution would have been some regulation of temperature ranges in the 3 tanks.

**System/Installation:**

The regulation system was disturbed when the pipework and storage tanks were insulated. The sensors for the control system were poorly calibrated. Collector insulation was changed from glass wool to polystyrene.

**Component Performance:** (Collector, Heat Exchangers, Storage, Pipework, Valves, Fittings, Pumps, Auxiliary)

Collector Performance. Results for a good, clear day during the heating season show an efficiency of 60-80%. During two months in the summer of 1978, collector performance was impaired because workmen had removed the top cover of the collector system, thus doubling the heat loss coefficient.

Storage efficiency has been good. From October to December 1978, 88% of stored energy was used.

**System Operation/Controls, Electronics:**

Some problems with electrical noise in the plant room caused by control switches.

Problems with the power supply for Pt. Resistance thermometers.

Sensor calibration had to be carefully checked.

There was a delay on the delivery of instrumentation, and there was a major problem with the interface for magnetic storage.

NB One of the Engineers lives in the building and is thus able to solve any problems as they occur. This also ensures continuity and reliability of data.

| PERFORMANCE EVALUATION/CONCLUSIONS | REF. | F | Bo | 78 | 79 | | 7 |
|---|---|---|---|---|---|---|---|

**Comparison of Measured with Predicted Performance**
(State reasons for any discrepancy and summarise the main factors affecting performance)

Predictions were made using both the manual ABACUS and the computer model (CLIM-MASOL), both prepared by E.D.F. There was some discrepancy between predicted and measured results, but the real meteorological data was not used for predictions. The program will be re-run in the near future using real met-data to test the model. It does seem that the biggest difference is caused by the occupants. System performance has been better than expected, due to low temperature working conditions of the collectors in the healing season. Storage losses are small.

**Modifications to System**

From 1/9/78 - 18/12/78 maximum 30m$^3$ storage was used. No modifications were carried out. From 18th Dec to end of heat. sea.only one tank was used (10m$^3$). This decreased solar contribution to space heating but increased it for D.H.w.

Extensive modifications were carried out in the other half of the building (East).

**Occupants/Response**

They cannot influence the behaviour of the system, but are conscious of auxiliary electricity consumption. Some flats are regularly below 20oC. The difference in auxiliary electricity consumption reaches a factor of 15 between flats. D.H.W. consumption is high on average.

**Conclusions**

After one year, the objectives of the project seem to be justified, and a considerable amount of useful information will be generated. There have been very few problems with the solar system, and the variation of system parameters is a practicable idea. The major problem is with the data storage system.

The delivery temperature for DHW could be reduced from 60oC to 50oC or 55°C to obtain better solar contribution.

**Future Work**

The experiments will continue for three years, modifying system parameters in different climatic conditions. Co-operation with other groups involved in model validation will be continued.

An experiment is also being carried out using Phase Change Material in one of the EAST tanks, for comparison with a WEST water tank.

## P1 MONTHLY PERFORMANCE

REF: F B0 78 79 7

| Month | Year | No Days Data | Average Temp °C: Internal | Average Temp °C: External | SOLAR ENERGY kWh: Global Irradiation On Collector Area | Solar Energy Collected | Solar Energy Used: Space Heating Useful Losses Incl. | Space Heating Useful Losses Excl. | Domestic Hot Water | Input To Heat | Heat Pump Total Output kWh: Space Heating | Domestic Hot Water | AUXILIARY ENERGY kWh: Heat Pump | Space Heating | Domestic Hot Water | Pumps & Fans | Free Heat Gains kWh: Internal | Direct Solar | SUMMARY kWh: Total Load | %: Percentage Solar | %: Syst. Eff. |
|---|---|---|---|---|---|---|---|---|---|---|---|---|---|---|---|---|---|---|---|---|---|
| Column No | | | 1 | 2 | 3 | 4 | 5 | 6 | 7 | 8 | 9 | 10 | 11 | 12 | 13 | 14 | 15 | 16 | 17 | 18 | 19 |
| J | 79 | 30 | 19E | 0.8 | 5452 | 1645 | (858) | 858 | 308 | | | | | 9000 | 2550 | 775 | 1100 | | 12716 | 9.1 | 21.4 |
| F | 79 | 27 | 19E | 4.9 | 4464 | 1242 | (685) | 685 | 268 | | | | | 6500 | 2550 | 825 | 1150 | | 10003 | 9.5 | 21.3 |
| M | 79 | 30.5 | 19E | 8.6 | 9355 | 3885 | (3229) | 3229 | 502 | | | | | 3000 | 1900 | 1200 | 1350 | | 8631 | 43.2 | 39.8 |
| A | 79 | 29.5 | 19E | 9.9 | 11626 | 4894 | (4074) | 4074 | 437 | | | | | 1200 | 1150 | 1175 | 1050 | | 6861 | 57.3 | 38.8 |
| M | 79 | 30.5 | 20E | 14.9 | 12049 | 3590 | (2315) | 2315 | 566 | | | | | 200 | 1800 | 1500 | 1000 | | 4881 | 59.0 | 23.9 |
| J | 79 | | 20 | | 11697 | 2781 | | | 738 | | | | | | 1050 | 1150 | 750 | | 1788 | 41.3 | 8.3 |
| J | | | | | | | | | | | | | | | | | | | - | | |
| A | | | | | | | | | | | | | | | | | | | - | | |
| S | 78 | 26 30 | - | 15.24 15.0 | 12329 | 4072 4451 | | | 548 622 | | | | | 0 | 805 | 1345 | 955 | | 1427 | 44 | 4.4 |
| O | 78 | 29.5 | 20E | 10.9 | 11843 | 5407 | (3748)+ | 3748+ | 426 | | | | | 600 | 1492 | 1225 | 900 | | 6266+ | 67 | 35 |
| N | 78 | 21 30 | 19E | 5.5 | 5578 | 2568 | (2506) | 2506 | 197 | | | | | 4380 | 2348 | 975 | 770 | | 9431+ | +29 | 48.5 |
| D | 78 | 27 31 | 19E | 7.4 | 5490 | 2889 | (2418) | 2418 | 275 | | | | | 5650 | 2180 | 830 | 650 | | 10523+ | +26 | 49.1 |
| TOTAL | | | — | — | 89853 | 33442 | (19833) | 19833 | 4339 | | | | | 30530 | 17825 | 11000= | 9675 | | 72527 | — | — |
| AVERAGE | | | 19E | 8.7 | | | | | | | | | | | | | | | | +32.9 | 26.9 |

TOTAL: 24172 kWh

E: Estimated; +: The solar energy used is underestimated since some days are missing; =: incl. fans for simple flux ventilation of the building.

| PROJECT DESCRIPTION DOURDAN | REF. | F | DO | 79 | 80 | 8 |
|---|---|---|---|---|---|---|

**Main Participants:**

Construction of the solar house and design of the solar system:

- Jacques MICHEL, architect (PARIS)
- Gerard GARY, physicist at CNRS (Lives in the house)
- Joel BRUCHE, engineer at "Société FORCLUM" (air-conditioning)

---

Monitoring of the data acquisition system and analysis of the data:

Groupe RAMSES (CNRS - PIRDES)

Laboratoire de l'Accélérateur Linéaire

91 405 - ORSAY Tel. 941 82 70

with the collaboration of: Centre de Recherches EDF (Dept ADE),
Les Renardièrs - Route de Sens,
77250 ECUELLES

**Project Description:**

The components of the solar heating system are :-

Air collectors / Rock Bed Storage / Electric Heat Pump (air/air). This system is quite new in France and the study of its performance could allow this solar technology to spread in the area of Paris. Air collectors with low thermal capacity have been chosen in order to insure a good efficiency of the energy collection, even with cloudy weather which is typical of the region of Paris. The photograph following shows a south-facing terrace at first floor level over the rock store. In front of the rock store are the collectors , $70m^2$.

**Project Objectives:**

One of the objectives of the experiment is to optimize the size of the storage in relation to the meteorological parameters of the region. The auxiliary system (heat pump) was chosen in order to minimize the needs of external energy. Another objective of this experiment is to test different schemes of regulation to define the most efficient means of coupling the solar system and the Heat Pump.

| SITE LOCATION MAP | DOURDAN | REF. | F | DO | 79 | 80 | 8 |
|---|---|---|---|---|---|---|---|

Distance to main city: 35 km from PARIS (South-west)

Please indicate height of nearest obstructions ; no obstruction shadowing the collectors.

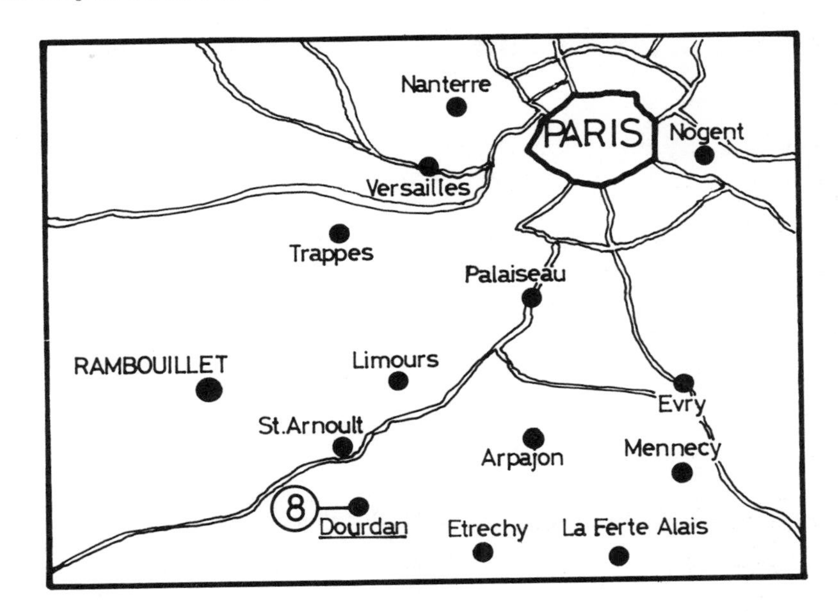


The photograph shows the south facing terrace at first floor level with the rock store below. In front of the the rock store are the collectors ($70m^2$)

Photograph of Project

**SUMMARY SHEET** DOURDAN **REF.** F | DO | 79 | 80 | **8**

**Design Data**

**A1 CLIMATE**

1) Source of Data: TRAPPES

2) Latitude: 48 ° 46 ′ Longitude: 2 ° 01 ′ Altitude: 167 metres

3) Global Irradiation (horiz plane): 1132 kWh/m²·year % Diffuse: 52%

4) Degree Days: 2917 Base Temp: 18 °C

5) Sunshine Hours: July: 230 January: 55 Annual: 1707

**B1 BUILDING**

1) Building Type: two-storey house No Occupants: 4 (2 ad. + 2 ch.)

2) Floor area: inhabited: 197 auxiliary: 124 m² Heated Volume: 534 m³

3) Design Temperature: External: -7 °C Internal: 20 °C

4) Ventilation Rate: 1 a.c.h. Vol. Heat Loss: 1.15 W/m³·K

5) Space Heat Load: kWh Hot Water Load: kWh

**C1 SYSTEM**

1) Absorber Type: Flat, black painted steel absorber.

2) Collector Area 62.5 m² (Aperture) Coolant: AIR

3) Orientation: South Tilt: 70° Glazing: Single: BSN Horti +

4) Storage Volume: 40 m³ Heat Emitters: Air ducts, air outlets

5) Auxiliary System: 4 kw heat pump, with 6 kw elect. resistance and 3 x 0.750 kw convectors.

**PERFORMANCE MONITORING**

1) Is there a Computer Model? One in operation (basic) one under development for large computer (CDC) 2) Start date for Monitoring Programme: early 1979

3) Period for which Results Available: a few days in 1979.

4) No of Measuring Points per house: 70

5) Data Acquisition System: Detailed follow-up of energy exchange between components of the heating system. All measuring points are recorded on a digital tape monitored by a SC/MP microprocessor.

6) Predicted Space Heating: 50% of 35,000 kWh.

| A2 LOCAL CLIMATE | DOURDAN | REF. | F | DO | 79 | 80 | 8 |
|---|---|---|---|---|---|---|---|

1 – Average Cloud cover: 4.9 Octas

2 – Average daily max temperatures (July) : 22 °C

Average daily min temperatures (Jan) : –3 °C

3 – Source of Weather Data: Meteorologie Nationale:
Centre de TRAPPES – 78 190

(TRAPPES is located at 25 kms at the north of DOURDAN; the town is underlined on the plan seen previously.)

4 – Micro Climate/Site Description: No particularities.
(The house is located in the countryside, among plain fields)

The data below has been averaged over five years (71–75)

| MONTH | YEAR | Irradiation on horizontal plane | | Degree Days | Precipitation | Average Ambient Temperature | Average Relative Humidity | Average Wind Speed | Prevailing Wind Direction |
|---|---|---|---|---|---|---|---|---|---|
| | | Global | Diffuse | Base: 18 °C | | | | | |
| JAN | | 24.5 | 17.5 | 431.5 | | 4.1 | 88.6 | 3.2 | |
| FEB | | 43.7 | 25.7 | 384.2 | | 4.3 | 84 | 3 | |
| MARCH | | 83.4 | 44.9 | 381.3 | | 5.7 | 76.2 | 2.8 | |
| APRIL | | 117 | 67.9 | 282.1 | | 8.9 | 73.4 | 3.2 | |
| MAY | | 141.7 | 80.9 | 177.4 | | 12.3 | 76 | 2.8 | |
| JUNE | | 158.1 | 86.5 | 88.8 | | 15 | 75.6 | 2.4 | |
| JULY | | 168.6 | 85.3 | 7.4 | | 17.8 | 74.6 | 2.2 | |
| AUGUST | | 172.7 | 72.9 | – | | 18.2 | 75.6 | 2 | |
| SEPT | | 103.8 | 52.4 | 108.6 | | 14.4 | 78.8 | 2.6 | |
| OCT | | 67 | 20.5 | 267.8 | | 9.4 | 83.8 | 2.2 | |
| NOV | | 30.6 | 19.8 | 355.8 | | 6.1 | 88.8 | 3 | |
| DEC | | 21.1 | 14.6 | 432.1 | | 4.1 | 89 | 2.8 | |
| Totals | | 1132.1 | 588.8 | 2917 | 650 | | | | |
| Averages | | | | | | $10^{O}$ | 80.4 | 2.7 | N & SW |
| Units | – | kWh/m²· month | | °C days | mm | °C | % | m/s | – |

| B2 BUILDING DESCRIPTION | REF. | F | DO | 79 | 80 | 8 |
|---|---|---|---|---|---|---|

**Design Concepts**
Outside insulation, active system, collectors due south with a 70° tilt.

**Construction**
Glazing:23% (59.7m²); 'U' values (W/m².K):3.2(glazing), 0.44(walls), 0.43(floor), 0.3(roof/loft); thermal mass:340 kg/m³; annual energy demand:35,000 kWh.

Wall:1cm plaster; 20cm concrete blocks; 7.5cm fibreglass; 1cm vacuum; 1cm plaster(on wire netting).
Glazing: N:15.65m², E:9.15m², W:17.60m², S:17.30m².

**Energy Conservation Measures**
Outside insulation, double-glazing.

## C2 SYSTEM DESCRIPTION DOURDAN

**Collector/Storage**

Primary circuit: air collectors ; insulation:6cm fibreglass (collector), polystyrene (storage wall), fibreglass (storage loft), 2.5cm fibreglass (pipework).

**Space Heating/Domestic Hot Water**
Auxiliary (SH):heat pump; heat emitters:air outlets.

**Control Strategy/Operating Modes**
Ti = inside temperature; Tc = collectors temp.; Tsh = hot side of storage temp.; Tsc = cold side of storage temp.

Five position:

| | |
|---|---|
| Direct | Ti < 20° and Tc > 20° |
| Release | Ti < 20° and Tc < 20° and Tsh > 20° |
| Auxiliary | Ti < 20° and Tc < 20° and Tsh < 20° |
| Storage | Ti > 20° and Tc > Tsc |
| Stop | Ti > 20° and Tc < Tsc |

**Diagram of System** Showing sensor locations, power ratings, storage volumes & flows.

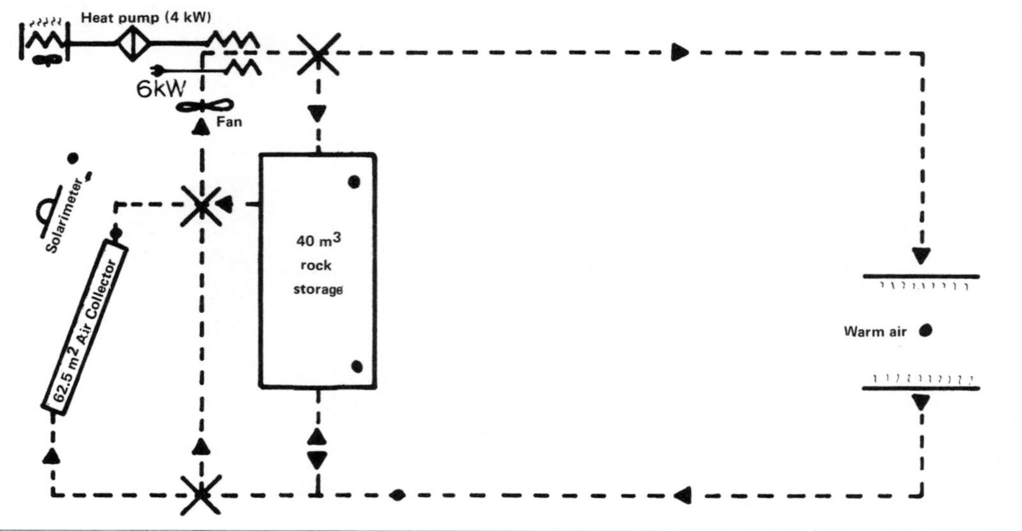

## TECHNICAL APPRAISAL/PRACTICAL EXPERIENCE

| REF. | F | DO | 79 | 80 | | 8 |
|---|---|---|---|---|---|---|

**System/Design:**

The size of collecting surface and of the rock-bed storage could be probably reduced.

**System/Installation:**

Difficulty in installing the rock-bed.

**Component Performance:** (Collector, Heat Exchangers, Storage, Pipework, Valves, Fittings, Pumps, Auxiliary)

- Special attention should be paid to the design of air-ducts and specially the air-valves which caused many difficulties.
- Considerable leaks from the collectors.
- Unpleasant smells came from collectors due to heat effects on wood.
- Heat-pump problems; cycle too quick, continual stop and go.

**System Operation/Controls, Electronics:**

- Monitoring too simple; the temperature of the air in the collectors is measured when there is no air circulating in the system. The system is monitored without knowledge of the temperature in the collectors once the system is functioning; hence a risk of stop and go oscillations exists. (This is a first stage monitoring system; it will be improved during the experiment.)

## PERFORMANCE EVALUATION/CONCLUSIONS

**Comparison of Measured with Predicted Performance**
(State reasons for any discrepancy and summarise the main factors affecting performance)

No data available yet, as the follow-up system was achieved in April 79. The analysis of a full year will be available in May 80.

**Modifications to System**

Leaks in the collectors have been repaired during summer 79.

The monitoring of the solar system will be improved.

**Occupants/Response**

The resident, who is research worker, has participated in the design of the solar system and in the preparation of this experiment; he contributes to the analysis of the experiment and to the improvement of the system.

**Future Work**

At spring 80 a new monitoring system, using a microprocessor (probably Z-80) will be installed in the house and connected to the heating system.

| PROJECT DESCRIPTION | LE HAVRE. FRANCE | REF. | F | LH | 77 | 78 | 9 |
|---|---|---|---|---|---|---|---|

**Main Participants:**

RESEARCH ENGINEERING MONITORING

ADE Research Department. Direction des Etudes et Recherches de Electricité de France (EDF) les Renardieres. Route de Sens. Ecuelles 77250 Moret-sur-Loing.

BUILDING MANAGEMENT

Societé Anonyme des HLM de l'Estuaire de la Seine. 16 rue Dupleix. 76000 Le Havre

FUNDING AUTHORITIES

Agence Nationale de Valorisation de la Recherche (ANVAR) 13 rue Madeleine Michelis, 92522 Neuilly Seine. Ministere de la qualité de la vie. EDF. ADE. CERCA. (Solar Collectors). Societé FBFC quartier des Berauts Zone Industrielle. 26100 Romans/Isere. Societe' des HLM de l'Estuaire de la Seine. Comité Interprofessionnel du logement de la Region Havraise.

**Project Description:**

A typical solar house in the Le Havre estate is a standard private house called 'Pavill on Normandie', several hundred of which have been built in the area since 1974 by the Societé des HLM de l'Estuaire de la Seine as 'all-electric' houses. The solar heating system consists of water cooled flat plate collectors, facing due south, water storage, and heat distribution with composite convectors (solar hot water coupled with electric resistance heaters) for space heating; Domestic hot water is solar pre-heated and electrically topped up.

**Project Objectives:**

1. To study the possibility of using solar energy to provide part of the space heating and hot water requirements of houses with normal insulation standards.

2. To find economical, optimum systems for the private housing market.

3. To find components and systems which are reliable and durable.

4. To obtain average measurements from the five solar houses and to compare these with forty three similar, all-electric reference houses in the neighbourhood.

5. To check the accuracy of the Abacus model used for calculating useful solar energy.

NOTE: For reference purposes ninety six all electric houses in the area with similar construction, were checked for energy consumption over a five year period. Two of the five solar houses were fully instrumented the remaining three being checked for auxiliary energy.

For this report, one of the instrumented houses has been chosen.

| SITE LOCATION MAP | LE HAVRE. FRANCE | REF. | F | LH | 77 | 78 | 9 |
|---|---|---|---|---|---|---|---|

– Distance to main city: 4 km from Le Havre (Town Centre)

– Please indicate height of nearest obstructions Flat plain, but 90m above Le Havre.

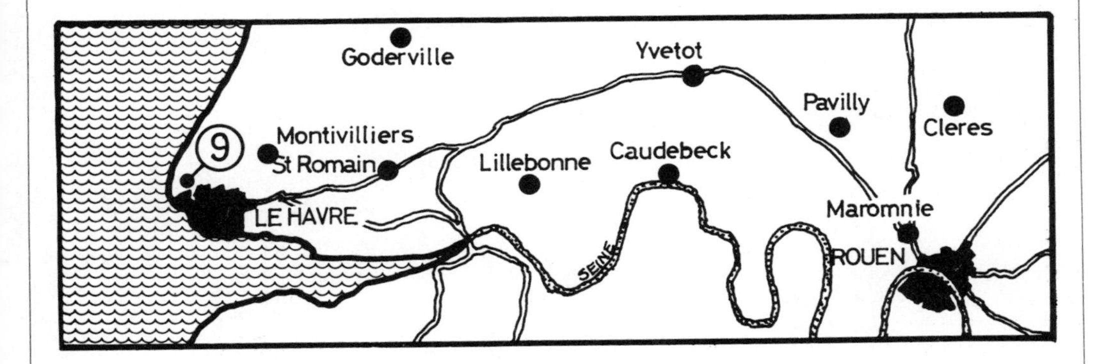

Photograph of Project

**SUMMARY SHEET** LE HAVRE . FRANCE **REF.** F | LH | 77 | 78 | **9**

**Design Data**

**A1 CLIMATE**

1) Source of Data: Met Station. Rouen. 72km from site for solar data cap de la Hève. 1km from site for temperature

2) Latitude: 49 ° 30 ′ Longitude: 0 ° 07 ′E Altitude: 92 metres

3) Global Irradiation (horiz plane): 1012 $kWh/m^2 \cdot year$ % Diffuse: 33

4) Degree Days: 2332 Base Temp: 18 °C

5) Sunshine Hours: July: 210.9 January: 49.6 Annual: 1602.6

**B1 BUILDING**

1) Building Type: detached house No Occupants: 4-5

2) Floor area: 116 $m^2$ Heated Volume: 290 $m^3$

3) Design Temperature: External: -7 °C Internal: 20°C °C

4) Ventilation Rate: 0.8 a.c.h. Vol. Heat Loss: 0.91 $W/m^3 \cdot K$

5) Space Heat Load: 12500 kWh Hot Water Load: 2000 kWh

**C1 SYSTEM**

1) Absorber Type: Aluminium Roll Bond. Black-Painted

2) Collector Area 39.5 $m^2$ (total) Coolant: Water & Additives

3) Orientation: 180° Tilt: 45° Glazing: Double-glazed

4) Storage Volume: 3 $m^3$ Heat Emitters: Composite

5) Auxiliary System: Electric Convectors Heat Pump:
SH. Off Peak electric DWH

**PERFORMANCE MONITORING**

1) Is there a Computer Model? Yes, two methods

2) Start date for Monitoring Programme: Winter 1975

3) Period for which Results Available: 1977/1978

4) No. of Measuring Points per house: 22

5) Data Acquisition System: 15 sensors, some connected to both a chart recorder and tape recording system, and 7 points connected to tape system only, including 2 electric meters.

| | Space Heating | Hot Water | Total |
|---|---|---|---|
| 6) Predicted | 43 % of 12500 kWh | 88 % of 2000 kWh | 49 % of 14500 kWh |
| 7) Measured 1977/78 | 31.4 % of 8875 kWh | 48 % of 4058 kWh | 36. % of 12933 kWh |

8) Solar Energy Used: Including useful losses ........ $kWh/m^2 \cdot year$
Excluding losses: 119.8 $kWh/m^2 \cdot year$

9) System Efficiency: $\frac{\text{Solar Energy Used}}{\text{Global Irradiation on collector}} \times 100 =$ 10.4 %

A2 LOCAL CLIMATE LE HAVRE. FRANCE REF. | F | LH | 77 | 78 | **9**

1 — Average Cloud cover: 6.2 Octas 1/7/77-30/6/78

2 — Average daily max temperatures (July): 19.8 °C

Average daily min temperatures (Jan): 3.1 °C

3 — Source of Weather Data:

For initial design work, mean ambient temperatures where taken from Cap de la Heve (Ref C) - averages from 1921-50. Global Irradiation was computed from Rouen data using sunshine hours (Ref R) averaged from 1950-75)

4 — Micro Climate/Site Description:

Meteorological data has been available from the Le Havre station since 23/9/76 (Ref H). The site was found not to have particular microclimatic characteristics except for a mean wind velocity markedly higher than at Cap de la Heve. There was good agreement for outside air temperature, but solar data from Le Havre is more reliable than the figures computed from Rouen data .

| | | Ref R | | | Ref H | Ref H | Ref R | Ref H | |
|---|---|---|---|---|---|---|---|---|---|
| MONTH | YEAR | Irradiation on horizontal plane: Global | Diffuse | Degree Days Base. 18 °C | Precipitation | Average Ambient Temperature | Average Relative Humidity | Average Wind Speed | Prevailing Wind Direction |
| JAN | 78 | 22.4 | | | 64 | 4.9 | 88.8 | 6.9 | |
| FEB | 78 | 37.3 | | | 74 | 4.0 | 80.4 | 5.7 | |
| MARCH | 78 | 76.2 | | | 97 | 7.4 | 79.8 | 6.5 | |
| APRIL | 78 | 109.3 | | | 73 | 7.7 | 78.5 | 4.3 | |
| MAY | 78 | 144.0 | | | 43 | 11.9 | 74.9 | 4.4 | |
| JUNE | 78 | 153.9 | | | 80 | 13.8 | 76.7 | 4.8 | |
| JULY | 77 | 151.0 | | | 18 | 16.7 | 77.2 | 4.5 | |
| AUGUST | 77 | 128.1 | | | 55 | 16.3 | 81.2 | 3.7 | |
| SEPT | 77 | 91,4 | | | 31 | 14.7 | 84.7 | 3.8 | |
| OCT | 77 | 54.9 | | | 62 | 13.9 | 83.9 | 5.2 | |
| NOV | 77 | 25.7 | | | 137 | 8.2 | 84.0 | 7.8 | |
| DEC | 77 | 17.7 | | | 58 | 7.2 | 87.9 | 6.0 | |
| Totals | | 1011.9 | | 2332 | 792 | | | | |
| Averages | | | 33%(R) | | | 10.4 | 81.5 | 5.3 | |
| Units | – | kWh/m²·month | | °C days | mm | °C | % | m/s | – |

**B2 BUILDING DESCRIPTION** LE HAVRE **REF.** F | LH | 77 | 78 | **9**

**Design Concepts**

The house is on two floors above a basement, and there are six main rooms. On the ground floor there is a living room, kitchen bedroom and utility room, which accomodates the solar equipment. On the first floor there are three bedrooms and a bathroom. In addition there is a garage outbuilding, and the solar storage tank is outside, below ground level.

**Construction**

1 – Glazing: 15 % Area: 18.92 $m^2$ 'U' value: 2.9 $W/m^2 \cdot K$

2 – 'U' values: Walls: 0.43 $W/m^2 \cdot K$ Floor: 1.05 $W/m^2 \cdot K$ Roof/loft: 0.41 $W/m^2 \cdot K$

3 – Thermal mass: 125 $kg/m^3$ Annual Energy Demand: 14500 kWh

Timber frame construction with masonry cladding. Walls are insulated with 8.5cm glass-wool. Windows are double glazed and include shutters on the ground floor.

**Energy Conservation Measures**

The house is insulated to approximately 30% better than minimum requirement for all-electric houses.

**Section Through House** Showing main solar components

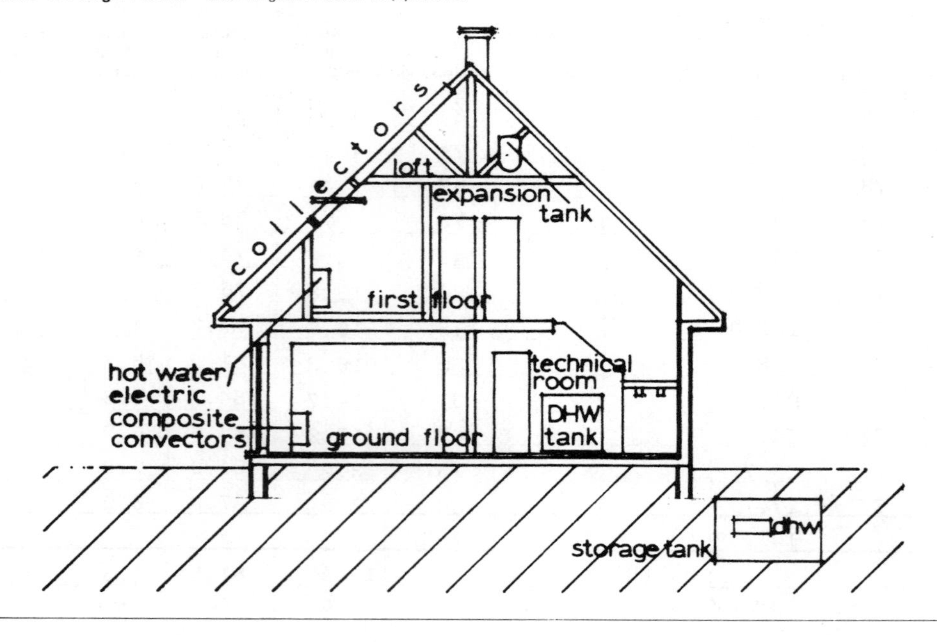

| C2 SYSTEM DESCRIPTION LE HAVRE. FRANCE | REF. | F | LH | 77 | 78 | 9 |
|---|---|---|---|---|---|---|

**Collector/Storage**

1 – Primary circuit. Water content: Primary 93 litres. Collectors 0.32 litres/m² collector area

2 – Insulation. Collector: 3cm Storage: 15cm Pipework: 3 or 4cm

3 – Frost Protection: 25% Ethylene Elycol

4 – Overheating Protection: Primary pump operated. Storage depleted if necessary.

5 – Corrosion Protection: 1-2% Inhibitor

**Space Heating/Domestic Hot Water**

1 – Auxiliary (S.H.): Electricity (DHW): 200 litre. 2.4kw immersion

2 – Heat Emitters: Composite convectors 8.5 kw Temp. Range: Adjustable

3 – DHW Insulation: 10cm Set Temperature: 60°C

**Other Points/** DWH pre-heated by 100 litre tank inside store. Collectors - 11 parallel sets of 2

**Control Strategy/Operating Modes**

A differential Thermostat operates the primary pump if DT > 3°C and stops it if DT ~ 0°C. A non-return valve prevents reverse thermocirculation at night. The primary heat exchanger is inside the storage tank. An adjustable aquastat (inside the store) permits solar heating if storage temperature exceeds 25°C. Heat distribution is ordered by a 2-level thermostate, 19.5°C - 20°C inside the living room. Solar energy if available is given priority at the composite convectors. If room temperature drops to below set temperature (-1°C), the electric convectors are operated.

**Diagram of System** Showing sensor locations, power ratings, storage volumes & flows.

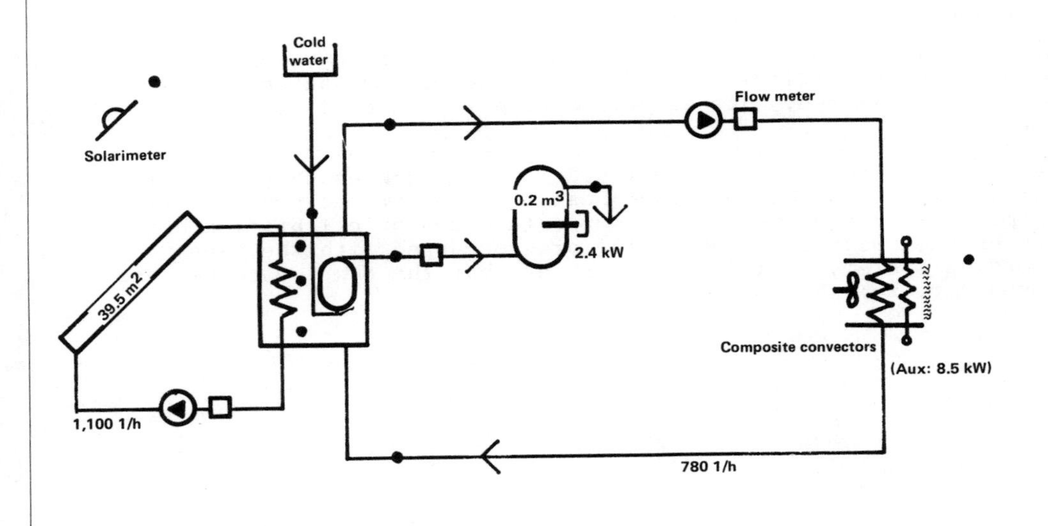

| TECHNICAL APPRAISAL/PRACTICAL EXPERIENCE | REF. | F | LH | 77 | 78 | 9 |
|---|---|---|---|---|---|---|

**System/Design:**

The initial design was computed using a simplified manual program, the Abacus method, for calculating useful solar energy. No useful measurements were obtained in 1976/77 due to problems with system operation.

**System/Installation:**

Collector leakage - expansion of rubber connection pipes and degradation of Aluminium panels. Mis-adjustment of non-return valve, the storage aquastat and 2-level thermostat caused deficient operation. The storage tank flooded, causing a four month delay.

**Component Performance:** (Collector, Heat Exchangers, Storage, Pipework, Valves, Fittings, Pumps, Auxiliary)

Useful data has been obtained during 1977-78. Heat losses have been much greater than anticipated. about 60% of stored energy, rather than 20%. Comparing solar energy stored, with the energy demand of the house, it was possible to calculate the non-useable energy ratio: approximately 50% of the total losses. Dynamic losses were increased due to malfunction of the non-return valve, and accounted for 30% of the total losses. If this had been avoided, global losses of stored energy would have been limited to 40%. Solar pre-heating of DHW was very inefficient, due to the use of a 100 litre tank located inside the main store.

**System Operation/Controls, Electronics:**

The 2 level thermostat and the aquastat ordering solar heating appeared to be difficult to adjust to take maximum advantage of solar energy stored. The aquastats were replaced after the store flooded.

The composite convectors were found to be strong enough, but they have made poor use of low grade heat. There were no problems with the monitoring equipment, but temperature sensor calibration has been checked regularly. Some failures occurred with the tape recording system, in an erratic manner, but the chart recorder has proved very reliable.

| PERFORMANCE EVALUATION/CONCLUSIONS | REF. | F | LH | 77 | 78 | 9 |
|---|---|---|---|---|---|---|

**Comparison of Measured with Predicted Performance**
(State reasons for any discrepancy and summarise the main factors affecting performance)

The initial estimates of heating loads were 18% above measured figures. This was because the ventilation rate had been assumed as 1.ach instead of the real 0.8, and room temperature assumed at 20°C instead of the real 19°C. When revised calculations were done, it was expected that solar energy would provide 45% of SH and 85% of DHW. Measured values were 31% and 38.7%. The main reasons for the shortfall were excessive heat losses. In a comparison with average all-electric houses, the average solar house had 26% less electricity consumption for space heating, but 31% in this particular house.

**Modifications to System**

- replacement of Aquastats
- adjustment of thermostatic controls
- Improvement of collector back insulation and changes of primary fluid contents and additive ratios

**Occupants/Response**

- they dis-adjusted non-return valves because of noise.
- they turned off the solar heating supply in mild weather
- there was no change in domestic habits due to solar
- electricity for DHW in solar houses was 34% greater than that for non-solar reference houses.

**Conclusions**

- engineering and measurements were constrained by HLM tenant rules which meant positioning the storage tank outside.
- lack of co-operation from component suppliers during failure periods
- the experiment has been very useful and many lessons have been learnt.

**Future Work**

Measurements stopped on June 30th 1978. Since then only auxiliary energy consumption for SH and DHW have been checked, on a 3 month basis.

**P1 MONTHLY PERFORMANCE** REF: F LH 76 77 9

| Month | Year | No Days Data | Average Temp °C Internal | External | SOLAR ENERGY kWh Global Irradiation On Collector Area | Solar Energy Collected | Solar Energy Used: Space Heating Useful Losses Incl. | Excl. | Domestic Hot Water | Input To Heat | Heat Pump Total Output kWh Space Heating | Domestic Hot Water | AUXILIARY ENERGY kWh Heat Pump | Space Heating | Domestic Hot Water | Pumps & Fans | Free Heat Gains kWh Internal | Direct Solar | SUMMARY kWh Total Load | % Percentage Solar | Syst. Eff. |
|---|---|---|---|---|---|---|---|---|---|---|---|---|---|---|---|---|---|---|---|---|---|
| Column No | | | 1 | 2 | 3 | 4 | 5 | 6 | 7 | 8 | 9 | 10 | 11 | 12 | 13 | 14 | 15 | 16 | 17 | 18 | 19 |
| 22/12-18/1 | 78 | 27 | 20.2 | 6.1 | 1025 | 470 | (42) | 42 | 79 | - | | | - | 1387 | 185 | | | | 1693 | 7 | 12 |
| 18/1-17/2 | 78 | 30 | 20.4 | 3.8 | 1887 | 875 | (299) | 299 | 120 | - | | | - | 1691 | 228 | | | | 2338 | 18 | 22 |
| 17/2-5/4 | 78 | 47 | 18.8 | 7.2 | 4368 | 1777 | (937) | 937 | 202 | - | | | - | 1233 | 348 | | | | 2720 | 42 | 26 |
| 5/4-2/5 | 78 | 27 | | 7.6 | 3470 | (1388) | (668) | 668 | 93 | - | | | - | 221 | 197 | | | | 1179 | 64 | 6 |
| M | 78 | 29 | 19.0 | 11.7 | 6545 | (2958) | (198) | 198 | 193 | - | | | - | 0 | 142 | | | | 553 | 73 | 4 |
| J | 78 | 29 | 19.2 | 13.4 | 4328 | (2054) | (24) | 24 | 243 | - | | | - | 1 | 119 | | | | 341 | 69 | 3 |
| J | 77 | 32 | 20.6 | 17 | 7051 | (3120) | 0 | 0 | 223 | - | | | - | 0 | 118 | | | | 327 | 64 | 4 |
| A | 77 | 31 | 20.7 | 15.5 | 5174 | (3245) | 0 | 0 | 210 | - | | | - | 0 | 117 | | | | 316 | 64 | 4 |
| S | 77 | 29 | 19.8 | 13.8 | 4911 | 2331 | 0 | 0 | 192 | - | | | - | 5 | 119 | | | | 509 | 61 | 8 |
| O | 77 | 34 | 20.1 | 13.2 | 3780 | 1986 | (135) | 135 | 182 | - | | | - | 9 | 183 | | | | 1473 | 62 | 26 |
| N | 77 | 29 | | 7.7 | 1626 | 757 | (301) | 301 | 131 | - | | | - | 826 | 215 | | | | 1116 | 29 | 19 |
| 2/12-22/12 | 77 | 20 | 19.3 | 7.0 | 1380 | 685 | (180) | 180 | 78 | - | | | - | 718 | 140 | | | | | 23 | |
| TOTAL | | | — | — | 45545 | (21646) | (2784) | 2784 | 1947 | - | | | - | 6091 | 2111 | | 8150 | | 12933 | — | — |
| AVERAGE | | | 19.5 | 10.4 | | | | | | | | | | | | | | | | 36 | 10 |

TOTAL: 4731 kWh

Annual averages for five solar houses: 6697 2421

(4): some figures are partly calculated due to tape recording failures; Primary pump duty time: 1659 hrs (813 hrs for S/H period); Storage temp (S/H period): top 41.3°C, bottom: 24.3°C

| PROJECT DESCRIPTION ODEILLO FRANCE | REF. | F | od | 74 | 75 | 10 |
|---|---|---|---|---|---|---|

Main Participants:

Laboratoire d'Energetique Solaire du Centre National de la Recherche Scientifique.

BP No.5 Odeillo 66120 Font Romeu, France
M Robert. Tel (68) 301024

Project Objectives:

1) To evaluate the amount of energy transmitted into the rooms by thermo-circulation, and by conduction through the wall.

2) To determine the average efficiency of the solar collector.

3) To evaluate the global thermal performance of the house.

Project Description:

This project is an example of the 'passive' approach to solar energy utilisation. It is the first occupied prototype solar house using the massive 'Trombe' wall system, and was built in 1967.

Photograph of Project

**SUMMARY SHEET** ODEILLO, FRANCE **REF.** F | od | 74 | 75 | **10**

**Design Data**

**A1 CLIMATE**

1) Source of Data: Odeillo
2) Latitude: 42 ° 29 ´ Longitude: 2 ° 1 ´ E Altitude: 1550 metres
3) Global Irradiation (horiz plane): 1605 kWh/m²·year % Diffuse: 34
4) Degree Days: 3942 Base Temp: 18 °C
5) Sunshine Hours: July: 320 January: 204 Annual: 2460

**B1 BUILDING**

1) Building Type: Single storey house No Occupants: 5
2) Floor area: 79.5 m² Heated Volume: 300 m³
3) Design Temperature: External: -8 °C Internal: 20 °C
4) Ventilation Rate: 1 a.c.h. Vol. Heat Loss: 1.67 W/m³·K
5) Space Heat Load: 48000 kWh Hot Water Load: - kWh

**C1 SYSTEM**

1) Absorber Type: Concrete Trombe Wall
2) Collector Area 55 m² (Aperture) Coolant: Air
3) Orientation: 180° Tilt: Glazing: Double
4) Storage Volume: 33 m³ Heat Emitters: Natural
5) Auxiliary System: Electrical Heat Pump:

**PERFORMANCE MONITORING**

1) Is there a Computer Model? Yes, after the experiments.
2) Start data for Monitoring Programme: November 1974.
3) Period for which Results Available: November 1974 to October 1975
4) No. of measuring points per house: 11 - 14
5) Data Acquisition System: 7 - 10 temperature measurements in system, outside air temp, anemometer, electricity consumption, 12 channel Chart Recorder.

| | Space Heating | Hot Water | Total |
|---|---|---|---|
| 6) Predicted | ... % of ... kWh | ... % of ... kWh | ... % of ... kWh |
| 7) Measured | 70 % of 27144 kWh | ... % of ... kWh | ... % of ... kWh |

8) Solar Energy Used: Including useful losses ... kWh/m²·year
Excluding losses: 347 kWh/m²·year

9) System Efficiency: $\frac{\text{Solar Energy Used}}{\text{Global Irradiation on collector}} \times 100 =$ 30 %

A2 LOCAL CLIMATE ODEILLO, FRANCE REF. | F | od | 74 | 75 | 10

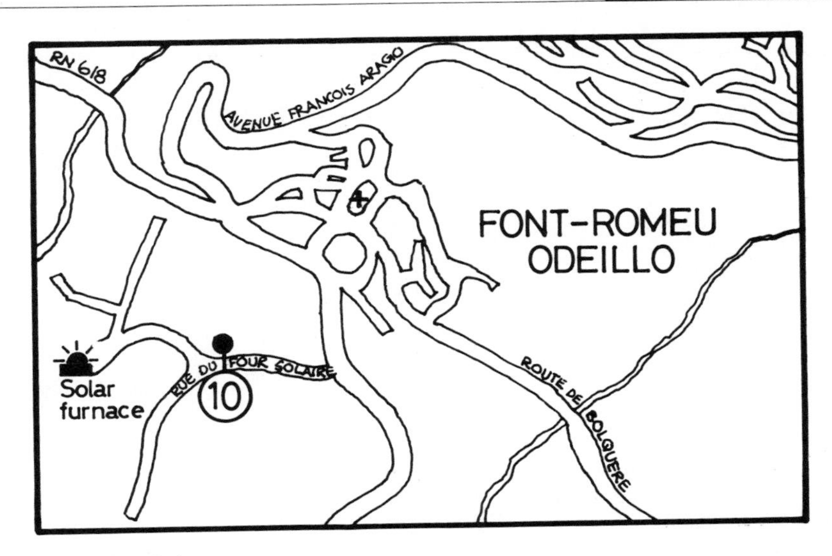


Micro Climate/Site Description:

The climate of Odeillo is a classical altitude climate with an important mediterranean influence. Odeillo is located on a south facing zone of the Cerdagne plain whose lower altitude is 1200 metres.

| MONTH | YEAR | Irradiation on horizontal plane | | Degree Days | Precipitation | Average Ambient Temperature | Average Relative Humidity | Average Wind Speed | Prevailing Wind Direction |
|---|---|---|---|---|---|---|---|---|---|
| | | Global | Diffuse | Base. 18°C | | | | | |
| JAN | 74 | 71.4 | 24.9 | 500 | 42 | 1.8 | 43 | | |
| FEB | 74 | 86.8 | 33.0 | 474 | 62 | 1.2 | 49 | | |
| MARCH | 74 | 129.9 | 45.9 | 484 | 59 | 2.3 | 45 | | |
| APRIL | 75 | 166.1 | 63.4 | 409 | 77 | 4.4 | 49 | | |
| MAY | 75 | 184.9 | 83.4 | 304 | 76 | 8.2 | 54 | | |
| JUNE | 75 | 189.7 | 63.5 | 180 | 67 | 12.0 | 57 | | |
| JULY | 75 | 211.0 | 56.0 | 90 | 108 | 15.1 | 54 | | |
| AUGUST | 75 | 177.4 | 52.0 | 95 | 96 | 15.0 | 56 | | |
| SEPT | 74 | 140.9 | 41.5 | 192 | 92 | 11.6 | 52 | | |
| OCT | 74 | 112.2 | 33.8 | 320 | 49 | 7.6 | 53 | | |
| NOV | 74 | 74.6 | 24.0 | 425 | 39 | 3.8 | 44 | | |
| DEC | 74 | 60.8 | 18.8 | 469 | 59 | 2.8 | 49 | | |
| Totals | | 1605.7 | 540.2 | 3942 | 896 | | | | |
| Averages | | | | | | 7.2 | 50.4 | 5 | 2850 |
| Units | – | kWh/m²·month | | °C days | mm | °C | % | m/s | – |

| B2 BUILDING DESCRIPTION | ODEILLO | REF. | F | od | 74 | 75 | 10 |
|---|---|---|---|---|---|---|---|

**Design Concepts**

The house is single storey, and detached, with four principal rooms.

**Construction**

1 – Glazing: ........ % Area: 15 $m^2$ 'U' value: ........ $W/m^2 \cdot K$

2 – 'U' values: Walls: 1.97 $W/m^2 \cdot K$ Floor: 0.65 $W/m^2 \cdot K$ Roof/loft: 0.95 $W/m^2 \cdot K$

3 – Thermal mass: 300 $kg/m^3$ Annual Energy Demand: 48.000* kWh

*excluding DHW

The south wall is a Trombe-wall consisting of 60cm concrete. The north east and west walls are made with hollow concrete blocks, 25cm thick, with timber cladding.

**Energy Conservation Measures**

Windows are single glazed but include timber shutters.

Free Heat Gains: Internal: 2500 kWh/year Direct Solar: 3300 kWh/year

**Section Through House** Showing main solar components

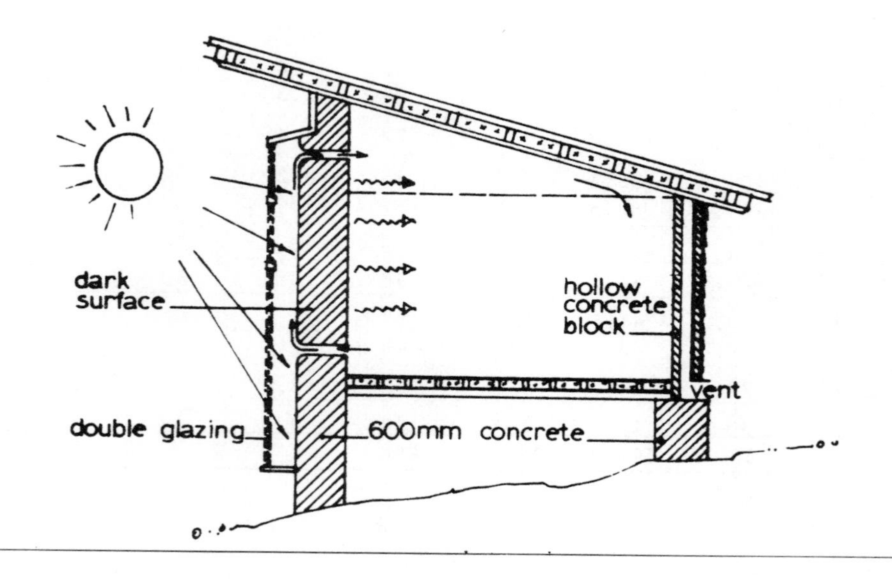

| C2 SYSTEM DESCRIPTION ODEILLO | REF. | F | od | 74 | 75 | 10 |
|---|---|---|---|---|---|---|

**Collector/Storage**

1 — Primary circuit. Water content: ........ litres/m$^2$ collector area

2 — Insulation. Collector: ........ Storage: ........ Pipework: ........

3 — Frost Protection: ........

4 — Overheating Protection: ........

5 — Corrosion Protection: ........

**Space Heating/Domestic Hot Water**

1 — Auxiliary (S.H.): Electric Convectors, 12KW (DHW): Electric

2 — Heat Emitters: ........ Temp. Range: ........

3 — DHW Insulation: ........ Set Temperature: ........

**Other Points/Heat Pump:**

**Control Strategy/Operating Modes**

Energy is collected between the double glazing and the dark-painted outside surface of the concrete wall. Cool air enters this space at low level and warmed air is transmitted into the room at high level by natural convective thermocirculation, control is by means of manual dampers. Energy is also transmitted through the wall over a 15 hour period in the case of a 60cm wall. There are no dampers to permit rejection of excess heat to outside during the summer.

**Diagram of System** Showing sensor locations, power ratings, storage volumes & flows.

Solarimeter

Temperature sensor

Cold water

55m$^2$ concrete wall

Electric convectors (12kW)

| TECHNICAL APPRAISAL/PRACTICAL EXPERIENCE | REF. | F | od | 74 | 75 | 10 |
|---|---|---|---|---|---|---|

System/Design:

The data collected are used to evaluate, on an hourly basis, the amounts of energy transmitted into the house by the collector, the house being maintained at 20oC by the electrical heating system and consequently, the air inlet temperature being constant.

1 - Thermocirculation of air

$$Qt = m\,Cp\ (Ts-Te)$$

m, the mass flow-rate is evaluated from the velocity measurement in the center of the lower vent by the determination of the average velocity v. This average velocity v was determined by the log-linear method for different flow rates.

2 - Restitution of energy by the wall

2-a-Radiation The energy restored by the thermal wall by radiation is evaluated by

$$Qr = \frac{\varepsilon}{2-\varepsilon}\,\sigma\,(Tm^4 - Tref^4)$$

$\varepsilon$ : emissivity of the walls of the room and of the solar wall
$\sigma$ : Stephan-Boltz-man constant
Tm : Temperature of the solar wall
Tref : Reference temperature of the other surfaces of the room.

Tref is an experimental average barycentric temperature corrected by the "shape factors" of the different walls versus the solar wall.

2-b-Convection. It is very difficult to know exactly the nature of the convection along a heated wall whose temperature is around 10°C above the room temperature (maximum). In addition, the nature of the flow can be modified by events occuring in a lived-in house : opening of doors and windows, passages of persons, draughts,.. So, the convection heat transfer was evaluated by averaging a turbulent and a laminar coefficient :

$$Q_{la} = 5,10\ (Tm-Ta)^{5/4}\quad (Kj/m^2.h)$$

$$Q_{tu} = 6,36\ (Tm-Ta)^{4/3}\quad (Kj/m^2.h)$$

3 - Evaluation of the efficiency of the collector

The amount of energy transmitted by the collector is compared for 24 hours or longer periods, with the incident solar energy on the collector.

Note 1 : The heat transmission through the storage wall was also evaluated by a finite difference method for the solution of the conduction heat transfer equation : the results are very similar of the results obtained by the method described above.

Note 2 : The precision of the evaluation of the heat balances and of the efficiencies are respectively 13% and 4%.

| PERFORMANCE EVALUATION/CONCLUSIONS | REF. | F | od | 74 | 75 | 10 |
|---|---|---|---|---|---|---|

**Comparison of Measured with Predicted Performance**
(State reasons for any discrepancy and summarise the main factors affecting performance)

From November 74 to October 75, the measured heating load was 27,144 kwh, or 56% of the calculated load. During this period the solar heat gain was 19094 kwh, or 70.3% of the load. During the past 9 years, the annual solar contribution has been an average of 60%.

System Efficiency during the monitoring period was 29.9% with a range from 16% in June to 36% in November.

**Modifications to System**

In subsequent test houses, the following modifications were made :-

- Reduction in thickness of collector to 37cm
- Air space reduced to 8cm
- Improved thermal insulation and reduction in collector area

**Conclusions**

The Trombe wall system operating in the climatic conditions at Odeillo can provide a significant part of the house heating load, excess heat is sometimes provided in sunny periods. Energy stored in the wall is generally not sufficient for the night and regulation is impossible.

**Future Work**

1) Reduction of Collector heat losses
2) Rejection of excess heat in summer
3) Passive Regulator to control air movement
4) Improve storage efficiency

## P1 MONTHLY PERFORMANCE

REF: F 0D 74 75 10

| Month | Year | No Days Data | Average Temp °C | | SOLAR ENERGY kWh | | | | | | Heat Pump Total Output kWh | | AUXILIARY ENERGY kWh | | | | Free Heat Gains kWh | | SUMMARY kWh | % | |
|---|---|---|---|---|---|---|---|---|---|---|---|---|---|---|---|---|---|---|---|---|---|
| | | | Internal | External | Global Irradiation On Collector Area | Solar Energy Collected | Solar Energy Used: Space Heating Useful Losses Incl. | Space Heating Useful Losses Excl. | Domestic Hot Water | Input To Heat | Space Heating | Domestic Hot Water | Heat Pump | Space Heating | Domestic Hot Water | Pumps & Fans | Internal | Direct Solar | Total Load | Percentage Solar | Syst. Eff. |
| Column No | | | 1 | 2 | 3 | 4 | 5 | 6 | 7 | 8 | 9 | 10 | 11 | 12 | 13 | 14 | 15 | 16 | 17 | 18 | 19 |
| J | 75 | 31 | 20 | 4.4 | 7432 | | 2654 | 2654 | - | - | | | - | 1350 | | | 725 | | 4004 | 66 | 36 |
| F | 75 | 27 | 20 | 2.4 | 6841 | | 2296 | 2296 | - | - | | | - | 710 | | | 725 | | 3006 | 76 | 33 |
| M | 75 | 21 | 20 | 0.9 | 2800 | | 1267 | 1267 | - | - | | | - | 1500 | | | 725 | | 2767 | 46 | 33 |
| A | 75 | 25 | 20 | 2.5 | 5741 | | 1504 | 1504 | - | - | | | - | 815 | | | 725 | | 2319 | 65 | 26 |
| M | 75 | 30 | 20 | 6.5 | 4944 | | 972 | 972 | - | - | | | - | 925 | | | 725 | | 1897 | 51 | 20 |
| J | 75 | 30 | 20 | 10.9 | 4740 | | 736 | 736 | - | - | | | - | 0 | | | - | | 736 | 100 | 15 |
| J | 75 | 3 | 22 | 16.2 | 538 | | 110 | 110 | - | - | | | - | 0 | | | - | | 110 | 100 | 20 |
| A | 75 | 20 | 21 | - | 3414 | | 901 | 901 | - | - | | | - | 0 | | | - | | 901 | 100 | 26 |
| S | 75 | 23 | 22 | 11.4 | 4744 | | 1283 | 1283 | - | - | | | - | 0 | | | - | | 1283 | 100 | 27 |
| O | 75 | 30 | 21 | 8.2 | 8703 | | 2909 | 2909 | - | - | | | - | 300 | | | 725 | | 3209 | 91 | 33 |
| N | 74 | 25 | 20 | 3.7 | 4761 | | 1594 | 1594 | - | - | | | - | 1250 | | | 725 | | 2844 | 56 | 33 |
| D | 74 | 31 | 20 | 5.4 | 8230 | | 2866 | 2866 | - | - | | | - | 1200 | | | 725 | | 4066 | 68 | 35 |
| TOTAL | | | — | — | 63892 | | 19094 | 19094 | - | - | | | - | 8050 | | | 5800 | | 27144 | — | — |
| AVERAGE | | | | 6.6 | | | | | | | | | | | | | | | | 70 | 30 |

TOTAL: 19094 kWh

*Domestic hot water not included.
(16) direct solar: 3300 kWh/year.

| PROJECT DESCRIPTION | AACHEN | REF. | D | Aa | 76 | 77 | 11 |
|---|---|---|---|---|---|---|---|

Main Participants:

Philips GmbH
Forschungslaboratorium Aachen,
Weisshausstra.. 5100 Aachen, West Germmany

Project Description:

The project consists of a one family, single house with attic and cellar, pre-fabricated "Streif-type", extra insulated.

Experimental house with the following features:

- SH with solar energy
- DHW with solar energy
- DHW from waste water enthalphy by a heat pump
- ventilation with heat recovery
- SH by heat pump coupled to an earth heat exchanger
- cooling via earth heat capacity.

Project Objectives:

To evaluate various measures of reducing the energy needs of buildings and using new sources.

Photograph of Project

| SUMMARY SHEET AACHEN | REF. | D | Aa | 76 | 77 | 11 |
|---|---|---|---|---|---|---|

**Design Data**

**A1 CLIMATE**

1) Source of Data: Site measurements
2) Latitude: 50 ° 76 ′ Longitude: 6 ° 09 Altitude: 195 metres
3) Global Irradiation (horiz plane): 1109 $kWh/m^2 \cdot year$ % Diffuse:
4) Degree Days: 3231 Base Temp: 15 °C
5) Sunshine Hours: July: 189 January: 50 Annual: 1,510

**B1 BUILDING**

1) Building Type: Single family house No Occupants: Simulated (4)
2) Floor area: 116 $m^2$ Heated Volume: 290 $m^3$
3) Design Temperature: External: 20 °C Internal: 14,16,20,22 °C dependent on room
4) Ventilation Rate: 1 (winter) 2(summer) a.c.h. Vol. Heat Loss: 0.03 $W/m^3 \cdot K$
5) Space Heat Load: 2,766 kWh (1976/77) Hot Water Load: 2,605 kWh

**C1 SYSTEM**

1) Absorber Type: Philips Mark I Blackened (Enamel 1 Metal Tubes
2) Collector Area 20.3 $m^2$ (Aperture) Coolant: Water
3) Orientation: 180° Tilt: Glazing: Evacuated glass tube
4) Storage Volume: 42 + 4 = 46 $m^3$ Heat Emitters: Radiators
5) Auxiliary System: None Heat Pump: Yes

**PERFORMANCE MONITORING**

1) Is there a Computer Model? Yes
2) Start Date for Monitoring Programme: 1975
3) Period for which Results Available: 1975 - 1977
4) No. of Measuring Points per house: 139
5) Data Acquisition System: P855 Mini Computer.

| | Space Heating | Hot Water | Total |
|---|---|---|---|
| 6) Predicted | % of kWh | % of kWh | % of kWh |
| 7) Measured | % of kWh | % of kWh | 50 % of 5448 kWh |

8) Solar Energy Used: Including useful losses $kWh/m^2 \cdot year$; Excluding losses: 134 $kWh/m^2 \cdot year$

9) System Efficiency: $\frac{\text{Solar Energy Used}}{\text{Global Irradiation on collector}} \times 100 =$ 12 %

| SITE LOCATION MAP AACHEN | REF. | D | AA | 76 | 77 | 11 |
|---|---|---|---|---|---|---|

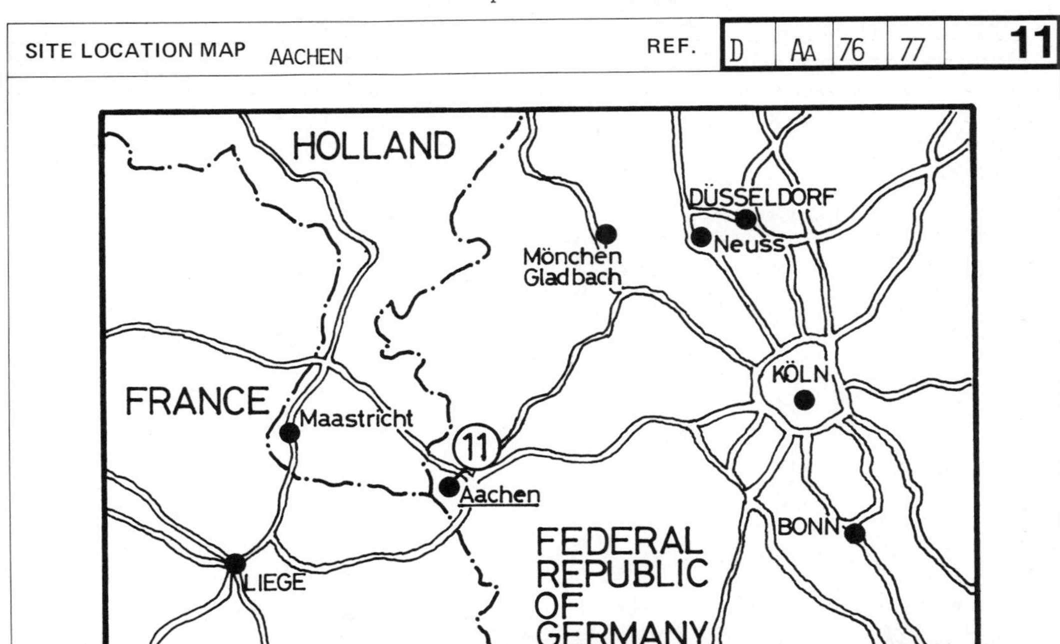


## A2 LOCAL CLIMATE

Average daily max temps (July): 26.7°C; average daily min temps (Jan): -2.7°C

* measured in 76/77 heating season on 48° tilted plan
** /m²
*** for 1969-1977

| MONTH | YEAR | Irradiation on horizontal plane | | Degree Days | ** | Average | Average | *** | Prevailing |
|---|---|---|---|---|---|---|---|---|---|
| | | * Global | Diffuse | Base: 15 °C | Precipitation | Ambient Temperature | Relative Humidity | Average Wind Speed | Wind Direction |
| JAN | 60-70 | 34.8 | | 536 | 64 | 1.7 | 84 | 4.1 | |
| FEB | " | 56.0 | | 459 | 59 | 2.6 | 82 | 3.7 | |
| MARCH | " | 96.6 | | 443 | 59 | 4.7 | 77 | 3.4 | |
| APRIL | " | 100.9 | | 279 | 76 | 9.7 | 74 | 3.5 | |
| MAY | " | 132.9 | | 206 | 68 | 12.4 | 72 | 2.8 | |
| JUNE | " | 154.8 | | – | 78 | 15.9 | 72 | 2.5 | |
| JULY | " | 146.6 | | – | 88 | 16.8 | 75 | 2.7 | |
| AUGUST | " | 150.6 | | – | 97 | 16.4 | 77 | 2.3 | |
| SEPT | " | 91.6 | | 129 | 53 | 14.7 | 78 | 2.7 | |
| OCT | " | 80.2 | | 245 | 60 | 11.1 | 81 | 2.9 | |
| NOV | " | 28.9 | | 396 | 75 | 5.8 | 83 | 4.5 | |
| DEC | " | 34.9 | | 539 | 79 | 1.6 | 85 | 3.8 | |
| Totals | | 1108.8 | | 3231 | 856 | | | | |
| Averages | | 92.4 | | | 71 | 9.4 | 78 | 3.2 | |
| Units | – | kWh/m²·month | | °C days | mm | °C | % | m/s | – |

| B2 BUILDING DESCRIPTION | AACHEN | REF. | D | Aa | 76 | 77 | 11 |
|---|---|---|---|---|---|---|---|

**Design Concepts**

Standard prefabricated "Streif-house" subdivided into the following areas :-

attic; solar collectors
upper floor; data acquisition, systems control and household simulation.
ground floor; experimental living area
cellar; heat storage, heat pump, air conditioning equipment
earth; heat exchanger under foundations

**Construction**

1 – Glazing: 10 % Area: 11.4 $m^2$ 'U' value: 1.9 $W/m^2 \cdot K$

2 – 'U' values: Walls: 0.17 $W/m^2 \cdot K$ Floor: 0.3 $W/m^2 \cdot K$ Roof/loft: 0.23 $W/m^2 \cdot K$

3 – Thermal mass: ........ $kg/m^3$ Annual Energy Demand: ........ kWh

**Energy Conservation Measures**

Extra insulation in walls, floors and ceilings; reduced uncontrolled ventilation losses; windows with low conductivity; heat recovery from exhaust air and waste water; use of heat pumps.

Free Heat Gains: Internal: 5,460 kWh/year Direct Solar: 1,800 kWh/year

**Section Through House** Showing main solar components

air outlet
air inlet
data aquisition and control
air inlet
cold
hot
water inlet
water outlet
earth heat exchanger

HEATING SYSTEM

air outlet
air inlet
radiator
hollow cinder brick wall
ventilator
filter
cooler
rotating heat exchanger econovent

VENTILATION SYSTEM

| C2 SYSTEM DESCRIPTION | AACHEN | REF. | D | Aa | 76 | 77 | 11 |
|---|---|---|---|---|---|---|---|

**Collector/Storage**

1 — Primary circuit. Water content: 8 litres/$m^2$ collector area

2 — Insulation. Collector: vacuum Storage: rockwool Pipework: rockwool

3 — Frost Protection: drainage of the system

4 — Overheating Protection: " " " "

5 — Corrosion Protection: none (essentially copper glass)

**Space Heating/Domestic Hot Water**

1 — Auxiliary (S.H.): Heat Pump (DHW): Heat Pump,

2 — Heat Emitters: 11 radiators with thermostat valves Temp. Range: 45°C inlet

3 — DHW Insulation: rock wool Set Temperature: 45-55°C

**Other Points/Heat Pump:**
An air-earth heat exchanger is formed by a hollow brick wall of $62.3m^3$, situated alongside the cellar walls. **Water and air heat recovery.**

**Control Strategy/Operating Modes**

1. Filling the system
2. Short circuit
3. Solar collector connected to seasonal storage tank
4. Solar collector connected to heating and hot water storage tank

**Diagram of System** Showing sensor locations, power ratings, storage volumes & flows.

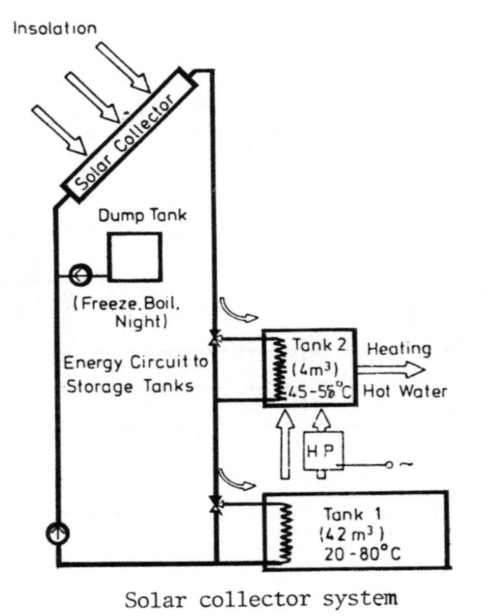


Solar collector system

## P1 MONTHLY PERFORMANCE

REF: D | AA | 76 | 77 | 11

| Month | Year | No Days Data | Average Temp °C Internal | Average Temp °C External | SOLAR ENERGY kWh: Global Irradiation On Collector Area | Solar Energy Collected | Solar Energy Used: Space Heating Useful Losses Incl. | Space Heating Useful Losses Excl. | Domestic Hot Water | Input To Heat | Heat Pump Total Output kWh: Space Heating | Domestic Hot Water | AUXILIARY ENERGY kWh: Heat Pump | Space Heating HP+rad +H.rec. | Domestic Hot Water HP+H.rec | Water H.recov | Free Heat Gains kWh: Internal | Direct Solar | SUMMARY* kWh: Total Load | %: Percentage Solar | Syst. Eff. |
|---|---|---|---|---|---|---|---|---|---|---|---|---|---|---|---|---|---|---|---|---|---|
| Column No | | | 1 | 2 | 3 | 4 | 5 | 6 | 7 | 8 | 9 | 10 | 11 | 12 | 13 | 14 | 15 | 16 | 17 | 18 | 19 |
| J | 77 | | | 2.5 | 706 | 237 | | 0 | 16 | 479 | | | heat pump consumption included in cols. 12 and 13 | 156 | 206 | 171 | 500 | | 857 | 58 | 70 |
| F | 77 | | | 4.9 | 1137 | 468 | | 100 | 30 | 337 | | | | 107 | 97 | 95 | 479 | | 671 | 70 | 41 |
| M | 77 | | | 7.5 | 1961 | 864 | | 44 | 130 | 199 | | | | 18 | 94 | 72 | 475 | | 485 | 77 | 19 |
| A | 77 | | | 6.2 | 2048 | 814 | | 48 | 125 | 299 | | | | 73 | 141 | 140 | 440 | | 647 | 67 | 21 |
| M | 76 | | | 11.5 | 2698 | 1182 | | 0 | 186 | 0 | | | | | 53 | 123 | 408 | | 239 | 78 | 7 |
| J | 76 | | | 19.0 | 3142 | 1332 | | 0 | 0 | 0 | | | | | 260 | 218 | 439 | | 260 | 0 | 0 |
| J | 76 | | | 20.6 | 2976 | 934 | | 0 | 0 | 0 | | | | | 168 | 141 | 337 | | 168 | 0 | 0 |
| A | 76 | | | 17.8 | 3057 | 1040 | | 0 | 0 | 0 | | | | | 199 | 163 | 465 | | 199 | 0 | 0 |
| S | 76 | | | 14.0 | 1860 | 511 | | 0 | 0 | 0 | | | | | 236 | 196 | 435 | | 236 | 0 | 0 |
| O | 76 | | | 11.1 | 1628 | 518 | | 0 | 0 | 0 | | | | | 275 | 236 | 438 | | 275 | 0 | 0 |
| N | 76 | | | 6.1 | 584 | 97 | | 509 | 0 | 0 | | | | | 255 | 174 | 447 | | 764 | 67 | 87 |
| D | 76 | | | 1.2 | 708 | 169 | | 182 | 0 | 73 | | | | 258 | 134 | 147 | 497 | | 647 | 39 | 36 |
| TOTAL | | | — | — | 22505 | 8166 | | 883 | 487 | 1348 | | | | 612 | 2118 | 1876 | 5460 | | 5448 | — | — |
| AVERAGE | | | | 10.2 | | | | | | | | | | | | | | | | 50 | 12 |

(8) :heat pump used for SH & DHW;

TOTAL: 2718 kWh | C.O.P =

(12)(13):includes heat recovery and heat pump; (17):includes waste water heat recovery; (14):water heat recovery used for SH and DHW.

* Solar Input To Heat Pump Included As Solar Contribution To Total Load

* Total Output From Heat Pump Included As Contribution To Total Load

| PROJECT DESCRIPTION | ESSEN | REF. | D | Es | 78 | 78 | 12 |
|---|---|---|---|---|---|---|---|

**Main Participants:**

Dornier/RWE Solar House Essen

Dornier System GmbH, Postfach 13 60, 7990 Friedrichshafen, FRG

RWE, Postfach 27, 4300 Essen 1, FRG

**Project Description:**

The project consists of a single house with eleven rooms on the two lower floors. It is of conventional construction. Dornier heat-pipe-collectors, 65m$^2$ effective area, provide thermal energy for SH, DHW and a 27.5m$^2$ indoor swimming pool.

**Project Objectives:**

- to obtain data for energy demand and distribution in a conventional one family house
- to obtain data for the distribution of directly used solar energy and indirectly used by way of a heat pump.
- to obtain data for the distribution of solar energy and auxiliary energy to the various heating demands.
- to obtain results on the efficiency of solar heat pipe collectors for both short-term and long-term conditions.
- the house is occupied by adults and children in order to provide realistic conditions.

| SITE LOCATION MAP | REF. | D | ES | 78 | 78 | 12 |
|---|---|---|---|---|---|---|

Distance to main city:12 km from Essen downtown.
Height of nearest obstructions: No shading by surrounding houses. Distant trees and multi-storey houses decrease irradiation in the morning and evening hours slightly

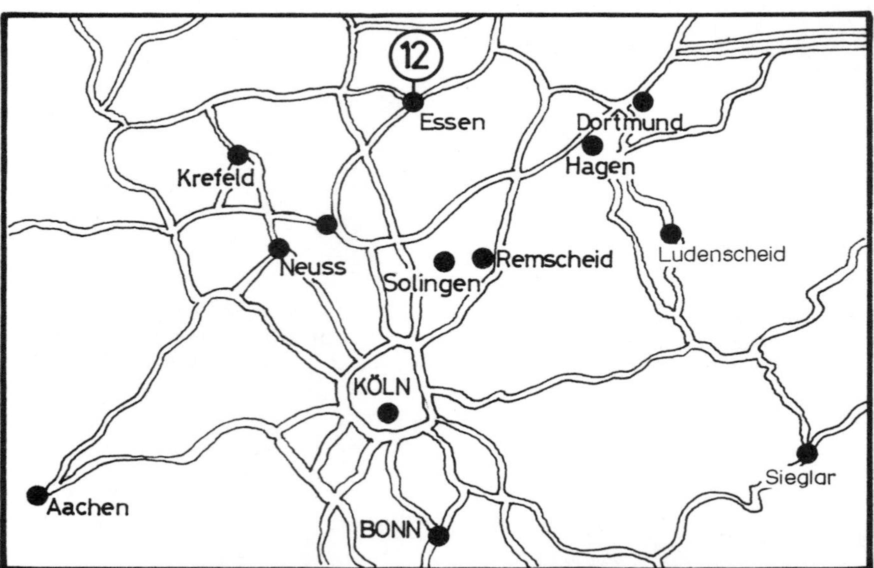

Photograph of Project

**SUMMARY SHEET** ESSEN **REF.** D | Es | 78 | 78 | **12**

**Design Data**

**A1 CLIMATE**

1) Source of Data: On site measurements

2) Latitude: 51 ° 27 ′ Longitude: 6 ° 57 ′ Altitude: 112 metres

3) +Global Irradiation (horiz plane): 932 kWh/m²·year % Diffuse:

4) +Degree Days: 3,881 Base Temp: 15 °C

5) +Sunshine Hours: July: 151 January: 49 Annual: 1,257

(+) 1978 data

**B1 BUILDING**

1) Building Type: Single family-house No Occupants: 3 adults and 1 child

2) Floor area: 200 m² Heated Volume: 552 m³

3) Design Temperature: External: 15 °C Internal: 21 °C

4) Ventilation Rate: a.c.h. Vol. Heat Loss: W/m³·K

5) Space Heat Load: kWh Hot Water Load: kWh

**C1 SYSTEM**

1) Absorber Type: Flat Plate. Heat Pipe. Aluminium Tube/Sheet

2) Collector Area 65 m² (Aperture) Coolant: Freon/Water

3) Orientation: 207° Tilt: 48° Glazing: Double Plexiglass

4) Storage Volume: 7.2 m³ Heat Emitters: Floor Heaters

5) Auxiliary System: Heat Pump: Yes

**PERFORMANCE MONITORING**

1) Is there a Computer Model? No

2) Start date for Monitoring Programme: March 1976

3) Period for which Results Available:

4) No of Measuring Points per house: 98

5) Data Acquisition System: Camac

| | Space Heating | Hot Water | Total |
|---|---|---|---|
| 6) Predicted | % of kWh | % of kWh | % of kWh |
| 7) Measured | % of kWh | % of kWh | 43 % of 25739 kWh |

8) Solar Energy Used: Including useful losses 173 kWh/m²·year

Excluding losses: kWh/m²·year

9) System Efficiency: Solar Energy Used / Global Irradiation on collector x 100 = 19 %

| A2 LOCAL CLIMATE | REF. | D | ES | 78 | 78 | 12 |
|---|---|---|---|---|---|---|

Average daily max temps(July):26°C(78),28.4°C(76),
Average daily min temps(Jan) :-1.7°C(78), -8.1°C(76).
Source of weather data:Bocholt 50km NW of Essen.
Micro climate/site description: Industrial region with decreased sunshine and equalized ambient temperatures.

| MONTH | YEAR | Irradiation on horizontal plane | | Degree Days | Precipitation | Average Ambient Temperature | Average Relative Humidity |
|---|---|---|---|---|---|---|---|
| | | Global | Diffuse | Base 15 °C | | | |
| JAN | 1978 | 17.7 | | 543.0 | | 2.5 | 84 |
| FEB | | 38.9 | | 539.6 | | 0.7 | 82 |
| MARCH | | 54.8 | | 430.7 | | 6.1 | 81 |
| APRIL | | 122.5 | | 365.0 | | 7.8 | 71 |
| MAY | | 132.7 | | 224.1 | | 12.8 | 74 |
| JUNE | | 142.9 | | 126.5 | | 15.2 | 72 |
| JULY | | 138.3 | | 120.2 | | 15.8 | 74 |
| AUGUST | | 122.6 | | 94.1 | | 15.5 | 74 |
| SEPT | | 71.8 | | 193.2 | | 13.0 | 83 |
| OCT | | 46.8 | | 263.8 | | 11.0 | 82 |
| NOV | | 26.9 | | 430.1 | | 5.7 | 79 |
| DEC | | 15.8 | | 550.3 | | 2.2 | 84 |
| Totals | | 931.7 | | 3880.6 | | 108.3 | |
| Averages | | 77.64 | | 323.38 | | 9.03 | 78.9 |
| Units | – | kWh/m²· month | | °C days | mm | °C | % |

## B2 BUILDING DESCRIPTION

**Design Concepts**

Conventional construction walls consisting of (from inside to outside):
11.5cm clinker (k=0.9 W/mK) roof consisting of tiles, 2cm air, 5cm glass wool(h=0.035),
17.5cm brick (h-0.4), 1.0cm plaster (h=0.17).

**Construction**

'U' values (W/m².K): 2.6/3.3 (glazing), 0.43 (walls), 0.29 (roof/loft).
Brick construction.

**Energy Conservation Measures**

Roof:insulation with 12cm "deltafoil" and wooden boards; windows:insulating glass, 2 sheets with 12mm air layer, at indoor swimming pool 3 sheets (h=2.3 */mK).

B2 BUILDING DESCRIPTION (CONT) REF. | D | Es | 78 | 78 | 12

**Section Through House** Showing main solar components

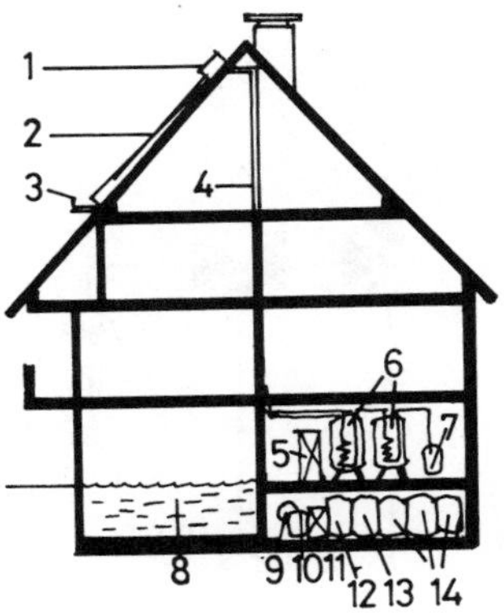

1. Heat exchanger
2. Solar collectors
3. Snow fence, walking board
4. Collector circuit
5. Control unit
6. Warmwater storage
7. Expansion vessel, collector circuit
8. Swimming pool
9. Expansion vessel, SH
10. Filter and solar afterheater
11. Heat pump
12. Heat storage (heat pump condensator)
13. Heat storage (heat pump evaporator)
14. Heat storage

## C2 SYSTEM DESCRIPTION

**Collector/Storage**

1 — Primary circuit. Water content: 0.4 litres/m$^2$ collector area

2 — Insulation. Collector: PU-foam Storage: Mineral wool Pipework: Armaflex

3 — Frost Protection: Glysantin

4 — Overheating Protection: expansion vessel;pressure valve, magnetic valve,heat pipe effect.

5 — Corrosion Protection: Antikorrosion fluid

**Space Heating/Domestic Hot Water**

1 — Auxiliary (S.H.): electric, heat pump (DHW): electric

2 — Heat Emitters: floor heating Temp. Range: 3K

3 — DHW Insulation: Set Temperature: 45°C

**Other Points/Heat Pump:** Indoor swimming pool (6.30 x 3.00m; 27.5m$^3$)

**Control Strategy/Operating Modes**

Base temperature for control is the temperature of one absorber panel(steering panel). Solar energy for DHW is collected when water temperature in the panels exceeds that in DHW storage by 5K. Priority is geven to DHW. When temperature in DHW storage reaches 45°C, solar energy is delivered to the storage tanks HB 1,2,3.
At small temperature differences across the collector, heat is delivered to the heat pump storage HB 4.

| C2 SYSTEM DESCRIPTION (CONT) | REF. | D | Es | 78 | 78 | 12 |
|---|---|---|---|---|---|---|

**Diagram of System** Showing sensor locations, power ratings, storage volumes & flows.

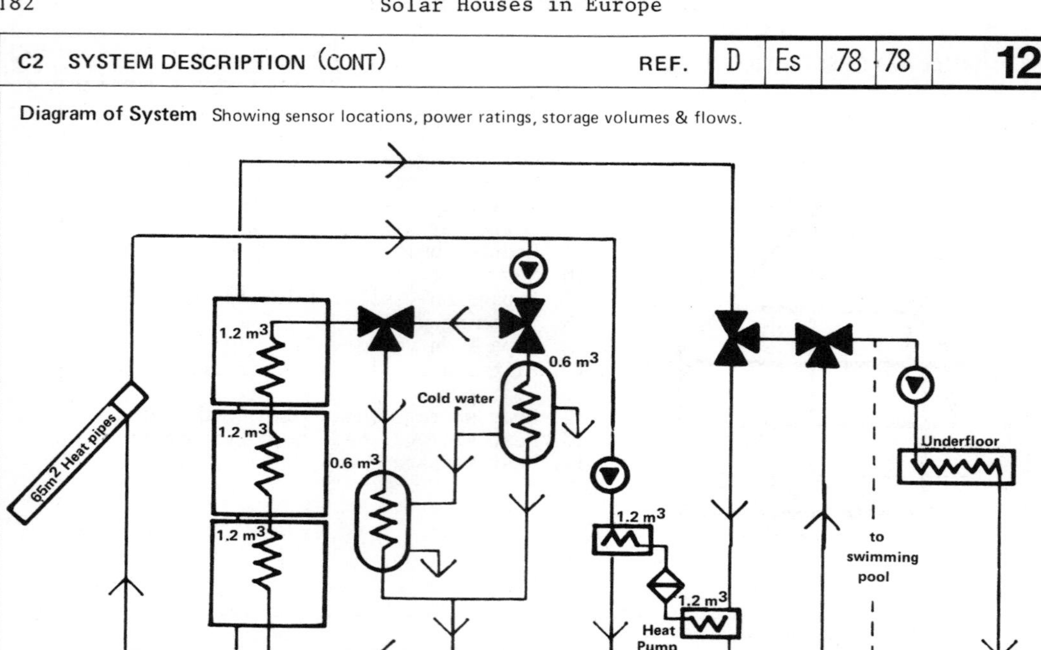


## PERFORMANCE EVALUATION/CONCLUSIONS

**Modifications to System**

Collector covering:originally double plexiglass plate (outside) + glass plate, then plexiglass double plate without glass plate (since 1977).

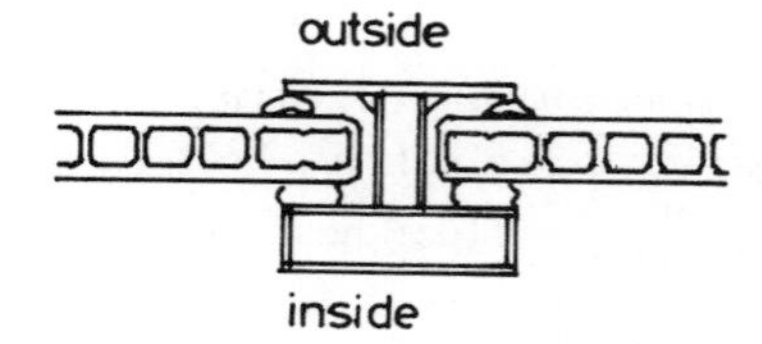


**Conclusions**

A relatively great amount (52%) of primary energy can be saved by this system. Considering prices for the solar system economic conditions are not yet reached.

**Future Work**

Absorber with selective coating ($\alpha^*/\varepsilon^* = 6.3$)

**P1 MONTHLY PERFORMANCE** REF: D ES 78 78 12

| Month | Year | No Days Data | Average Temp °C | | SOLAR ENERGY kWh | | | | | | Heat Pump Total Output kWh | | AUXILIARY ENERGY kWh | | | | Free Heat Gains kWh | | SUMMARY * | | |
|---|---|---|---|---|---|---|---|---|---|---|---|---|---|---|---|---|---|---|---|---|---|
| | | | | | | | Solar Energy Used | | | | | | | | | | | | kWh | % | |
| | | | | | | | Space Heating Useful Losses | | | | | | | | | | | | | | |
| | | | Internal | External | Global Irradiation On Collector Area | Solar Energy Collected | Incl. | Excl. | Domestic Hot Water | Swim Pool | Space Heating | Domestic Hot Water | Heat Pump | Space Heating | Domestic Hot Water | Pumps & Fans | Internal | Direct Solar | Total Load | Percentage Solar | Syst. Eff. |
| Column No | | | 1 | 2 | 3 | 4 | 5 | 6 | 7 | 8 | 9 | 10 | 11 | 12 | 13 | 14 | 15 | 16 | 17 | 18 | 19 |
| J | 78 | | | 2.5 | 1987 | 755 | 495 | | 89 | 171 | Solar input to heat pump included in column 5 | 0 | 835 | 2518 | 473 | | | | 3475 | 17 | 29 |
| F | 78 | | | 0.7 | 3870 | 1045 | 603 | | 85 | 357 | | 0 | 1055 | 1779 | 514 | | | | 2981 | 23 | 18 |
| M | 78 | | | 6.1 | 4033 | 1815 | 1006 | | 284 | 525 | | 0 | 890 | 1706 | 557 | | | | 3553 | 36 | 32 |
| A | 78 | | | 7.8 | 7878 | 2521 | 1109 | | 777 | 635 | | 0 | 371 | 649 | 269 | | | | 2804 | 67 | 24 |
| M | 78 | | | 12.8 | 4782 | 1674 | 449 | | 813 | 412 | | 0 | 117 | 128 | 134 | | | | 1524 | 83 | 26 |
| J | 78 | | | 15.2 | 8477 | 1526 | 115 | | 631 | 780 | | 0 | 89 | 34 | 62 | | | | 842 | 88 | 9 |
| J | 78 | | | 15.8 | 7264 | 1816 | 340 | | 663 | 813 | | 0 | 125 | 100 | 87 | | | | 1190 | 84 | 14 |
| A | 78 | | | 15.5 | 7133 | 1070 | 27 | | 604 | 439 | | 0 | 3 | 1 | 32 | | | | 664 | 95 | 9 |
| S | 78 | | | 13.0 | 4774 | 1098 | 124 | | 465 | 509 | | 0 | 192 | 81 | 166 | | | | 836 | 70 | 12 |
| O | 78 | | | 11.0 | 2847 | 1395 | 639 | | 362 | 394 | | 0 | 380 | 566 | 283 | | | | 1850 | 54 | 35 |
| N | 78 | | | 5.7 | 3154 | 1293 | 700 | | 241 | 352 | | 0 | 704 | 1437 | 340 | | | | 2718 | 35 | 30 |
| D | 78 | | | 2.2 | 1676 | 754 | 591 | | 20 | 153 | | 0 | 921 | 2361 | 340 | | | | 3302 | 18 | 36 |
| TOTAL | | | — | — | 57875 | 16762 | 6188 | | 5034 | 5540 | | 0 | 5682 | 11260 | 3257 | | | | 25739 | — | — |
| AVERAGE | | | | | | | | | | | | | | | | | | | | 43 | 19 |

TOTAL: kWh C.O.P = (3)

* Solar Input To Heat Pump Included As Solar Contribution To Total Load

* Total Output From Heat Pump Included As Contribution To Total Load

Excluding swimming pool

*Excluding swimming pool and pumps. (12):Heat pump consumption included.

| SUMMARY SHEET | FREIBURG | REF. | D | Fr | | | 13 |
|---|---|---|---|---|---|---|---|

**Design Data**

**A1 CLIMATE**

1) Source of Data:
2) Latitude: (48) ° ′ Longitude: ° ′ Altitude: metres
3) Global Irradiation (horiz plane): (1,180) kWh/m$^2$·year % Diffuse:
4) Degree Days: (3,147) Base Temp: 12 °C
5) Sunshine Hours: July: (226) January: (37) Annual: (1,460)

**B1 BUILDING**

1) Building Type: 3 storey, 6 flats No Occupants:
2) Floor area: 641 m$^2$ Heated Volume: 4,284 m$^3$
3) Design Temperature: External: -12 °C Internal: °C
4) Ventilation Rate: pre-heated, forced vent a.c.h. Mean U Value: 0.50 W/m$^3$·K
5) Space Heat Power @ -12°C: 62 kW Hot Water Load: kWh

**C1 SYSTEM**

1) Absorber Type: Evacuated, high efficiency glass tube collectors.
2) Collector Area 26.6 + 28 m$^2$ (Aperture) Coolant: Water & anti-freeze
3) Orientation: Tilt: 55° Glazing:
4) Storage Volume: 22.5 m$^3$ Heat Emitters: Radiators
5) Auxiliary System: 72kW oil burner (SH, DHW) Electric (DHW) Heat Pump:

**PERFORMANCE MONITORING**

1) Is there a Computer Model? Yes
2) No of Measuring Points per house: 180
3) Data Acquisition System: All sensor scanned at few seconds interval. Averages recorded on magnetic tape at 5 min. interval.

| | Space Heating | Hot Water | Total |
|---|---|---|---|
| 6) Predicted | % of kWh | % of kWh | % of kWh |
| 7) Measured | % of kWh | % of kWh | % of kWh |

8) Solar Energy Used: Including useful losses kWh/m$^2$·year
Excluding losses: kWh/m$^2$·year

9) System Efficiency: $\frac{\text{Solar Energy Used}}{\text{Global Irradiation on collector}} \times 100 =$ %

40

| PROJECT DESCRIPTION | WALLDORF | REF. | D | Wa | 76 | 77 | 14 |
|---|---|---|---|---|---|---|---|

**Main Participants:**

BROWN, BOVERI & CIE AG,
Zentrales Forschungslabor,
Eppelheimer Str. 82,
D-6900 Heidelberg.

BROWN, BOVERI & CIE AG,
Spezialbereich Solarenergie,
Impexstrasse 9,
D-6907 Walldorf

**Project Description:**

The BBC-Solarhouse is a prefabricated single family house which is used as an office. The characteristics of the heat pump assisted energy system are :

- 71.5m$^2$ collector area
- 8m$^3$ heat storage volume of water
- electrical compressor heat pump of 2.25kW$_{el}$
- oilburner of 16.5kW$_{th}$

**Project Objectives:**

The objective of the project was to investigate the basic technology for the development of an economic solar system for room-heating and the preparation of domestic hot water in a typical single family house under central European weather conditions.

Steps in this investigation were:
- to test the performance of solar components under realistic conditions
- to gain experience in the operation of the solar system with different control stratergies.
- to study the longtime behaviour of the system and its components.

The experimental data was taken to prove the validity of a computer programme for sizing solar systems.

Photograph of Project

**SUMMARY SHEET** WALLDORF **REF.** D | Wa | 76 | 77 | **14**

**Design Data**

**A1 CLIMATE**

1) Source of Data: Meteorological Station - Heidelberg

2) Latitude: 49 ° 4 ´ Longitude: 8 ° 7 ´ Altitude: 112 metres

3) +Global Irradiation (horiz plane): 952 kWh/m²·year % Diffuse:

4) +Degree Days: 3,050 Base Temp: 12 °C

5) +Sunshine Hours: July: 220 January: 45 Annual: 1540

(+) (1976/77 data)

**B1 BUILDING**

1) Building Type: Single family house No Occupants: 6 (office)

2) Floor area: 176 m² Heated Volume: m³

3) Design Temperature: External: -12 °C Internal: 20 °C

4) Ventilation Rate: a.c.h. Vol. Heat Loss: W/m³·K

5) Space Heat Load: 13kW at -12°C kWh Hot Water Load: kWh

**C1 SYSTEM**

1) Absorber Type: Aluminium Roll Bond - Non selective

2) Collector Area 71.5 m² (Aperture) Coolant: Water-Glycol

3) Orientation: 180° Tilt: Glazing: Single

4) Storage Volume: 8 m³ Heat Emitters: Underfloor

5) Auxiliary System: Oil Boiler 16.5kW Heat Pump: 2.25kW elec.

**PERFORMANCE MONITORING**

1) Is there a Computer Model? Yes 2) Start Date for Monitoring Programme: August 1979

3) Period for which Results Available: 1st September 1976 to 31 August 1977

4) No. of Measuring Points per house: 88

5) Data Acquisition System: BBC Metrawatt Metramatic F4, 48 temperatures, 2 Solar Radiation, 13 Volume Flow, 13 Heat Meters, 4 Electric Meters, 10 running hours.

| | Space Heating | Hot Water | Total |
|---|---|---|---|
| 6) Predicted | % of kWh | % of kWh | % of kWh |
| 7) Measured | % of kWh | % of kWh | 58 % of 24743 kWh |

8) Solar Energy Used: Including useful losses 199 kWh/m²·year

Excluding losses: kWh/m²·year

9) System Efficiency: $\frac{\text{Solar Energy Used}}{\text{Global Irradiation on collector}} \times 100 =$ 20 %

## A2 LOCAL CLIMATE WALLDORF

REF. | D | Wa | 76 | 77 | 14

Source of weather data:measurements in the BBC Solar House

Micro climate/site description:radiation and temperatures in summer are below the longtime average.

| MONTH | YEAR | Irradiation on horizontal plane Global | Diffuse | Degree Days Base 12 °C | Precipitation | Average Ambient Temperature | Average Relative Humidity | Average Wind Speed | Prevailing Wind Direction |
|---|---|---|---|---|---|---|---|---|---|
| JAN | 1977 | 16.4 | | 509 | | 2.8 | | * | |
| FEB | " | 34.7 | | 369 | | 6.5 | | 2.1 | |
| MARCH | " | 75.0 | | 314 | | 9.5 | | 2.9 | |
| APRIL | " | 108.0 | | 308 | | 9.0 | | 1.8 | |
| MAY | " | 148.2 | | 182 | | 12.8 | | 0.7 | |
| JUNE | " | 135.0 | | – | | 14.0 | | 0.4 | |
| JULY | " | 137.3 | | – | | 16.7 | | 0.3 | |
| AUGUST | " | 119.7 | | – | | 16.3 | | 0.4 | |
| SEPT | 1976 | 84.0 | | 155 | | 13.8 | | * | |
| OCT | " | 54.3 | | 247 | | 11.3 | | * | |
| NOV | " | 22.2 | | 405 | | 6.1 | | * | |
| DEC | " | 18.0 | | 561 | | 1.7 | | * | |
| Totals | | 952.8 | | 3050 | | | | | |
| Averages | | 79.4 | | | | 10.0 | | | |
| Units | – | kWh/m²·month | | °C days | mm | °C | % | m/s | – |

## B2 BUILDING DESCRIPTION

A prefabricated house.

Glazing:24.5m$^2$ ; 'U' values (W/m$^2$.K):3.0 (glazing), 0.51 (walls), 0.38 (roof/loft)
Annual energy demand:24743 kWh; Thermal load at -12°C:13 kW.

## C2 SYSTEM DESCRIPTION

**Collector/Storage**

Primary circuit, heat carrier content:1.6 litres/m$^2$ collector area ;
Insulation:60mm PU-foam (collector), 150mm mineral wool (storage), 20mm mineral wool (pipework) ; frost protection:50% ethylene glycol - 50% water mixture ; no overheating protection; corrosion protection:yes.

**Space Heating/Domestic Hot Water**

Auxiliary (SH and DHW):HP-2.25kW (elec) and oilburner 16.5 (th);
heat emitters:floor heating system; temp range: max. 45°C;
DHW insulation:5mm glasswool, set temp:55°C.

**Control Strategy/Operating Modes**

ΔT-control between absorber and storage tank.
Priority SH and DHW: 1. Solar direct from storage 2. Solar heat pump) alternately
3. Oilburner ) operated

| C2 SYSTEM DESCRIPTION (cont) | WALLDORF | REF. | D | Wa | 76 | 77 | 14 |
|---|---|---|---|---|---|---|---|

**Diagram of System** Showing sensor locations, power ratings, storage volumes & flows.

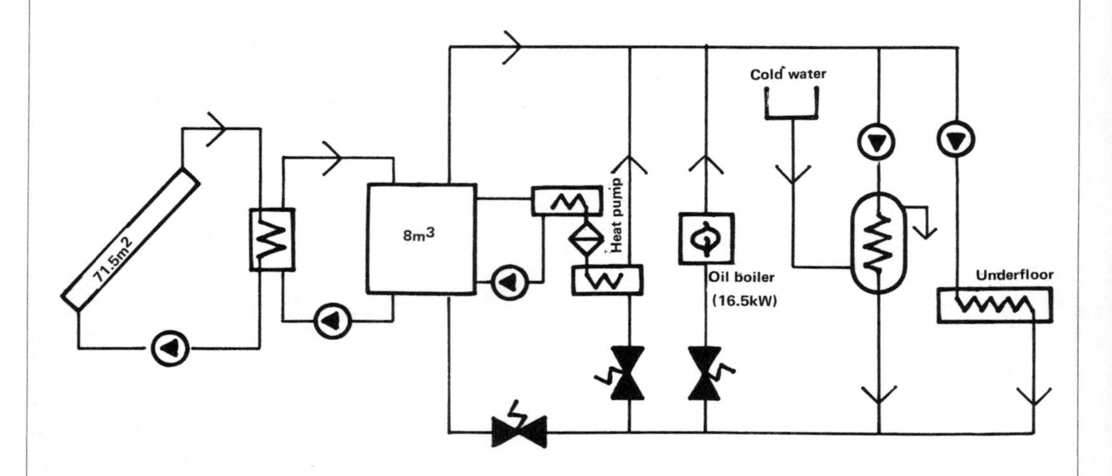


PERFORMANCE EVALUATION/CONCLUSIONS

**Comparison of Measured with Predicted Performance**
(State reasons for any discrepancy and summarise the main factors affecting performance)

During the test phase the system was operated with different collector area (33 - 71.5m$^2$) and storage volumes (4 - 8m$^3$) The hot water consumption was simulated (200 l/d). The computer simulation of the system with actual data for solar radiation, ambient temperature and heat load (hourly and/or daily basis) showed very good agreement within the accuracy of measurement (2 - 5%)

## P1 MONTHLY PERFORMANCE

REF: D | WA | 76 | 77 | 14

| Month | Year | No Days Data | Average Temp °C: Internal (1) | Average Temp °C: External (2) | SOLAR ENERGY kWh: Global Irradiation On Collector Area (3) | Solar Energy Collected (4) | Solar Energy Used: Space Heating Useful Losses Incl. (5) | Solar Energy Used: Space Heating Useful Losses Excl. (6) + Domestic Hot Water (7) + Input To Heat (8) | Heat Pump Total Output kWh: Space Heating (9) | Domestic Hot Water (10) | AUXILIARY ENERGY kWh: Heat Pump (11) | + Space Heating (12) + | Domestic Hot Water (13) | Pumps & Fans (14) | Free Heat Gains kWh: Internal (15) | Direct Solar (16) | SUMMARY * kWh: Total Load (17) | %: Percentage Solar (18) | Syst. Eff. (19) |
|---|---|---|---|---|---|---|---|---|---|---|---|---|---|---|---|---|---|---|---|
| J | 77 | 31 | 20 | 2.8 | 1509 | 747 | | 811 | | | elec. 433 | + | gas 2867 | | | | 4111 | 20 | 54 |
| F | 77 | 28 | 20 | 6.5 | 3403 | 1907 | | 1712 | | | 615 | + | 645 | | | | 2975 | 58 | 50 |
| M | 77 | 31 | 20 | 9.5 | 6850 | 2491 | | 2338 | | | 300 | + | 0 | | | | 2638 | 89 | 34 |
| A | 77 | 30 | 20 | 9.0 | 6950 | 2594 | | 2534 | | | 434 | + | 47 | | | | 3015 | 84 | 36 |
| M | 77 | 31 | 20 | 12.8 | 10282 | 1997 | | 1232 | | | 14 | + | 0 | | | | 1246 | 89 | 12 |
| J | 77 | 30 | 20 | 14.0 | 8687 | 565 | | 362 | | | 0 | + | 0 | | | | 362 | 100 | 4 |
| J | 77 | 31 | 20 | 16.7 | 9002 | 741 | | 374 | | | 0 | + | 0 | | | | 374 | 100 | 4 |
| A | 77 | 31 | 20 | 16.3 | 8759 | 792 | | 375 | | | 0 | + | 0 | | | | 375 | 100 | 4 |
| S | 77 | 30 | 20 | 13.8 | 7250 | 1024 | | 897 | | | 1 | + | 0 | | | | 898 | 100 | 12 |
| O | 77 | 31 | 20 | 11.3 | 5534 | 1036 | | 1270 | | | 98 | + | 0 | | | | 1368 | 93 | 23 |
| N | 77 | 30 | 20 | 6.1 | 2231 | 1223 | | 1312 | | | 520 | + | 952 | | | | 2784 | 47 | 59 |
| D | 77 | 31 | 20 | 1.7 | 2152 | 1126 | | 1016 | | | 472 | + | 3110 | | | | 4597 | 22 | 47 |
| TOTAL | | | — | — | 72609 | 16243 | | 14233 | | | 2886 | + | 7624 | | | | 24743 | — | — |
| AVERAGE | | | 20 | 10. | | | | | | | | | | | | | | 58 | 20 |

total: 10510

TOTAL: 14233 kWh

(1): min temp 16 C (between 22 & 6 hrs.)

(4): output from collectors;
(11)(12)(13): electricity and gas for space heating and domestic hot water; (6)(7)(8): total solar energy used, including input to heat pump

* Solar Input To Heat Pump Included As Solar Contribution To Total Load

* Total Output From Heat Pump Included As Contribution To Total Load

| PROJECT DESCRIPTION | OTTERFING | REF. | D | OTT | 77 | 78 | 15 |
|---|---|---|---|---|---|---|---|

**Main Participants:**

Messerschmitt-Bölkow-Blohm Gmbh, Space Division RX 15, PO Box 60 11 79
D-8000 München 80 (Ottobrunn)

**Project Description:**

The Solar house "Otterfing" is a non-occupied one family house(Knodler type) pre-alpine house style, fitted with 80m2 effective absorber area. The house is used daily for sales' consultation and for demonstration to visitors.

**Project Objectives:**

To study the system installation parameters; to demonstrate and to measure the solar heat input and utilization; to test the system operation under long term conditions.

**SITE LOCATION MAP**

Situated 30km south of Munich

Nearest obstructions: the building is free from essential obstructions related to wind and sunshine.

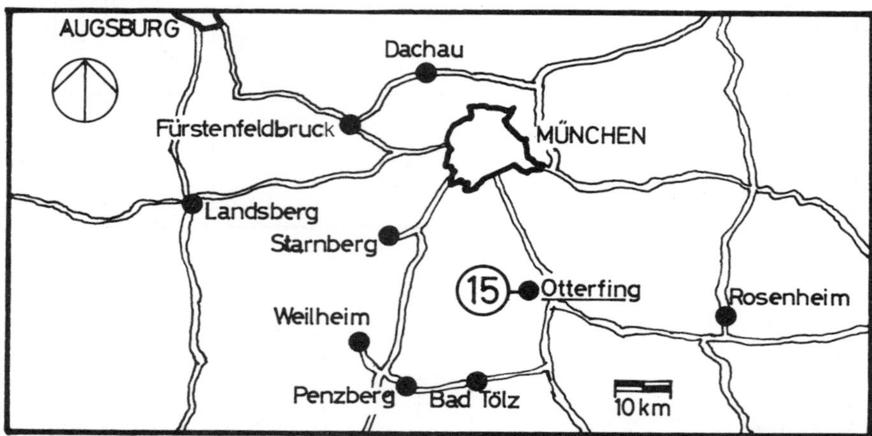

Photograph of Project

SUMMARY SHEET OTTERFING REF. D OTT 77 78 **15**

**Design Data**

**A1 CLIMATE**

1) Source of Data: Published by weather bureau, Holzkirchen

2) Latitude: 48 ° ´ Longitude: 11 ° 6 ´ Altitude: 600 metres

3) +Global Irradiation (horiz plane): 1093 kWh/m²·year % Diffuse:

4) +Degree Days: 4,045 Base Temp: 20 °C

5) +Sunshine Hours: (1977) July: 255 January: 38 Annual: 1,567

(+): (77/78) data

**B1 BUILDING**

1) Building Type: Prefabricated, single house No Occupants: none

2) Floor area: 145 m² Heated Volume: 900 m³

3) Design Temperature: External: °C Internal: °C

4) Ventilation Rate: a.c.h. Vol. Heat Loss: W/m³·K

5) Space Heat Load: 21,000 kWh Hot Water Load: 3,400 kWh

**C1 SYSTEM**

1) Absorber Type: Aluminium Roll bond, normal black colour

2) Collector Area 80 m² (Aperture) Coolant: water + antifrogen N

3) Orientation: south Tilt: 30 Glazing: double

4) Storage Volume: 8; 105m² (soil) m³ Heat Emitters: floor heat. system

5) Auxiliary System: Oil heater (20kW); electric heater (2.5 kW; DHW)

**PERFORMANCE MONITORING**

1) Is there a Computer Model? Yes 2) Start Date for Monitoring Programme: November 1976

3) Period for which Results Available: February 1977 to January 1978

4) No. of Measuring Points per house: 50 to 70

5) Data Acquisition System: daily manual registration.

| | Space Heating | Hot Water | Total |
|---|---|---|---|
| 6) Predicted | 32 % of 31500 kWh | 50 % of 6400 kWh | 35 % of 37900 kWh |
| 7) Measured | 16 % of 29,720 kWh | 90 % of 2,880 kWh | 23 % of 32,600 kWh |

8) Solar Energy Used: Including useful losses kWh/m²·year; Excluding losses: 93 kWh/m²·year

9) System Efficiency: Solar Energy Used / Global Irradiation on collector x 100 = 8 %

| A2 LOCAL CLIMATE | REF. | D | OTT | 77 | 78 | 15 |
|---|---|---|---|---|---|---|

1. Average cloud cover: 5.6 octas
2. Average daily max temps (July):17.7°C, average daily min temps (Jan): -1.6°C.
3. Source of weather data: weather bureau Munchen.

4. Micro climate/site description: Pre-alpine climate; land climate with weather influence from the Bay of Biscay. Sky conditions are sometimes heavily influenced by warm winds falling down from the north side of the Alps. Negligible influence from air pollution by traffic or industry.

Table below: all data derived from official Munich weather data ;
*on 30° tilted area, calculated from mean daily insolation;+degree days:3730 long term.

| MONTH | YEAR | Irradiation on horizontal plane | | Degree Days | Precipitation | Average Ambient Temperature | Average Relative Humidity | Average Wind Speed | Prevailing Wind Direction |
|---|---|---|---|---|---|---|---|---|---|
| | | Global * | Diffuse | Base: 20°C | | | | | |
| JAN | 1977 | 37.51 | | 638 | | -0.6 | | - | |
| FEB | | 60.76 | | 466 | | 3.3 | | - | |
| MARCH | | 99.82 | | 418 | | 6.5 | | 3.4 | |
| APRIL | | 108.3 | | 426 | | 5.8 | | 3.8 | |
| MAY | | 147.56 | | 253 | | 11.8 | | 3.2 | |
| JUNE | | 165.0 | | 91 | | 15.8 | | 3.0 | |
| JULY | | 174.84 | | 35 | | 17.1 | | 2.6 | |
| AUGUST | | 128.96 | | 87 | | 16.0 | | 2.3 | |
| SEPT | | 98.40 | | 216 | | 12.0 | | 2.1 | |
| OCT | | 80.91 | | 325 | | 9.5 | | 1.9 | |
| NOV | | 57.6 | | 470 | | 4.3 | | 4.2 | |
| DEC | | 33.79 | | 620 | | 0 | | 2.6 | |
| Totals | | 1193.45 | | 4045 + | | | | | |
| Averages | | 99.45 | | 337 | | 8.46 | | 2.9 | |
| Units | – | kWh/m²·month | | °C days | mm | °C | % | m/s | – |

## B2 BUILDING DESCRIPTION

### Design Concepts

Standard pre-alpine house style. The house is dominated by a living hall with veranda. The upper floor is constructed like a gallery. The ridge of the main roof is set down against the north side to obtain additional light into the hall.

### Construction

Glazing: 15% (23m²) ; 'U' values(W/m².K): 2.6 (glazing), 0.40 (walls), 0.58 (floor), 0.32 (roof/loft); Annual energy demand: 24,400 kWh.

B2 BUILDING DESCRIPTION (CONT) REF. D OTT 77 78 15

**Section Through House** Showing main solar components

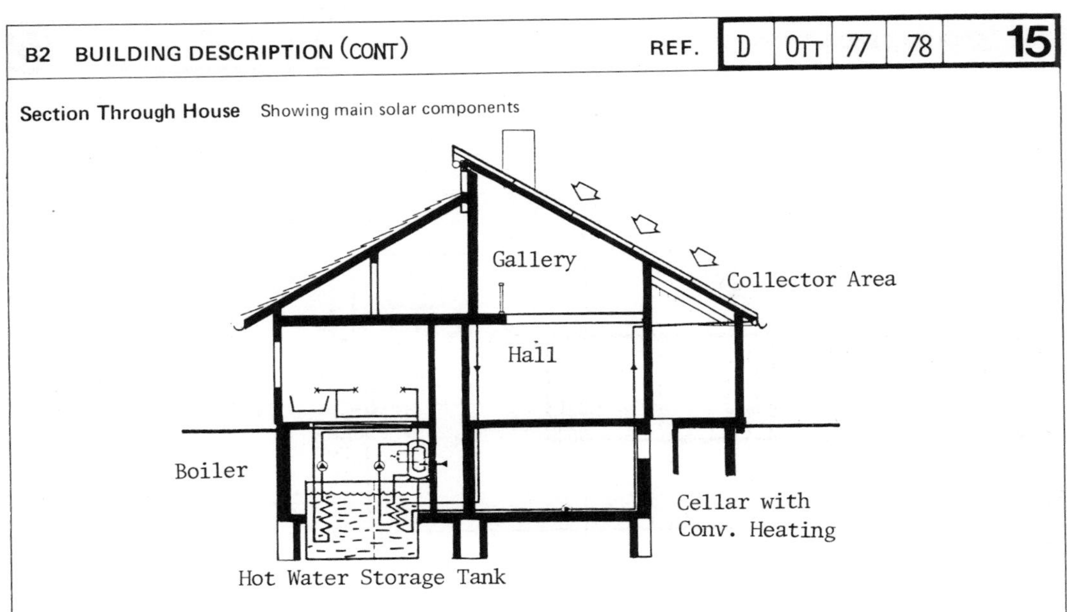


## C2 SYSTEM DESCRIPTION

**Collector/Storage**

Primary circuit, water content: 2.6 litres/$m^2$ collector area;
Insulation (W/m2.K):collector 'U' value 4, storage 'U' value 0.1, pipework 1.1 ;
Frost protection: antifrogen-N ; Overheating protection: overpressure valve empties collectors at 97°C ; corrosion protection:Antifrogen-N.

**Space Heating/Domestic Hot Water**

Auxiliary(SH):oil heater 20 kW, (DHW):2.5 kW immersion heater (electric);
DHW:insulation:0.32 W/$m^2$.K , set temperature: boiling point.

**Other Points/Heat Pump:**

The cascaded heat storage system includes an 8$m^3$ hot water storage tank with two compartments and 3 soil storage compartments.

**Control Strategy/Operating Modes**

Operating modes are: pure solar heating, alternative conventional space heating, and additional electric boiler heating respectively; heat storage of excess solar heat in the soil storage compartments. Operating modes are controlled automatically by an electronic two point control system. This contains sub-systems for: collector loop control, safety device operation, heat output from each storage tank into conventional energy loops, heating control of conventional energy loops.

| C2 SYSTEM DESCRIPTION (CONT) | REF. | D | OTT | 77 | 78 | 15 |
|---|---|---|---|---|---|---|

**Diagram of System** Showing sensor locations, power ratings, storage volumes & flows.

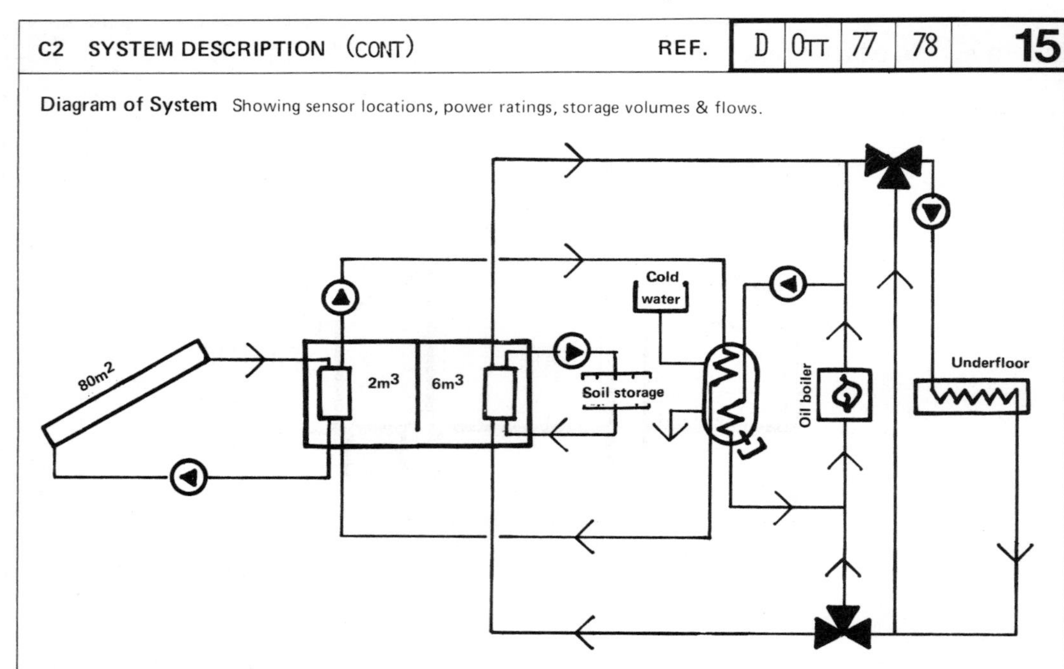


## TECHNICAL APPRAISAL/PRACTICAL EXPERIENCE

**System/Design:**
In the first year of operation the predicted savings were not completely achieved due to departures in weather and operating conditions. Only 80% of the predicted annual oil savings were obtained.

**Component Performance:** (Collector, Heat Exchangers, Storage, Pipework, Valves, Fittings, Pumps, Auxiliary)
The adjustment of hardware was not optimal:controller, circulating pumps and measurement equipment had to be improved. In the conventional heating system some departures from design characteristics were found. The mass flow of the working fluid in the collector loop was approximately 50% of the design value. Due to low storage temperature (35°C) no use of solar energy stored in the soil could be made. The heat transfer from the dry plaster floor heating system and the circulation line of the boiler seem to be questionable due to increased operating temperatures.

**System Operation/Controls, Electronics:**
Teh operation costs for supplemental energy of circulating pumps, control etc. were less than 160 DM/year. Other operational costs were negligible. Several interruptions due to special system tests, reconstruction and adjustment of control units. This solar heating system is flaxible to test all relevant operation modes.

| PERFORMANCE EVALUATION/CONCLUSIONS | REF. | D | OTT | 77 | 78 | 15 |
|---|---|---|---|---|---|---|

### Comparison of Measured with Predicted Performance
(State reasons for any discrepancy and summarise the main factors affecting performance)

April peak: snowfall;
June peak: garden sprinkling with hot water

actual energy demand 32.6 MWh/year
predicted energy demand Model 1 24.5 MWh/year
predicted energy demand Model 2 21.4 MWh/year



### Modifications to System

The soil storage volume must be reduced in order to obtain higher storage temperatures. Apart from the solar heating system function, some improvements in building insulation and conventional heating system should be carried out.
The amount of measured values can be decreased.
With respect to commercial utilization it is recommended to reduce effective collector area to about 50m$^2$ per 1 family house and not to instal soil storage compartments.

### Conclusions

For commercial utilization it is important to be aware of: simple and safe elements, automated operation, minimum installation effort at the building site, critical selection of mass produced parts, stagnation temperature capability, pre-installed compact modules and safety device for each mode of operation.

### Future Work

The influence of bad weather conditions and other departures from design has to be analysed by means of a post computer calculation at MBB.

## P1 MONTHLY PERFORMANCE

REF: D OTT 77 78 15

| Month | Year | No Days Data | Average Temp °C: Internal | Average Temp °C: External | SOLAR ENERGY kWh: Global Irradiation On Collector Area | SOLAR ENERGY kWh: Solar Energy Collected | Solar Energy Used: Space Heating Useful Losses Incl. | Solar Energy Used: Space Heating Useful Losses Excl. | Solar Energy Used: Domestic Hot Water | Solar Energy Used: Input To Heat | Heat Pump Total Output kWh: Space Heating | Heat Pump Total Output kWh: Domestic Hot Water | AUXILIARY ENERGY kWh: Heat Pump | AUXILIARY ENERGY kWh: Space Heating | AUXILIARY ENERGY kWh: Domestic Hot Water | AUXILIARY ENERGY kWh: Pumps & Fans | Free Heat Gains kWh: Internal | Free Heat Gains kWh: Direct Solar | SUMMARY kWh: Total Load | SUMMARY %: Percent-age Solar | SUMMARY %: Syst. Eff. |
|---|---|---|---|---|---|---|---|---|---|---|---|---|---|---|---|---|---|---|---|---|---|
| Column No | | | 1 | 2 | 3 | 4 | 5 | 6 | 7 | 8 | 9 | 10 | 11 | 12 | 13 | 14 | 15 | 16 | 17 | 18 | 19 |
| J | 78 | | | | 2000 | 350 | | 400 | 0 | - | - | - | - | 5800 | 150 | | | | 6350 | 6 | 20 |
| F | 77 | | | | 4800 | 650 | | 0 | 250 | - | | | - | 3000 | 0 | | | | 3250 | 8 | 5 |
| M | 77 | | | | 8000 | 1700 | | 800 | 200 | - | | | - | 2200 | 0 | | | | 3200 | 31 | 12 |
| A | 77 | | | | 8900 | 1100 | | 750 | 150 | - | | | - | 2600 | 0 | | | | 3500 | 26 | 10 |
| M | 77 | | | | 10000 | 1600 | | 550 | 150 | - | | | - | 450 | 0 | | | | 1150 | 61 | 7 |
| J | 77 | | | | 11200 | 2300 | | 400 | 1150 (+) | - | | | - | 50 | 0 | | | | 1600 | 97 | 14 |
| J | 77 | | | | 11800 | 2250 | | 400 | 150 | - | | | - | 50 | 0 | | | | 600 | 92 | 5 |
| A | 77 | | | | 10400 | 1700 | | 450 | 0 | - | | | - | 200 | 130 | | | | 780 | 58 | 5 |
| S | 77 | | | | 9200 | 1600 | | 550 | 150 | - | | | - | 600 | 0 | | | | 1300 | 54 | 8 |
| O | 77 | | | | 6500 | 1300 | | 400 | 150 | - | | | - | 1200 | 0 | | | | 1750 | 31 | 8 |
| N | 77 | | | | 4700 | 450 | | 50 | 150 | - | | | - | 3630 | 0 | | | | 3830 | 5 | 4 |
| D | 77 | | | | 2600 | 400 | | 100 | 100 | - | | | - | 5090 | 0 | | | | 5290 | 4 | 7 |
| TOTAL | | | — | — | 90100 | 15400 | | 4850 | 2600 | - | | | - | 24870 | 280 | 1061 | | | 32600 | — | — |
| AVERAGE | | | | | | | | | | | | | | | | | | | | 23 | 8 |

TOTAL: 7450 kWh

(+): garden sprinkling with hot water.

PROJECT DESCRIPTION WERNAU REF. D WE 77 78 **16**

**Main Participants:**

Robert Bosch GmbH, Geschaftsbereich Junkers
Postfach 13 09
D-7314 Wernau F.R.G.

**Project Description:**

The project consists of a one family, terraced house with basement and attic, built in a wood-skeleton construction.
The heating systen (Tri-therm) consists of a combination of three subsystems:

- a solar heating system with collector and storage tank
- a heat pump with an air heat exchanger or a ground heat exchanger
- a gas fired auxiliary heater

**Project Objectives:**

- To develop and test non-conventional heating systems especially to provide for the further development of the so-called Tri-therm system for energy saving
- To test operation parameters for performance improvement
- To test various ways of heating by radiators and floor heating
- To investigate possibilities of saving resources by extraction of heat from waste water and recycling this water
- The house is used as a demonstration object to convey the possible kinds of future heating systems and their effectiveness. It is not occupied. Domestic water consumption is simulated at a rate of about 200 l/d at 50$^{o}$C.

Photograph of Project

| SUMMARY SHEET | WERNAU | REF. | D | We | 79 | 78 | 16 |
|---|---|---|---|---|---|---|---|

**Design Data**

**A1 CLIMATE**

1) Source of Data: Stuttgart-Hohenheim Met Station 15km
2) Latitude: 48° 42′ Longitude: 9° 12′ Altitude: 400 metres
3) Global Irradiation (horiz plane): 962 $kWh/m^2 \cdot year$ % Diffuse:
4) Degree Days: ∾ 3,100 Base Temp: 12 °C
5) Sunshine Hours: July: 240 January: 47 Annual: 1763

**B1 BUILDING**

1) Building Type: Single family house No Occupants: None
2) Floor area: 174 $m^2$ Heated Volume: 1400* $m^3$
3) Design Temperature: External: -15 °C Internal: °C
4) Ventilation Rate: a.c.h. Vol. Heat Loss: $W/m^3 \cdot K$
5) Space Heat Load: 36,228 kWh Hot Water Load: 3372 kWh

* inc. basement and attic

**C1 SYSTEM**

1) Absorber Type: Copper tube with aluminium sheet, non-selective.
2) Collector Area 38.4 $m^2$ (Aperture) Coolant: Water/Glycol
3) Orientation: 180° Tilt: Glazing: Double
4) Storage Volume: 4.7 $m^3$ Heat Emitters: Radiator/U floor
5) Auxiliary System: Gas boiler 20kW Heat Pump: 12kW

**PERFORMANCE MONITORING**

1) Is there a Computer Model? Yes
2) Start Date for Monitoring Programme: June 1977
3) Period for which Results Available: June 1977 to May 1978

| | Space Heating | Hot Water | Total |
|---|---|---|---|
| 6) Predicted | % of kWh | % of kWh | 38 % of 39600 kWh |
| 7) Measured | % of kWh | % of kWh | 26 % of 37750 kWh |

8) Solar Energy Used: Including useful losses 260 $kWh/m^2 \cdot year$
Excluding losses: $kWh/m^2 \cdot year$

9) System Efficiency: $\frac{\text{Solar Energy Used}}{\text{Global Irradiation on collector}} \times 100 =$ 26 %

| SITE LOCATION MAP | WERNAU | REF. | D | WE | 77 | 78 | 16 |
|---|---|---|---|---|---|---|---|

2 — Distance to main city: 35 km from Stuttgart

Gunzburg
AUGSBURG
Wernau
ULM
16
Biberach
Mindelheim
Landsberg
Memmingen
10 km

## A2 LOCAL CLIMATE

Average daily temps (July):17.6°C ; average daily temps (Jan): – 0.8°C.
Source of weather data: Meteorological station Stuttgart-Hohenheim

| MONTH | YEAR | Irradiation on horizontal plane | | Degree Days | Precipitation | Average Ambient Temperature | Average Relative Humidity | Average Wind Speed | Prevailing Wind Direction |
|---|---|---|---|---|---|---|---|---|---|
| | | Global | Diffuse | Base: 12 °C | | | | | |
| JAN | 1978 | 25 | | | | 0.7 | 87 | 2.8 | |
| FEB | | 42 | | | | -0.3 | 85 | 2.3 | |
| MARCH | | 80 | | | | 5.7 | 78 | 3.8 | |
| APRIL | | 109 | | | | 8.0 | 73 | 3.3 | |
| MAY | | 104 | | | | 11.5 | 64 | 2.5 | |
| JUNE | 1977 | 138 | | | | 15.8 | 71 | 2.6 | |
| JULY | | 154 | | | | 17.2 | 63 | 2.34 | |
| AUGUST | | 112 | | | | 16.2 | 801 | 1.7 | |
| SEPT | | 95 | | | | 12.4 | 87 | 1.8 | |
| OCT | | 49 | | | | 10.5 | 83 | 1.8 | |
| NOV | | 28 | | | | 5.0 | 81 | 4.3 | |
| DEC | | 26 | | | | 1.8 | 88 | 2.4 | |
| Totals | | 962 | | 3,100 | | – | – | – | |
| Averages | | 80 | | | | 8.7 | 78 | 2.6 | |
| Units | – | kWh/m$^2$·month | | °C days | mm | °C | % | m/s | – |

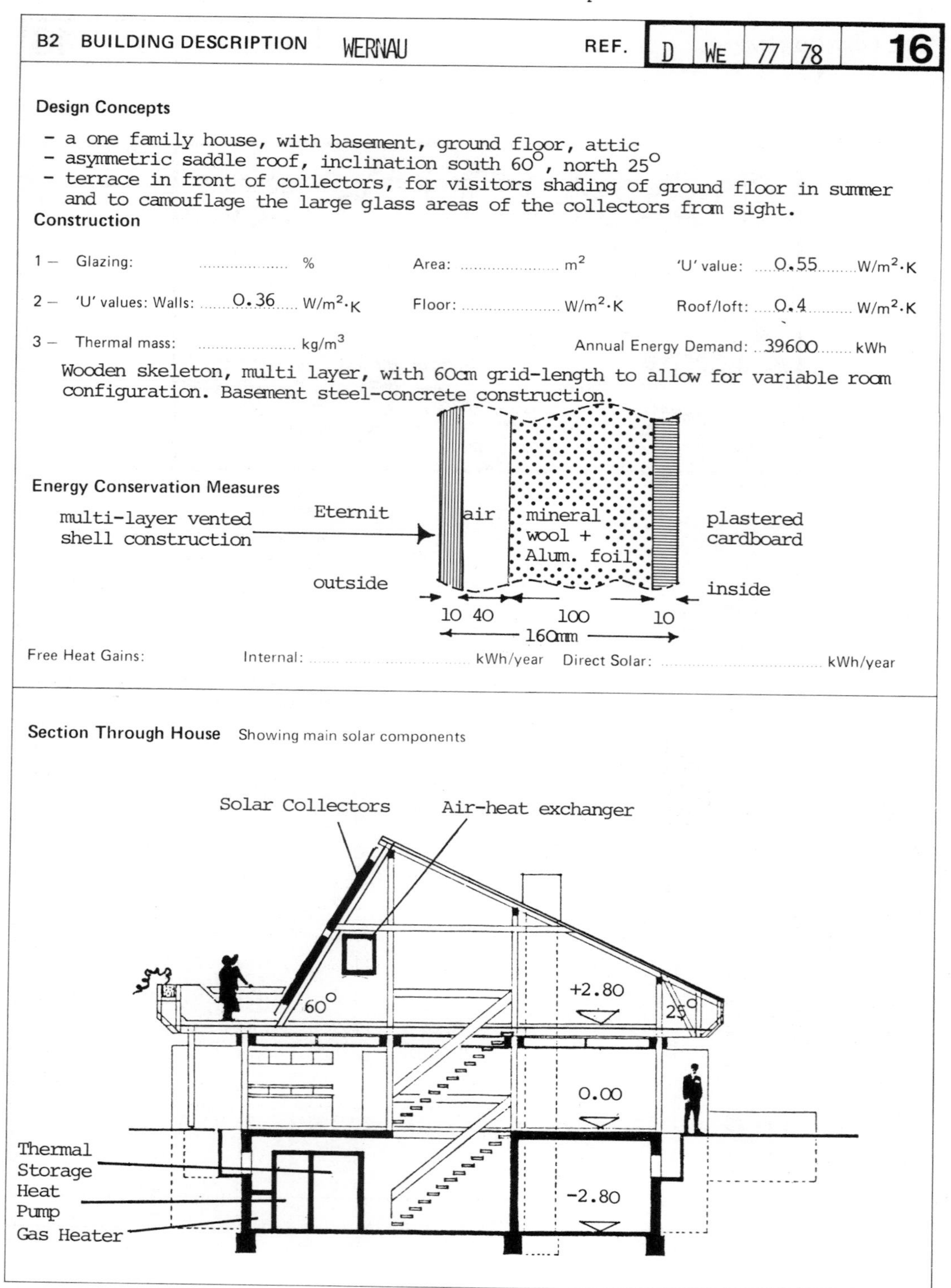

**B2 BUILDING DESCRIPTION** WERNAU **REF.** D | WE | 77 | 78 | **16**

**Design Concepts**

- a one family house, with basement, ground floor, attic
- asymmetric saddle roof, inclination south 60°, north 25°
- terrace in front of collectors, for visitors shading of ground floor in summer and to camouflage the large glass areas of the collectors from sight.

**Construction**

1 — Glazing: .................... % Area: .................... m² 'U' value: 0.55 W/m²·K

2 — 'U' values: Walls: 0.36 W/m²·K Floor: .................... W/m²·K Roof/loft: 0.4 W/m²·K

3 — Thermal mass: .................... kg/m³ Annual Energy Demand: 39600 kWh

Wooden skeleton, multi layer, with 60cm grid-length to allow for variable room configuration. Basement steel-concrete construction.

**Energy Conservation Measures**

multi-layer vented shell construction



Free Heat Gains: Internal: .................... kWh/year Direct Solar: .................... kWh/year

**Section Through House** Showing main solar components

| C2 SYSTEM DESCRIPTION | WERNAU | REF. | D | WE | 77 | 78 | 16 |
|---|---|---|---|---|---|---|---|

## Collector/Storage

1 – Primary circuit. Water content: 1.0 litres/m$^2$ collector area

2 – Insulation. Collector: glass wool Storage: styrodur 300mm Pipework:

3 – Frost Protection: 40% ethylen-glykol water mixture

4 – Overheating Protection: expansion vessels can take the entire liquid from collectors

5 – Corrosion Protection:

## Space Heating/Domestic Hot Water

1 – Auxiliary (S.H.): heat pump 12kW gas heater 20kW (DHW): heat pump, gas heater

2 – Heat Emitters: radiators or floor heating Temp. Range: inlet 60°*

3 – DHW Insulation: Set Temperature: 50°C

*for maximum demand

**Other Points/Heat Pump:** warm-water floor heating alternatively to radiators. Soil heat exchanger to support air heat exchanger.

**Control Strategy/Operating Modes** S.H. is covered as far as possible by solar energy either directly from collectors or by the heat store. If necessary the heat pump is used additionally with the heat gains from: collector, heat store, air heat exchanger. The auxiliary gas heater works in parallel to the heat pump at ambient temperatures below 0°C, for space heating only. For operation control one collector panel is used under no-load conditions.

**Diagram of System** Showing sensor locations, power ratings, storage volumes & flows.

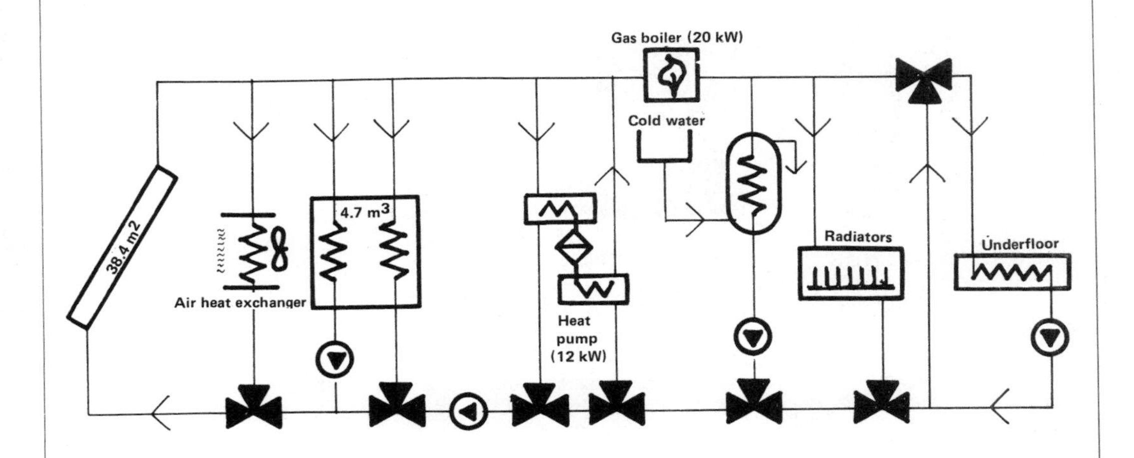


G We

| TECHNICAL APPRAISAL/PRACTICAL EXPERIENCE | REF. | D | WE | 77 | 78 | 16 |
|---|---|---|---|---|---|---|

**Component Performance:** (Collector, Heat Exchangers, Storage, Pipework, Valves, Fittings, Pumps, Auxiliary)

Heat Exchangers: Heat transfer from the glykol-water mixture to the tubes proved to be considerably temperature dependent due to viscosity, so that turbulent flow may turn into laminar flow at temperatures below 0°C.

Storage: Insulation was not satisfactory; thermal bridges had to be eliminated.

Valves: Motoric three path valves presented leakage to the by-pass tubes after a short time of operation, requiring a replacement by differently constructed valves.

**System Operation/Controls, Electronics:**

Better overall efficiency is expected when, from February until October, the collector and the air heat exchanger only is used as heat source for the heat pump so that the heat storage tank can be discharged at a higher temperature directly for S.H.

The operation strategy of the Tri-therm heating uses primarily direct (non-delayed) solar heating; the large capacity thermal store could be replaced by a smaller pressurized store without heat exchangers.

The use of the house as a demonstration object causes unpredictable deviations from a normally occupied home, temperature rises and un-controlled ventilation increase the heat demand.

## PERFORMANCE EVALUATION/CONCLUSIONS

**Comparison of Measured with Predicted Performance**
(State reasons for any discrepancy and summarise the main factors affecting performance)

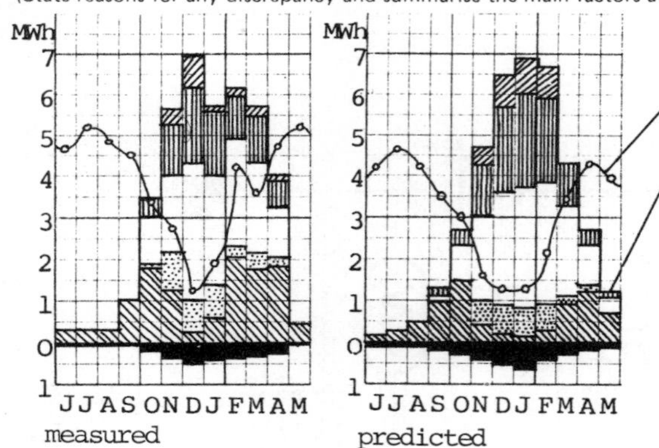


Global irradiance
Heat demand
Auxiliary Heating
Elect. Energy for Heat Pump
Air Heat for Heat Pump
Solar Heat for Heat Pump
Solar Heat direct
Auxiliary Energy (elect.)

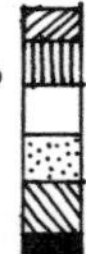

Discrepancies are due to significant differences in sunshine hours and global insolation; both proved to be 20% lower in the measuring period than in the prediction period 1975. For the heat pump the use of a heat carrier with a strongly temperature dependent viscosity caused lower outputs than predicted. This again caused the auxiliary heating to double its real output.

**Conclusions** The coverage of energy demand for SH and DHW cannot be provided by pure solar heating system. The use of the Tri-therm heating system may decrease the fuel demand to less than 10% of the present. The investment costs, however, still seem to be too high for a wider application.

**Future Work** Improvements and simplifications of the system to make it more economic.

## P1 MONTHLY PERFORMANCE

REF: D | WE | 77 | 78 | 16

| Month | Year | No Days Data | Average Temp °C: Internal | Average Temp °C: External | SOLAR ENERGY kWh: Global Irradiation On Collector Area | Solar Energy Collected | Solar Energy Used: Space Heating Useful Losses Incl. | Space Heating Useful Losses Excl. | + Domestic Hot Water | + Input To Heat | Heat Pump Total Output kWh: Space Heating | Domestic Hot Water | AUXILIARY ENERGY kWh: Heat Pump | + Space Heating | + Domestic Hot Water | Pumps & Fans | Free Heat Gains kWh: Internal | Direct Solar | SUMMARY* kWh: Total Load | %: Percentage Solar | Syst. Eff. |
|---|---|---|---|---|---|---|---|---|---|---|---|---|---|---|---|---|---|---|---|---|---|
| Column No | | | 1 | 2 | 3 | 4 | 5 | 6 | 7 | 8 | 9 | 10 | 11 | 12 | 13 | 14 | 15 | 16 | 17 | 18 | 19 |
| J | 78 | | | 0.7 | 1380 | 600 | | | 790 | | | | | 6210 | | | | | 7000 | 11 | 57 |
| F | 78 | | | -0.3 | 2260 | 1090 | | | 900 | | | | | 5880 | | | | | 6780 | 13 | 40 |
| M | 78 | | | 5.7 | 3220 | 1440 | | | 1100 | | | | | 3200 | | | | | 4310 | 26 | 34 |
| A | 78 | | | 8.0 | 4280 | 1850 | | | 1380 | | | | | 1180 | | | | | 2560 | 54 | 32 |
| M | 78 | | | 11.5 | 3880 | 1400 | | | 620 | | | | | 630 | | | | | 1250 | 50 | 16 |
| J | 77 | | | 15.8 | 4250 | 920 | | | 250 | | | | | 0 | | | | | 250 | 100 | 6 |
| J | 77 | | | 17.2 | 4600 | 580 | | | 290 | | | | | 0 | | | | | 290 | 100 | 6 |
| A | 77 | | | 16.2 | 4150 | 630 | | | 340 | | | | | 0 | | | | | 340 | 100 | 8 |
| S | 77 | | | 12.4 | 3740 | 1150 | | | 960 | | | | | 220 | | | | | 1180 | 81 | 26 |
| O | 77 | | | 10.5 | 3010 | 1390 | | | 1500 | | | | | 1110 | | | | | 2610 | 58 | 50 |
| N | 77 | | | 5.0 | 1760 | 780 | | | 980 | | | | | 3740 | | | | | 4720 | 21 | 56 |
| D | 77 | | | 1.8 | 1370 | 730 | | | 870 | | | | | 5590 | | | | | 6460 | 14 | 63 |
| TOTAL | | | — | — | 37900 | 12620 | | | 9990 | | | | | 27760 | | 3390 | | | 37750 | — | — |
| AVERAGE | | | | 8.7 | | | | | | | | | | | | | | | | 26 | 26 |

TOTAL: 9990 kWh | C.O.P = 3.3

* Solar Input To Heat Pump Included As Solar Contribution To Total Load

Additionally: air heat exchanger.

* Total Output From Heat Pump Included As Contribution To Total Load

| PROJECT DESCRIPTION | FIUME VENETO | REF. | I | FV | 78 | 79 | 17 |
|---|---|---|---|---|---|---|---|

**Main Participants:**

| | | |
|---|---|---|
| Building Construction | : | I.A.C.P. (Autonomous Institute for Popular Houses ) - Pordenone |
| Solar System Designer and Manufacturer | : | Sun Life S.p.A. - Pordenone |
| Monitoring | : | Industrie Zanussi - Pordenone |
| Data Acquisition System Manufacturer | : | Industrie Micros - Conegliano |
| Financial Support | : | C.N.R. - National Council of Research |

**Project Description:** This experiment consists of a 12 apartment building, designed in 1975 and built between 1975 and 1976 by Autonomous Institute for Popular Houses, using, above all, the criteria of economy and simplicity. The thermal insulation does not satisfy the new Italian law regarding energy conservation in buildings. The solar system is very simple and consists of 129 $m^2$ of solar collectors mounted on the roof, 60° inclined.

**Project Objectives:**

Evaluation of the possibility of energy conservation by means of a solar heating system

Evaluation of the impact of the solar energy use on inhabitants behaviour

Evaluation of the problems concerning the durability of solar components

Measurement and evaluation of the energy balances

**SITE LOCATION MAP**

Distance to main city:15km from Pordenone.

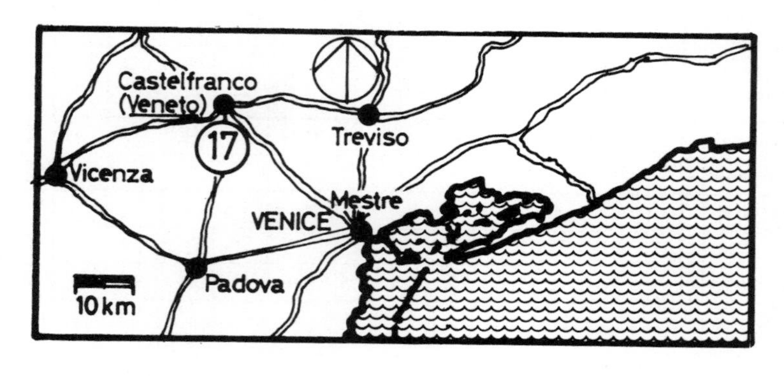

**SUMMARY SHEET** FIUME VENETO REF. I FV 78 79 **17**

**Design Data**

**A1 CLIMATE**

1) Source of Data: Conigliano (35 km), Udine (45 km)
2) Latitude: 46 ° 03 ′ Longitude: ° ′ Altitude: 119 metres
3) Global Irradiation (horiz plane): 1,159 kWh/m²·year % Diffuse: 41
4) Degree Days: 2,386 Base Temp: 19 °C
5) Sunshine Hours: July: January: Annual: 2,012

**B1 BUILDING**

1) Building Type: 12 apartment building No Occupants:
2) Floor area: 1,011.3 m² Heated Volume: m³
3) Design Temperature: External: - 5 °C Internal: 20 °C
4) Ventilation Rate: . 0.5 a.c.h. Vol. Heat Loss: 1.82 W/m³·K
5) Space Heat Load: 377,045 kWh Hot Water Load: 43,953 kWh

**C1 SYSTEM**

1) Absorber Type: flat AL-Roll bond, black paint
2) Collector Area m² (Aperture) Coolant: water & monoeth.glycol
3) Orientation: 180° Tilt: 60° Glazing: single
4) Storage Volume: 3 m³ Heat Emitters: fan coil & radiators
5) Auxiliary System: gas fired boiler Heat Pump:

**PERFORMANCE MONITORING**

1) Is there a Computer Model? No
2) Start Date for Monitoring Programme: March 1978
3) No. of Measuring Points per house: 21
4) Data Acquisition System: micro-processor controlled data logging system.

| | Space Heating | Hot Water | Total |
|---|---|---|---|
| 6) Predicted | 30 % of ... kWh | 80* % of ... kWh | % of ... kWh |
| 7) Measured | % of ... kWh | % of ... kWh | % of ... kWh |

8) Solar Energy Used: Including useful losses ... kWh/m²·year
Excluding losses: ... kWh/m²·year

9) System Efficiency: $\frac{\text{Solar Energy Used}}{\text{Global Irradiation on collector}} \times 100 =$ %

* outside heating season

## A2 LOCAL CLIMATE

REF. I | FV | 78 | 79 | **17**

Average daily max temps(July):+26.6 C; average daily min temps(Jan): -1.2 C.

Source of Weather Data: Udine 40km, Conegliano 30km.

| MONTH | YEAR | Irradiation on horizontal plane | | Degree Days | Precipitation | Average Ambient Temperature | Average Relative Humidity | Average Wind Speed | Prevailing Wind Direction |
|---|---|---|---|---|---|---|---|---|---|
| | | Global | Diffuse | Base: 19 °C | | | | | |
| JAN | 1958/69 | 40.03 | 18.1 | 536 | | + 2.7 | | | |
| FEB | " | 52.9 | 24.3 | 434 | | + 4.5 | | | |
| MARCH | " | 92.4 | 42.5 | 378 | | + 7.8 | | | |
| APRIL | " | 106.2 | 46.7 | 66 | | + 12.7 | | | |
| MAY | " | 151.0 | 63.4 | – | | + 16.5 | | | |
| JUNE | " | 150.6 | 63.2 | – | | + 20.5 | | | |
| JULY | " | 165.5 | 62.3 | – | | + 23.4 | | | |
| AUGUST | " | 141.7 | 54.5 | – | | + 22.1 | | | |
| SEPT | " | 108.6 | 43.4 | – | | + 18.9 | | | |
| OCT | " | 79.7 | 32.9 | 95 | | + 13.2 | | | |
| NOV | " | 37.2 | 18.5 | 375 | | + 7.5 | | | |
| DEC | " | 32.9 | 15.2 | 502 | | + 3.8 | | | |
| Totals | | 1159 | 485 | 2386 | | | | | |
| Averages | | 96.5 | 40.4 | | | 12.8 | | | |
| Units | – | kWh/m$^2$·month | | °C days | mm | °C | % | m/s | – |

## B2 BUILDING DESCRIPTION

**Design Concepts** The design and construction are very traditional. The external walls consist of a double layer of hollow bricks separated by a cavity. The solar and auxiliary systems are placed in a room under the roof.

**Construction**

1 – Glazing: 14 % Area: 160 m$^2$ 'U' value: 6 W/m$^2$·K

2 – 'U' values: Walls: 1.7 W/m$^2$·K Floor: 1.2 W/m$^2$·K Roof/loft: 1.5 W/m$^2$·K

3 – Thermal mass: 500 kg/m$^3$ Annual Energy Demand: 335,000 kWh

**Energy Conservation Measures**

Individual thermostat controls in each room.

| C2 SYSTEM DESCRIPTION | REF. | I | FV | 78 | 79 | **17** |
|---|---|---|---|---|---|---|

**Collector/Storage**

1 – Primary circuit. Water content: ........ litres/$m^2$ collector area

2 – Insulation. Collector: 50mm minl. wool Storage: 70mm minl. wool Pipework: 50/30mm minl. wool.

3 – Frost Protection: Glycol 30%

4 – Overheating Protection: None

5 – Corrosion Protection: Inhibitors

**Space Heating/Domestic Hot Water**

1 – Auxiliary (S.H.): Gas boiler (DHW): Gas boiler

2 – Heat Emitters: Fan coils Temp. Range: 48°C Solar 80°C Auxiliary

3 – DHW Insulation: 70mm minl. wool. Set Temperature: 55 C°

**Control Strategy/Operating Modes**

Direct use of the solar heat in the distribution system. The thermal storage is at the return of the ambient fan coil and collects the excess energy.

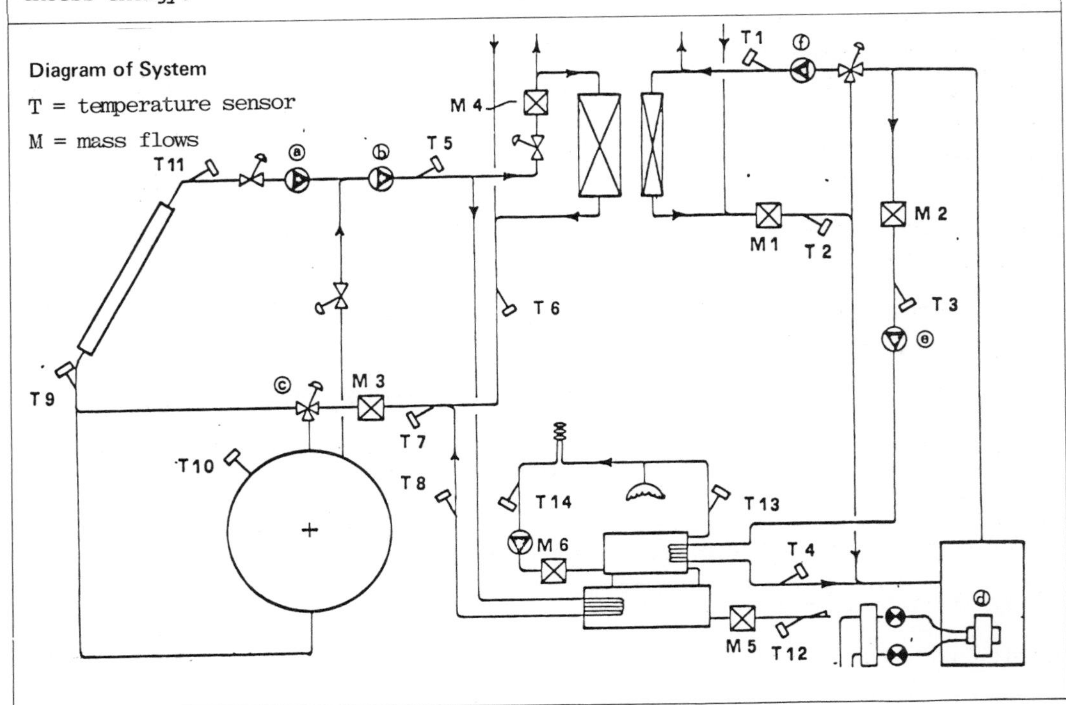

| TECHNICAL APPRAISAL/PRACTICAL EXPERIENCE | REF. | I | FV | 78 | 79 | 17 |
|---|---|---|---|---|---|---|

**System/Design:** It seems that the solutions chosen of using solar energy for heating directly during the day can be in accord with the reduced value of the ratio of collector area to volume heated. Some problems were noted in the heat distribution concerning a heat exchange between the solar and the auxiliary fan coil. A reverse circulation was noted between the storage and solar collectors during the night. A considerable thermal inertia was noted in the system: in fact a certain delay was noted in delivering energy early in the morning and an extension in energy delivery in the afternoon (approx. 1 hour).

**System/Installation:**

There were no useful losses due to the position of the heating plants (inside a room under the roof).

**Component Performance:** (Collector, Heat Exchangers, Storage, Pipework, Valves, Fittings, Pumps, Auxiliary)

Collector performances did not decrease during a period of 16 months. It was noted that the collectors were subject to overheating conditions during 3 months in Summer, because of system problems. Some leaks were noted in the collector absorber (Aluminium Roll Bond), but after some time, these holes were plugged by the limestone contained in the water.

**System Operation/Controls, Electronics:**

The most important failures have been noted in this part of the system:

- Mass flow transducers have excessibe loss of pressure (replaced).
- Human errors in connecting the new transducers to the data acquisition system.
- Human errors in system maintenance; a valve between the storage and the distribution was left closed and the storage was not able to deliver the energy stored in the heating system.
- Due to the length of time taken for the processing of data, failures in the system were noted at a late stage, with a consequent delay in their repair.
- Failures in the data logging system due to line break down and to a bad line up of recorder magnetic heads.

PERFORMANCE EVALUATION/CONCLUSIONS REF. I FV 78 79 17

### Comparison of Measured with Predicted Performance

(State reasons for any discrepancy and summarise the main factors affecting performance)

It is difficult to make a significant comparison between the measured and predicted performances because of the imcomplete data. However, two considerations can be outlined:

- The calculated performances of 30% of total load for solar energy seems to be very optimistic. A more careful calculation gives 21% of total load.
- An evaluation of solar energy used extrapolated from existing data gives a solar energy percentage of 16.7% of solar total load, 180,000 kWh.

### Modifications to System

It has been necessary to change the mass flow transducers because of too large pressure losses:

- a non-return valve has been added between the storage and the collectors in order to avoid the reverse circulation during the night.
- the time interval of change of magnetic tape has been reduced from two months to 15 days.

### Occupants/Response

It is perhaps too soon to evaluate this, but some data collected shows that the DHW consumption of this solar building has been about 15 - 20% greater than another solar building built by I.A.C.P., having about the same volume and number of occupants.

### Conclusions

This first year of experiment gave much information about the main problems of a solar system having certain dimensions. It seems necessary to study the influence of system design to obtain a good efficiency in heat distribution, taking adbantage of useful losses from the system. Further information is necessary concerning the influence of thermal inertia and the monitoring and maintenance problems.

### Future Work

The monitoring program will continue until the summer of 1981. The main object of the work will be the continuous measurement of energy balance and the improvement of system operation.

## P1 MONTHLY PERFORMANCE

REF: I FV 78 79 17

| Month | Year | No Days Data | Average Temp °C | | SOLAR ENERGY kWh | | | | | | Heat Pump Total Output kWh | | AUXILIARY ENERGY kWh | | | | Free Heat Gains kWh | | SUMMARY | | |
|---|---|---|---|---|---|---|---|---|---|---|---|---|---|---|---|---|---|---|---|---|---|
| | | | | | | | Solar Energy Used | | | | | | | | | | | | kWh | % | |
| | | | | | | | Space Heating Useful Losses | | | | | | | | | | | | | | |
| | | | Internal | External | Global Irradiation On Collector Area | Solar Energy Collected | Incl. | Excl. | Domestic Hot Water | Input To Heat | Space Heating | Domestic Hot Water | Heat Pump | Space Heating | Domestic Hot Water | Pumps & Fans | Internal | Direct Solar | Total Load | Percent-age Solar | Syst. Eff. |
| Column No | | | 1 | 2 | 3 | 4 | 5 | 6 | 7 | 8 | 9 | 10 | 11 | 12 | 13 | 14 | 15 | 16 | 17 | 18 | 19 |
| J | | | | | | | | | | | | | | | | | | | | | |
| F | 79 | 15 | | 3.9 | 5289 | 1462 | | 530 | 331 | | | | | 15486 | 294 | | | | 16641 | 5.2 | 16.3 |
| M | 79 | 18 | | 8.1 | 3999 | 1284 | | 235 | 364 | | | | | 10596 | 244 | | | | 11439 | 5.2 | 15.0 |
| A | 79 | 8 | | 12.5 | 2709 | 1105 | | 342 | 279 | | | | | 15 | 165 | | | | 800 | 77.6 | 22.9 |
| M | | | | | | | | | | | | | | | | | | | | | |
| J | | | | | | | | | | | | | | | | | | | | | |
| J | | | | | | | | | | | | | | | | | | | | | |
| A | | | | | | | | | | | | | | | | | | | | | |
| S | | | | | | | | | | | | | | | | | | | | | |
| O | | | | | | | | | | | | | | | | | | | | | |
| N | 78 | 9 | | 8.3 | 3096 | 992 | | 466 | 264 | | | | | 4824 | 135 | | | | 5689 | 12.8 | 22.9 |
| D | 78 | 12 | | 4.1 | 4257 | 872 | | 545 | 201 | | | | | 13982 | 314 | | | | 15042 | 5.0 | 17.5 |
| TOTAL | | | — | — | | | | | | | | | | | | | | | | — | — |
| AVERAGE | | | | | | | | | | | | | | | | | | | | | |

TOTAL: kWh

| PROJECT DESCRIPTION | FIRENZE | REF. | I | Fi | 77 | 78 | 18 |
|---|---|---|---|---|---|---|---|

**Main Participants:**

C.E.C. Solar Energy R & D Programme
(Contract No. 121-76-ESI)

Contractor: Nouvo Pignone SpA
Via F. Mattencci,
Firenze,
Italy.

**Project Description:**

This project was carried out within the C.E.C. Solar Energy R & D Programme (Project A). The building provides $260m^2$ of office accommodation and has been equipped with solar-assisted air conditioning.

**Project Objectives:**

1. To check both the availability of a perpetual non-polluting energy source, such as solar, and the economics of application.
2. To collect experimental data on global irradiation and other weather conditions.
3. To test the performance of individual components, and the whole system, with the purpose of obtaining data on the characteristics of the collector, the absorption/refridgeration machine, the heat storage tank, the auxiliary heater and the control system.
4. To check operating and maintenance expenses.
5. To process and analyse collected data.
6. Economic optimisation of design.

| SITE LOCATION | FIRENZE | REF. | I | Fi | 77 | 78 | 18 |
|---|---|---|---|---|---|---|---|

Situated 1km from Florence

Photograph of Project

## A2 LOCAL CLIMATE

Average cloud cover:5 octas ;
Average daily max temps(July):27°C ; average daily min temps(Jan):2°C
Source of weather data: Meteorolical station PP Ximeniani, Florence

| MONTH | YEAR | Irradiation on Collector Plane | | Degree Days | Precipitation | Average Ambient Temperature | Average Relative Humidity | Average Wind Speed | Prevailing Wind Direction |
|---|---|---|---|---|---|---|---|---|---|
| | | Global | Diffuse | Base. 19 °C | | | | | |
| JAN | | 72.6 | 15.4 | | 93.2 | 7.0 | 67 | | |
| FEB | | 98.5 | 34.7 | | 99.6 | 7.9 | 66 | | |
| MARCH | | 125.5 | 41.8 | | 87.6 | 10.9 | 58 | | |
| APRIL | | 142.3 | 50.6 | | 172.8 | 12.0 | 63 | | |
| MAY | | 164.0 | 63.8 | | 104.0 | 16.7 | 55 | | |
| JUNE | | 159.5 | 61.7 | | 77.8 | 19.7 | 54 | | |
| JULY | | 175.9 | 61.6 | | 53.8 | 23.2 | 49 | | |
| AUGUST | | 175.7 | 54.0 | | 47.2 | 23.1 | 54 | | |
| SEPT | | 152.2 | 43.6 | | 34.6 | 20.1 | 52 | | |
| OCT | | 120.5 | 36.4 | | 81.4 | 15.0 | 61 | | |
| NOV | | 92.1 | 25.2 | | 42.2 | 7.7 | 64 | | |
| DEC | | 60.2 | 19.0 | | 50.4 | 6.8 | 66 | | |
| Totals | | 1539 | 507.8 | 2144 | | | | | |
| Averages | | | | | 78.7 | 14.1 | 59 | | |
| Units | – | kWh/m²· month | | °C days | mm | °C | % | m/s | – |

| SUMMARY SHEET | FIRENZE | REF. | I | Fi | 77 | 78 | 18 |
|---|---|---|---|---|---|---|---|

**Design Data**

**A1 CLIMATE**

1) Source of Data: Meteorological Station 2.5km from site
2) Latitude: 43 ° 46′ Longitude: 11 ° 15′ Altitude: 49 metres
3) Global Irradiation (horiz plane): 1539 kWh/m²·year % Diffuse: 33
4) Degree Days: 2144 Base Temp: 19°C
5) Sunshine Hours: July: 10 January: 2.9 Annual: 2213

**B1 BUILDING**

1) Building Type: Office Building No Occupants: 22
2) Floor area: 260 m² Heated Volume: 750 m³
3) Design Temperature: External: -2 °C Internal: 20 °C
4) Ventilation Rate: 1 a.c.h. Vol. Heat Loss: 0.824 W/m³·K
5) Space Heat Load: 9476 kWh Hot Water Load: 1566 kWh

**C1 SYSTEM**

1) Absorber Type: A1 Roll Bond. Black Paint
2) Collector Area 110 m² (Aperture) Coolant: Water.
3) Orientation: 50% S. 50% SW Tilt: 48° Glazing: Double
4) Storage Volume: Winter 16.5 Summer 5.5 m³ Heat Emitters: Warm Air
5) Auxiliary System Winter Gas Boiler Summer Elec.Refrigerator Heat Pump: No

**PERFORMANCE MONITORING**

1) Is there a Computer Model? Yes (own software)
2) Start Date for Monitoring Programme: 10th December 1977
3) Period for which Results Available: 12 months
4) Data Acquisition System: Schlumberger Data Logger with punched tape and printed paper output.

| | Space Heating | Hot Water | Total |
|---|---|---|---|
| 6) Predicted | 70 % of 9476 kWh | 80 % of 1566 kWh | % of kWh |
| 7) Measured | 64 % of 8348 kWh | 87 % of 1473 kWh | 68 % of 9821 kWh |

space cooling: 77% of 9697 kWh

8) Solar Energy Used: Including useful cooling 128.4 kWh/m²·year
Excluding cooling 60.6 kWh/m²·year

9) System Efficiency: $\frac{\text{Solar Energy Used}}{\text{Global Irradiation on collector}} \times 100 =$ 12 %

B2 BUILDING DESCRIPTION FIRENZE REF. | I | Fi | 77 | 78 | **18**

**Design Concepts**

The space behind the solar wall is not heated.

**Construction**

1 – Glazing: 21 % Area: m² 'U' value: 3.3 W/m²·K

2 – 'U' values: Walls: 0.438 W/m²·K Floor: 1.18 W/m²·K Roof/loft: 0.438 W/m²·K

3 – Thermal mass: kg/m³ Annual Energy Demand: 20.736 kWh

External walls: 100mm lightweight, prefabricated with 90mm polyurethane.

Roof construction: Corrugated sheet iron with lightweight concrete infill and two 25mm polyurethane layers.

False ceiling height varies from 1.6 - 2.3m

**Energy Conservation Measures**

External windows and doors, without thermal bridges.
Inner door at the entrance.
Heat recovery from ventilation air

Free Heat Gains: Internal: 1980 kWh/year Direct Solar: 3141 kWh/year

**Section Through House** Showing main solar components

**usable surface area: 260 m²**

**total surface area: 370 m²**

**basement volume: 295 m³**

**solar panel surface area: 110 m²**

| C2 SYSTEM DESCRIPTION | FIRENZE | REF. | I | Fi | 77 | 78 | 18 |
|---|---|---|---|---|---|---|---|

Primary circuit:water content:16,500 litres, 150 litres/m2collector area;
Insulation:100mm(collector), 300mm(storage), 200mm(pipework);
Frost protection:drain down at +6°C; overheating protection:overpressure valves;
corrosion protection: inhibitors.

Auxiliary(SH)+(DHW):gas boiler ; heat emitters:warm air, temp range 22-28°C ;
set temp(DHW):45°C ; insulation(DHW):100mm

**Diagram of System** Showing sensor locations, power ratings, storage volumes & flows.

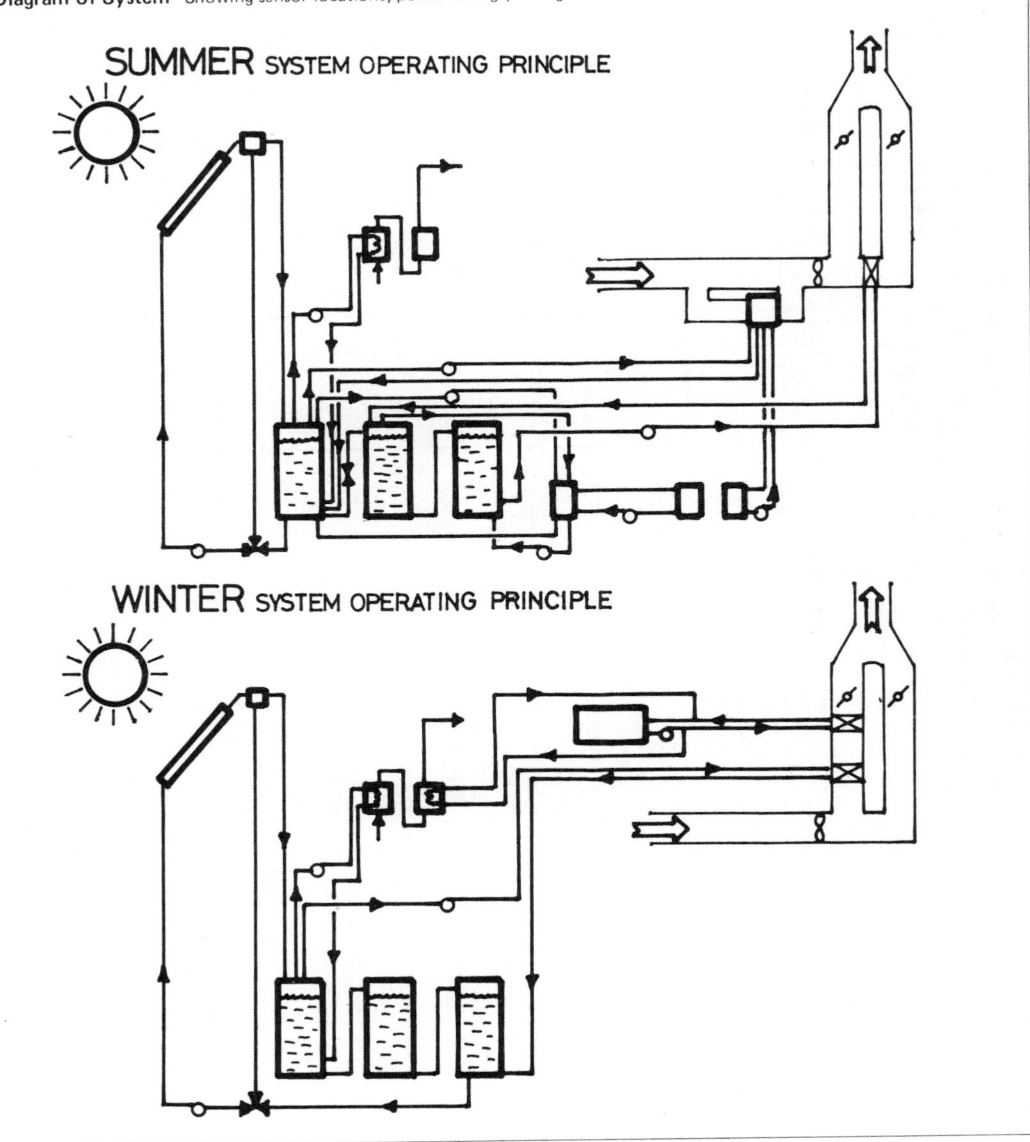

## P1 MONTHLY PERFORMANCE

REF: I | FI | 77 | 78 | 18

| Month | Year | No Days Data | Average Temp °C Internal | Average Temp °C External | SOLAR ENERGY kWh: Global Irradiation On Collector Area | Solar Energy Collected | Solar Energy Used: Space Heating Useful Losses Incl. | Space Heating Useful Losses Excl. | Domestic Hot Water | Space Cooling | Heat Pump Total Output kWh: %Space Cooling | %Solar Domestic Hot Water | AUXILIARY ENERGY kWh: Space Cooling | Space Heating | Domestic Hot Water | Pumps & Fans | Free Heat Gains kWh: Internal | Direct Solar | SUMMARY ** kWh: Total Load | %: Percent-age Solar | Syst. Eff. |
|---|---|---|---|---|---|---|---|---|---|---|---|---|---|---|---|---|---|---|---|---|---|
| Column No | | | 1 | 2 | 3 | 4 | 5 | 6 | 7 | | | | | 12 | 13 | 14 | 15 | 16 | 17 | 18 | 19 |
| J | 78 | 31 | 21.3 | 4.9 | 7778 | 2341 | 1218 | | 79 | - | | | - | 1142 | 58 | | | | 2497 | 52 | 16 |
| F | 78 | 28 | 19.8 | 7.6 | 6333 | 1333 | 942 | | 78 | - | | | - | 1265 | 69 | | | | 2354 | 43 | 16 |
| M | 78 | 31 | 19. | 12.1 | 11200 | 3320 | 1314 | | 116 | - | | | - | 69 | 15 | | | | 1514 | 94 | 12 |
| A | 78 | 21 | 22.9 | 13.0 | 6270 | 987 | 778 | | 100 | - | | | - | | | | | | 870 | 100 | 14 |
| M | | | | | | | | | 160 | 733* | 100 | 100 | 0+ | 0 | 0 | | | | 893 | 100 | |
| J | 78 | 30 | 21.8 | 20.5 | 14411 | 2794 | 0 | | 158 | 1288* | 70 | 100 | 552+ | 0 | 0 | | | | 1999 | 72 | 10 |
| J | 78 | 31 | 22. | 23.2 | 18758 | 3320 | 0 | | 148 | 1520* | 61 | 100 | 972+ | 0 | 0 | | | | 2640 | 63 | 8 |
| A | 78 | 31 | 21.9 | 23.0 | 17251 | 3571 | 0 | | 93 | 1643* | 75 | 100 | 562+ | 0 | 0 | | | | 2298 | 75 | 10 |
| S | 78 | 30 | 21.6 | 20.1 | 15486 | 3650 | 0 | | 154 | 1611* | 91 | 100 | 154+ | 0 | 0 | | | | 1918 | 91 | 11 |
| O | 78 | 31 | 19.9 | 16.5 | 14086 | 2652 | 0 | | 144 | 622* | 100 | 100 | 0+ | 0 | 0 | | | | 805 | 100 | 5 |
| N | | | | | | | | | | - | | | | | | | | | | | |
| D | 77 | 21 | 16.0 | 4.5 | 4874 | 1552 | 1129 | | 53 | - | | | | 491 | 48 | | | | 1721 | 68 | 24 |
| TOTAL | | | — | — | 116446 x | 25520 x | 5381 | | 1283 | 7457 | | | | 2967 | 190 x | | | | 19518 | — | — |
| AVERAGE | | | | 14.5 | SH:36455 | | | | SH: 426 | | | | | | | | | | SH: 8964 | 72 | 12 x |
| | | | | | | | | | | | | | | | | | | | | SH:65 | SH:16 |

TOTAL: kWh

* space cooling refrigeration units;
x 10 month data; + cooling load.

** includes cooling

| PROJECT DESCRIPTION ROSSANO CALABRO | REF. | I | RC | 79 | 80 | 19 |
|---|---|---|---|---|---|---|

**Main Participants:**

Consruction: E.N.E.L. Natonal Electricity Board
System and Monitoring: C.R.T.N./E.N.E.L. Thermal and Nuclear research Centre.
Funding: C.N.R. National Council of Research.

**Project Description:**

Consists of two 6 apartment buildings, one of which incorporates a solar heating system. The other is electrically heated and is used for comparison. The building design and construction are traditional, but roof insulation has been improved. 50% of tho collector area is vertical (on external walls) and 50% roof mounted.

**Project Objectives:**

- Performance monitoring and technical evaluation of the selected design solution including solar and auxiliary installations.
- Energy balance measurements for the two buildings, and comparison.
- Evaluation of the selected design solutions in terms of architectural integration.
- Testing and validation of computer model.

**SITE LOCATION MAP**

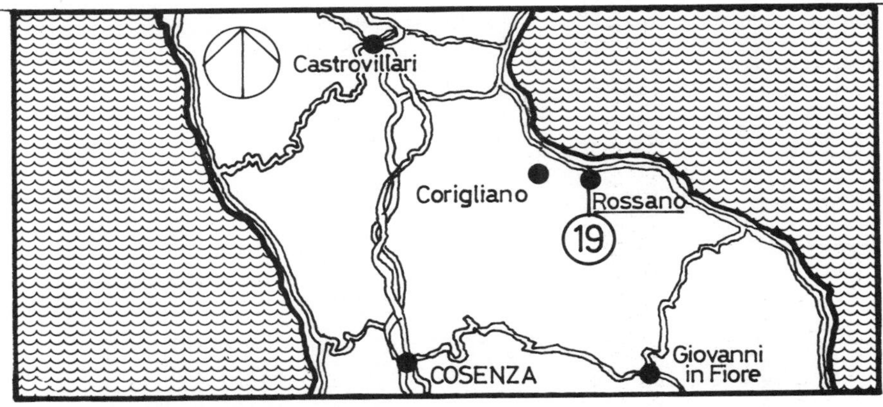


Situated 2km from Rossano Calabro.

Photograph of Project

| SUMMARY SHEET ROSSANO CALABRO | REF. | I | RC 79 | 80 | 19 |
|---|---|---|---|---|---|

**Design Data**

**A1 CLIMATE**

1) Source of Data: Air Force Weather Station, CROTONE
2) Latitude: 39 ° 04 ′ Longitude: 17 ° 04 ′ Altitude: 158 metres
3) Global Irradiation (horiz plane): 1,382 kWh/m²·year % Diffuse:
4) Degree Days: 1297 Base Temp: 19 °C
5) Sunshine Hours: July: 347 January: 158 Annual: 2200

**B1 BUILDING**

1) Building Type: 6 Apartments No Occupants: 4 x 6
2) Floor area: 780 m² Heated Volume: 3300 m³
3) Design Temperature: External: +2 °C Internal: 20 °C
4) Ventilation Rate: 05 a.c.h. Vol. Heat Loss: 1.51 W/m³·K
5) Space Heat Load: 50,000 kWh Hot Water Load: 13,400 kWh

**C1 SYSTEM**

1) Absorber Type: 50% Al Roll Bond. 50% Extrud Aluminium
2) Collector Area 120 m² (Aperture) Coolant: Water/Glycol
3) Orientation: 195° Tilt: 90° 25° Glazing: Single
4) Storage Volume: 7.5 m³ Heat Emitters: Fan Coil
5) Auxiliary System: Electric Night Storage Heaters Heat Pump: No

**PERFORMANCE MONITORING**

1) Is there a Computer Model? Yes
2) Start Date for Monitoring Programme: April 1979
3) No. of Measuring Points per house: 40
4) Data Acquisition System: Hewlwtt Packard, 100 channels.

| | Space Heating | Hot Water | Total |
|---|---|---|---|
| 6) Predicted | 45 % of 50,000 kWh | 85 % of 13,400 kWh | 53 % of 63400 kWh |
| 7) Measured | % of kWh | % of kWh | % of kWh |

8) Solar Energy Used: Including useful losses kWh/m²·year
Excluding losses: kWh/m²·year

9) System Efficiency: $\frac{\text{Solar Energy Used}}{\text{Global Irradiation on collector}} \times 100 =$ %

| A2 LOCAL CLIMATE ROSSANO CALABRO | REF. | I | RC | 79 | 80 | 19 |
|---|---|---|---|---|---|---|

Average daily max temps(July):29.5 C; average daily min temps(Jan):5.7 C.

Source of Weather Data: Crotona Air Force weather station.

| MONTH | YEAR | Irradiation on horizontal plane Global | Diffuse | Degree Days Base: °C | Sunshine Hours (daily) | Average Ambient Temperature | Average Relative Humidity | Average Wind Speed | Prevailing Wind Direction |
|---|---|---|---|---|---|---|---|---|---|
| JAN | 58-69 | 52.6 | | | 4.9 | 8.4 | 71.5 | 5.8 | |
| FEB | | 73.5 | | | 4.4 | 8.5 | 74.4 | 5.9 | |
| MARCH | | 101. | | | 6.5 | 8.6 | 66.1 | 5.6 | |
| APRIL | | 132.2 | | | 6.3 | 13. | 80.0 | 3.2 | |
| MAY | | 167.2 | | | 7.1 | 16.8 | 76.2 | 2.9 | |
| JUNE | | 177.2 | | | 10.3 | 21.8 | 70.7 | 2.3 | |
| JULY | | 189.2 | | | 11.3 | 25.2 | 53.8 | 3.0 | |
| AUGUST | | 170. | | | 11. | 25.3 | 64.9 | 1.9 | |
| SEPT | | 128.3 | | | 8.3 | 26.9 | 69.9 | 3.5 | |
| OCT | | 89.4 | | | 7.1 | 17.2 | 68.8 | 4.6 | |
| NOV | | 55.1 | | | 4.3 | 9.3 | 81.6 | 3.4 | |
| DEC | | 46.5 | | | 4.7 | 10.1 | 75.2 | 6.4 | |
| Totals | | 1,381.9 | | | | | | | |
| Averages | | | | | | | | | |
| Units | – | kWh/m²·month | | °C days | mm | °C | % | m/s | – |

## B2 BUILDING DESCRIPTION

### Design Concepts

The design and construction are traditional. The external walls consist of double hollow brick work (8cm. 5cm cavity 12cm). The roof is insulated with 15cm light weight concrete.

### Construction

Glazing:10%(area 74m$^2$); 'U' values (W/m$^2$.K):3.3(glazing), 1.3(walls), 1.8(floor), 1.0(roof/loft) ; thermal mass:500kg/m$^3$ ; annual energy demand:63,400kWh.

### Energy Conservation Measures

There are individual thermostat controls in each room.
Free heat gains: 21,000kW/year (internal).

| C2 SYSTEM DESCRIPTION ROSSANO CALABRO | REF. | I | RC | 79 | 80 | 19 |
|---|---|---|---|---|---|---|

**Collector/Storage**

1 – Primary circuit. Water content: 2000 litres ... 16.7 litres/m$^2$ collector area

2 – Insulation. Collector: 50/100mm Storage: 100mm Pipework: 40mm

3 – Frost Protection: 39 Ethylene Glycol

4 – Overheating Protection: None

5 – Corrosion Protection: Inhibitors

**Space Heating/Domestic Hot Water**

1 – Auxiliary (S.H.): 10m$^3$ Elec. Storage (DHW): Electric auxiliary

2 – Heat Emitters: Fan Coil Units Temp. Range: 40 – 45°C

3 – DHW Insulation: 100mm Set Temperature: 48°C

**Other Points/Heat Pump:** All insulation – mineral wool

**Control Strategy/Operating Modes**

| | | |
|---|---|---|
| DAY | – Solar System On | Store if T 40°C |
| | – Heat distribution | Auxiliary if T 40°C |
| | – Auxiliary off | |
| NIGHT | – Heat distribution | Ditto |
| | – Auxiliary on (until night storage = 85°C) | |

**Diagram of System** Showing sensor locations, power ratings, storage volumes & flows.

| PROJECT DESCRIPTION SEQUALS | REF. | I | Se | 79 | 80 | 20 |
|---|---|---|---|---|---|---|

**Main Participants:**

Owner: A.N.A. (Associazione Nazionale Alpini)

Building: Industrie Zanussi-Farsura - Spilimbergo

Solar System Designer and Manufacturer: Industrie Zanussi - Pordenone

Monitoring: Industrie Zanussi - Pordenone

Data Acquisition System Manufacturer: Industrie Micros - Conegliano

Funding Authority: Ministry of Industry, Commerce and Small Industry

**Project Description:**

This experiment consists of 4 one family solar houses, built in Sequals, a small town near Udine (35 Km). The building construction system is a prefabricated by means of tridimensional components. The external building envelope has a thermal transmittance value that meets the new italian law concerning energy conservation in building. Two of three houses are provided with a heat pump which operates in the solar heating system, increasing the thermal level of fluid when the solar energy contribution is too low. The construction was financially sponsored by the italian Ministry of Industry and Commerce.

**Project Objectives:**

1) Performance evaluation of a solar system in relation to its use in a type of economic house;

2) Evaluation of the particular solutions selected in the field of architectural integration of solar collectors, and in relation to the solar heating and storage systems;

3) Comparison of performance between the houses provided with solar heating system only and the houses provided with a solar heating system plus heat pump;

4) Evaluation of the external envelope behaviour and construction system design improvement from point of view of energy conservation requirements;

5) Evaluation of the solar application impact on the inhabitants' usage pattern.

| SUMMARY SHEET SEQUALS | REF. | I | Se | 79 | 80 | 20 |
|---|---|---|---|---|---|---|

**Design Data**

**A1 CLIMATE**

1) Source of Data: Udine (35km) Conegliano (50Km)
2) Latitude: 46 ° 05′ Longitude: 13 ° Altitude: 139 metres
3) Global Irradiation (horiz plane): 1159 kWh/m$^2$·year % Diffuse: 42
4) Degree Days: 2386 Base Temp: 19 °C
5) Sunshine Hours: July: 268 January: 105 Annual: 2007

**B1 BUILDING**

1) Building Type: Prefabricated house No Occupants:
2) Floor area: 95 m$^2$ Heated Volume: 332 m$^3$
3) Design Temperature: External: -5 °C Internal: 20 °C
4) Ventilation Rate: 0.5 a.c.h. Vol. Heat Loss: 1.035 W/m$^3$·K
5) Space Heat Load: 13,848 kWh Hot Water Load: 2665 kWh

**C1 SYSTEM**

1) Absorber Type: Aluminium Roll Bond
2) Collector Area 33 m$^2$ (Aperture) Coolant: Water/Glycol
3) Orientation: 180 Tilt: 90° 45° Glazing: Single
4) Storage Volume: .5 + .5 m$^3$ Heat Emitters: Warm air
5) Auxiliary System: Gas Boiler Heat Pump: In 2 houses

**PERFORMANCE MONITORING**

1) Is there a Computer Model? No
2) Start Date for Monitoring Programme: 27th January 1979
3) No. of Measuring Points per house: 59
4) Data Acquisition System: Microprocessor Controlled data logging system, collecting data from 2 houses and a weather station (6 variables).

| | Space Heating | Hot Water | Total |
|---|---|---|---|
| 6) Predicted | % of kWh | % of kWh | 42 % of 16513 kWh |
| 7) Measured | % of kWh | % of kWh | % of kWh |

8) Solar Energy Used: Including useful losses kWh/m$^2$·year
Excluding losses: kWh/m$^2$·year

9) System Efficiency: $\frac{\text{Solar Energy Used}}{\text{Global Irradiation on collector}} \times 100 =$ %

| SITE LOCATION MAP | SEQUALS | REF. | 1 | Se | 79 | 80 | 20 |
|---|---|---|---|---|---|---|---|

Distance to main city: 35km from Udine

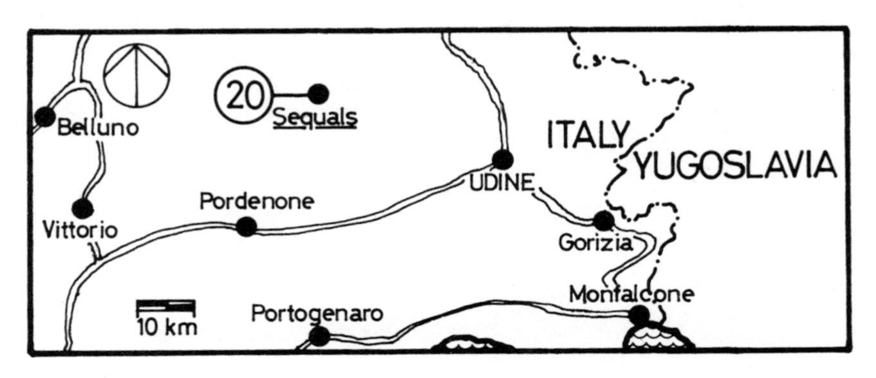


## A2 LOCAL CLIMATE SEQUALS

Source of Weather Data: Udine (35km) Air Force Weather Station, Conegliano (50km).

| MONTH | YEAR | Irradiation on horizontal plane Global | Diffuse | Degree Days Base: 19 °C | Precipitation | Average Ambient Temperature |
|---|---|---|---|---|---|---|
| JAN | 1958-59 | 40,03 | 18.1 | 536 | | +2,7 |
| FEB | | 52,9 | 24,3 | 434 | | +4,5 |
| MARCH | | 92,4 | 42,5 | 378 | | +7,8 |
| APRIL | | 106,2 | 46,7 | 66 | | +12,7 |
| MAY | | 151,0 | 63,4 | - | | +16,5 |
| JUNE | | 150,6 | 63,2 | - | | +20,5 |
| JULY | | 165,5 | 62,3 | - | | +23,4 |
| AUGUST | | 141,7 | 54,5 | - | | +22,1 |
| SEPT | | 108,6 | 43,4 | - | | +18,9 |
| OCT | | 79,7 | 32,9 | 95 | | +13,2 |
| NOV | | 37,2 | 18,5 | 375 | | +7,5 |
| DEC | | 32,9 | 15,2 | 502 | | +3,8 |
| Totals | | 1159 | 485,0 | 2,386 | | +12.8 |
| Averages | | | | | | |
| Units | – | kWh/m²·month | | °C days | mm | °C |

## B2 BUILDING DESCRIPTION SEQUALS

REF. | I | Se | 79 | 80 | **20**

**Design Concepts** The house is prefabricated by means of three dimensional components, generally parallel piped boxes. These components are completely finished at the factory before being assembled on site. The wall insulation is obtained by means of a polyurethane layer placed between two structural concrete layers. The structural performance of the house meet Italian law concerning earthquake prevention.

**Construction**

1 — Glazing: 9 % Area: 15.7 $m^2$ 'U' value: 3.3 $W/m^2 \cdot K$

2 — 'U' values: Walls: 0.71 $W/m^2 \cdot K$ Floor: 0.58 $W/m^2 \cdot K$ Roof/loft: 0.71 $W/m^2 \cdot K$

3 — Thermal mass: 380 $kg/m^3$ Annual Energy Demand: 16513 kWh

**Energy Conservation Measures**

Insulated Shutters

Free Heat Gains: Internal: 1000 kWh/year Direct Solar: 1850 kWh/year

## C2 SYSTEM DESCRIPTION SEQUALS

**Collector/Storage**

1 — Primary circuit. Water content: (Excluding storage) 1.9 litres/$m^2$ collector area

2 — Insulation. Collector: 50mm Storage: 50mm Pipework: 42mm

3 — Frost Protection: Polypropylene Glycol - 40%

4 — Overheating Protection: None

5 — Corrosion Protection: Distilled Primary fluid with inhibitors

**Space Heating/Domestic Hot Water**

1 — Auxiliary (S.H.): Gas Boiler (DHW): Gas Boiler

2 — Heat Emitters: Warm Air Temp. Range: 20-35°C

3 — DHW Insulation: Polyurethane Set Temperature: 55°C

**Control Strategy/Operating Modes**

| Winter | Collector Output | Storage | Heating | |
|---|---|---|---|---|
| 1 | 30°C | - | Yes | Direct Mode |
| 2 | 5 Bottom Store | - | No | To Storage |
| 3 | 30°C | 30°C | Yes | Direct Mode |
| 4 | 30°C but gaining | - | Yes | Heat Pump/Coll. |
| 5 | Not Gaining | 30°C 0°C | Yes | Heat Pump/St. |
| Summer | | | | |
| 1 | Gaining | | | To Storage |
| 2 | Not Gaining | | | To DHW |

C2 SYSTEM DESCRIPTION (CONT) REF. I Se 79 80 **20**

**Diagram of System** Showing sensor locations, power ratings, storage volumes & flows.

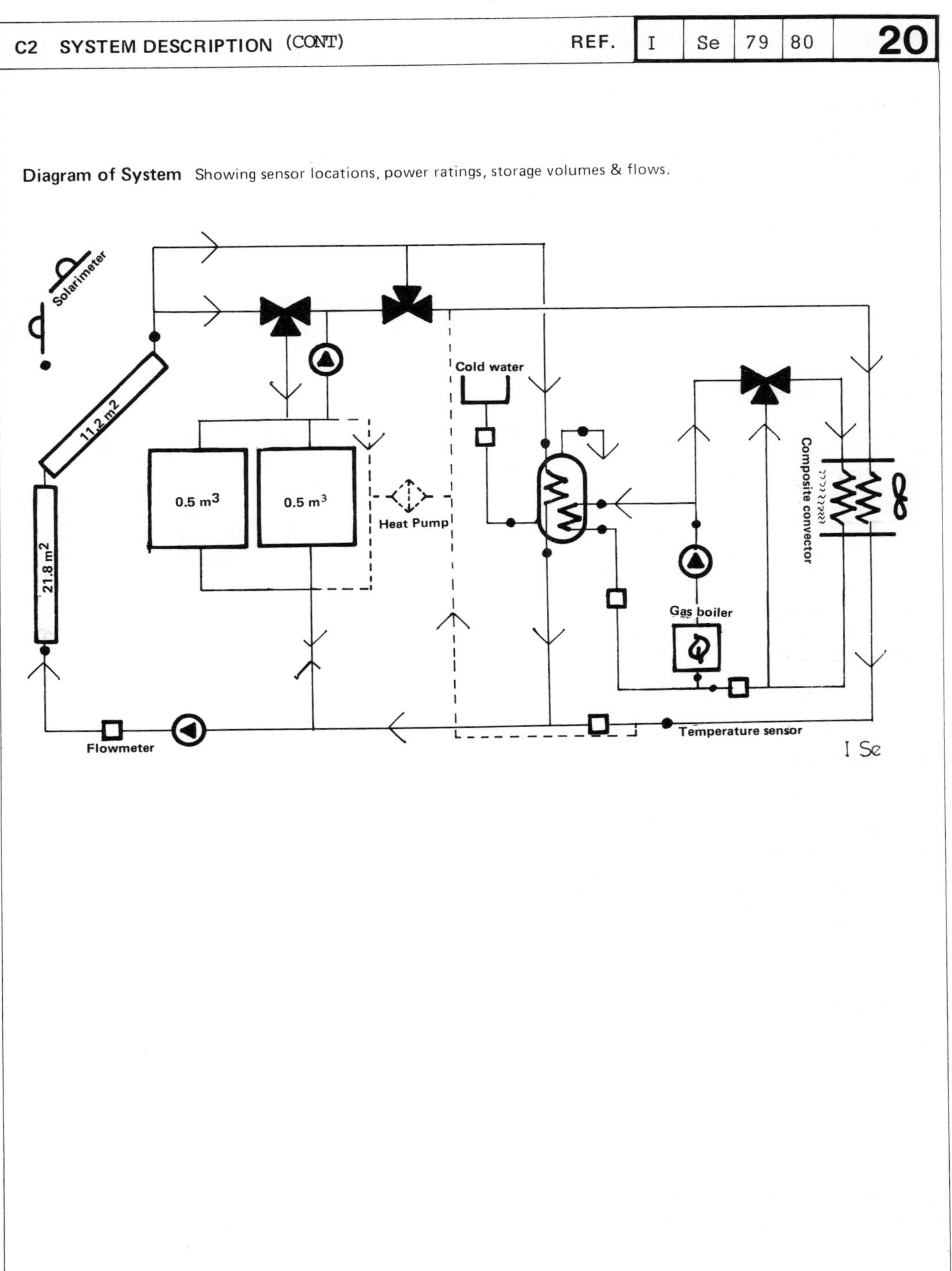

| PROJECT DESCRIPTION | EINDHOVEN | REF. | NL | EI | 77 | 78 | 21 |
|---|---|---|---|---|---|---|---|

**Main Participants**:

Coordination, engineering, monitoring: Eindhoven University of Technology (section Heat and Flow Technology and section Physical Engineering in Relation to Building Design), P.O. Box 513 Eindhoven

Architects: H.C.A. Hoekstra, Parkstraat 54, Nuenen

Financial support: house privately owned. Solar installation and monitoring subsidized by Dutch Government.

**Project Description:**

The solar house of the EUT, a rather spacious, detached house, is situated in the outskirts of Eindhoven. In the design of the house both the architectural and the mechanical engineering implications of the utilization of solar energy were taken into consideration. The solar roof, tilted at an angle of $48^o$ to the South, embodies $51m^2$ of aluminium finned tube absorber plates, coated with black chrome. The cooling medium is water. In the $4.1m^3$ solar heat storage tank the advantages of thermal stratification were exploited to the limits of their potential. The solar system serves both space heating and domestic hot water supply. The performance of the solar system is monitored continuously, readings of 30 measuring points being taken each minute and recording executed each half hour.

**Project Objectives**:

- to gain expeience with the design, construction and operation of a solar heating system.
- to identify the problems which need further attention in the development of solar systems.
- to help establish a well founded design method for such systems on the basis of performance measurments.

Project schedule:
house construction started May 1975
house completed in November 1976
monitoring commenced in April 1977
monitoring completed in April 1980

| SUMMARY SHEET | EINDHOVEN | REF. | NL | Ei | 77 | 78 | 21 |
|---|---|---|---|---|---|---|---|

**Design Data**

**A1 CLIMATE**

1) Source of Data: Eindhoven airport (5 km) and average of RDMI in De Bilt (75 km N) and Zuid Limburg airport (60 km SSE).

2) Latitude: 51 °28 ′ N Longitude: 5 ° 30 ′ E Altitude: 16 metres

3) Global Irradiation (horiz plane): 1284 $kWh/m^2 \cdot year$ % Diffuse: 40% summer 60% winter

4) Degree Days: 2,785 Base Temp: 18 °C

5) Sunshine Hours: July: 225 January: 73 Annual: 1,400

**B1 BUILDING**

1) Building Type: ........ No Occupants: 4

2) Floor area: 220 $m^2$ Heated Volume: 650 $m^3$

3) Design Temperature: External: -10 °C Internal: +20 °C

4) Ventilation Rate: 0.75 a.c.h. Vol. Heat Loss: 0.92 $W/m^3 \cdot K$

5) Space Heat Load: 34,440 kWh Hot Water Load: 2,000 kWh

**C1 SYSTEM**

1) Absorber Type: Selective, finned tube type, extruded aluminium

2) Collector Area 51. $m^2$ (Aperture) Coolant: water

3) Orientation: 187.° Tilt: 42.° Glazing: single, 4mm.

4) Storage Volume: 4.1 $m^3$ Heat Emitters: warm air

5) Auxiliary System: Gas fired air heater Heat Pump: -

**PERFORMANCE MONITORING**

1) Is there a Computer Model? No

2) Start Date for Monitoring Programme: April 1979

3) Period for which Results Available: 16th May 1977 to 15th May 1978

4) No. of Measuring Points per house: 30

5) Data Acquisition System: Munster and Diehl data logger, microprocessor and teletype.

| | Space Heating | Hot Water | Total |
|---|---|---|---|
| 6) Predicted | ... % of ... kWh | ... % of ... kWh | 61 % of 21,000 kWh |
| 7) Measured | ... % of ... kWh | ... % of ... kWh | 41 % of 30505 kWh |

8) Solar Energy Used: Including useful losses 245 $kWh/m^2 \cdot year$; Excluding losses: 206 $kWh/m^2 \cdot year$

9) System Efficiency: $\frac{\text{Solar Energy Used}}{\text{Global Irradiation on collector}} \times 100 =$ 27 %

| SITE LOCATION | REF. | NL | EI | 77 | 78 | 21 |
|---|---|---|---|---|---|---|

Situated about 2km from Eindhoven town centre

Photograph of Project

## A2 LOCAL CLIMATE

1 – Average Cloud cover: 5.5 Octas

2 – Average daily max temperatures (July) : 20.8 °C

Average daily min temperatures (Jan) : -1.3 °C

3 – Source of Weather Data:

The local data in the table is taken from the weather records of the Welschap airport, 12 km SW of the house. Some of the data is the arithmetical average of the records of the Royal Dutch Meteorological Institute (75km NNW) in De Bilt and the records at Zuid Limburg airport (60km SSE).

4 – Micro Climate/Site Description:

The Dutch climate is characterised by rather a large diversity. The differences between the continental and the maritime climate are noticeable in a relatively small area. In the western part of the country the climate has a strong maritime tendency with a moderate character (rater low temperatures in summer and relatively high temperatures in winter and also rather strong westerly winds). Somewhat up -country the air soon gets a more continental character (less wind, higher temperature differences between summer and winter and even during a day).

A2 LOCAL CLIMATE (CONT) REF. NL EI 77 78 21

| MONTH | Irradiation on horizontal plane | | Degree Days | Precipitation | Average Ambient Temperature | Average Relative Humidity | Average Wind Speed | Prevailing Wind Direction |
|---|---|---|---|---|---|---|---|---|
| | Global | Diffuse | Base: 18 °C | | | | | |
| JAN | 25 | winter | 495 | 72.9 | 1.6 | 85.5 | 4.1 | s.w. |
| FEB | 49 | ca. 60% | 412 | 54.2 | 1.9 | 82.2 | 4.1 | n.e. |
| MARCH | 87 | | 391 | 40.9 | 5.1 | 73.6 | 3.5 | n.n.e. |
| APRIL | 146 | | 286 | 36.2 | 8.5 | 65.7 | 3.7 | s.w. |
| MAY | 188 | | 163 | 50.1 | 12.4 | 63.2 | 3.0 | s.w. |
| JUNE | 212 | summer | 0 | 58.7 | 15.5 | 64.5 | 2.9 | s.w. |
| JULY | 183 | ca. 40% | 0 | 71.1 | 17.1 | 70.1 | 3.0 | w.n.w. |
| AUGUST | 162 | | 0 | 91.5 | 16.9 | 72.3 | 3.2 | s.w. |
| SEPT | 116 | | 0 | 68.2 | 14.3 | 72.8 | 3.0 | s.w. |
| OCT | 63 | | 232 | 63.3 | 9.9 | 78.9 | 3.3 | s.s.w. |
| NOV | 32 | | 367 | 53.9 | 5.8 | 84.7 | 3.3 | e.n.e. |
| DEC | 21 | | 439 | 61.9 | 2.9 | 88.3 | 3.6 | e.n.e. |
| Totals | 1284 | | 2785 | 722.9 | | | | |
| Averages | 107 | | | 60.2 | 9.3 | 75.1 | 3.4 | s.w. |
| Units | kWh/m²·month | | °C days | mm | °C | % | m/s | – |

B2 BUILDING DESCRIPTION

**Design Concepts**

The heated floor area amounts to 220m². An air heating system provides the heat for living spaces (living room, study, kitchen, pantry, 5 bedrooms, 2 bathrooms). The garage is heated by waste ventilation air. The total outside roof area - about 141m² - with a forced ventilation rate of 200m³ hr⁻¹ and an estimated unintentional ventilation of 100m³ hr⁻¹. The total design heat load amounts to 0.39kW°C⁻¹.

**Construction**

1 – Glazing: 28 % Area: 40 m² 'U' value: 3.2 W/m²·K

2 – 'U' values: Walls: 0.4 W/m²·K Floor: 0.5 W/m²·K Roof/loft: 0.4 W/m²·K

3 – Thermal mass: 294 kg/m³ Annual Energy Demand: 21,000 kWh (calculated)

Insulation: ground floor - 4cm above clearing space (120m², k = 0.5Wm⁻² °C⁻¹)
walls - 7.5cm glass wool (172m², k = 0.4Wm⁻² °C⁻¹)
roof - 5cm foam plastic (141m², k = 0.4Wm⁻² °C⁻¹)

**Energy Conservation Measures**

Both roofs and walls have a theoretical heat transmission coefficient of 0.4Wm⁻² °C⁻¹ - a very low value compared to conventional Dutch standards. In order to achieve a low unintentional ventilation, numerous measures had to be taken to eliminate air leaks in the envelope of the house. All movable elements in the facades have been provided with weather stripping.

Free Heat Gains: Internal: 7300 (7 months) kWh/year Direct Solar: 4700 (7 months) kWh/year

B2 BUILDING DESCRIPTION (CONT) REF. NL EI 77 78 21

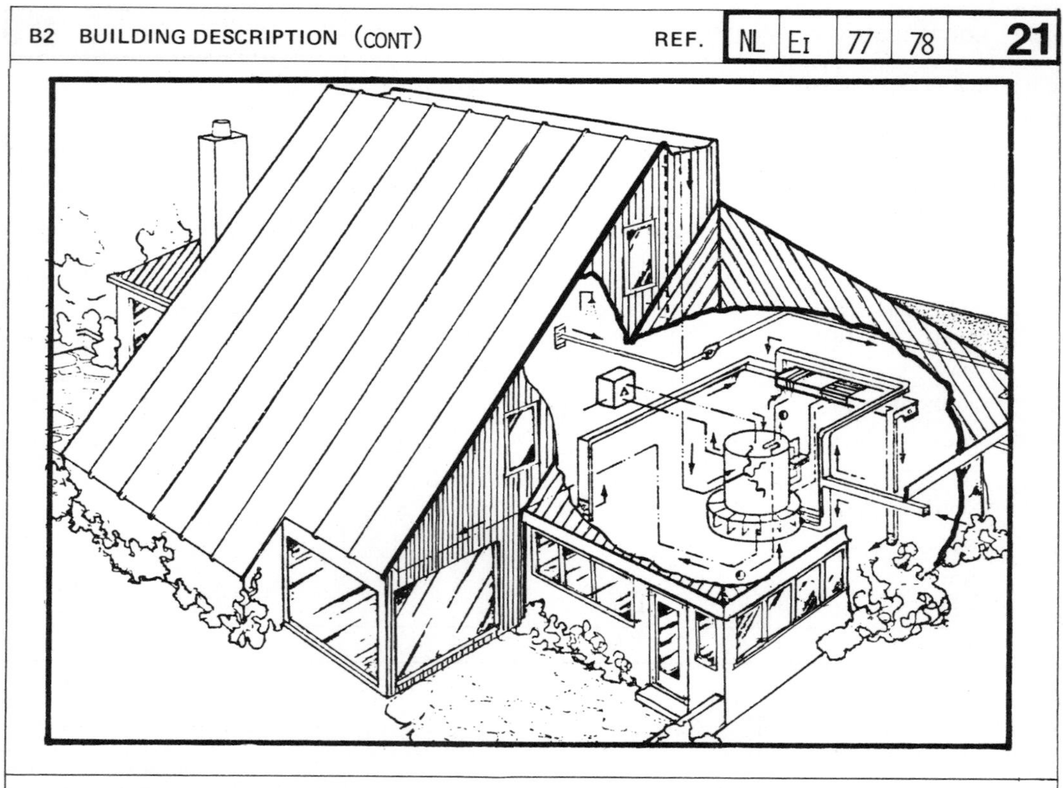

C2 SYSTEM DESCRIPTION

**Collector/Storage**

Insulation:4cm glass wool+8cm polyurethane foam (collector) ;
frost protetion: mono-ethylene glycol filled expansion tubes (collectors);
overheating protection:overload heat-exchanger in soil.

**Space Heating/Domestic Hot Water**

Auxiliary (SH):gas fired boiler ; heat emitters:air heating ;
Auxiliary (DHW): nothing special ; set temp:55°C.

**Control Strategy/Operating Modes** : control of the solar and auxiliary heating system is fully automatic. The storage collector system pump is started when the difference between the temp. of the upper portion of the absorber plates and the water at 0.25m above the lowest point of the storage vessel when the temp. is at 0.6m below the highest point of the storage vessel sinks below the (adjustable) value of 55°C. It stops when the temp at the same point has increased 5°C. The control of the delivery of the heat of the house has been described above. All control systems act mutually independently, so all six possible combinations do actually occur. The maximum temp. in the storage is set at 80° C. When this temp. is exceeded (summer) the flow from the collectors is automatically diverted from the storage to a 40m long copper tube running through the ground around the house.

C2 SYSTEM DESCRIPTION (CONT) REF. NL EI 77 78 21

**Diagram of System** Showing sensor locations, power ratings, storage volumes & flows.

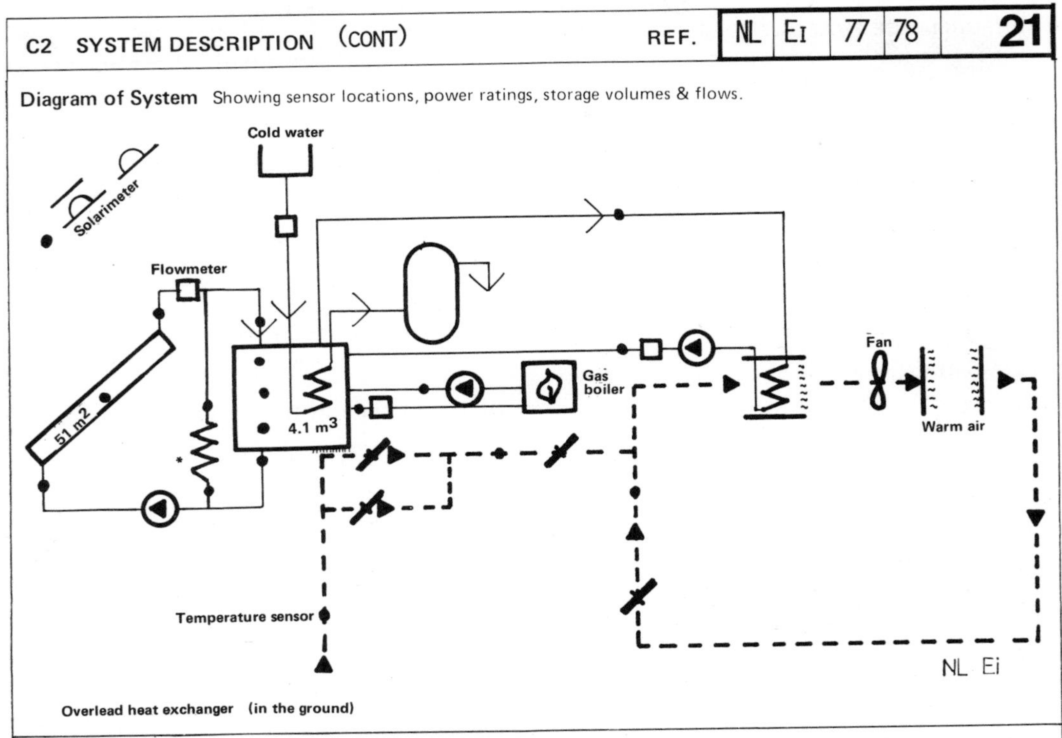


## TECHNICAL APPRAISAL/PRACTICAL EXPERIENCE

**System/Design:**

The optimal stratification improves the efficiency of the collector. This stratification is obtained by four items:
- a newly developed device - floating inlet. This device automatically moves to that level in the tank where the temp. is equal to the exit temp. of collectors.
- the fresh air intake for ventilating the house passes along the bottom part of the tank, thereby cooling the water in that part and becoming pre-heated itself.
- domestic hot water is prepared by passing cold water from the mains through a pipe running from the bottom to the top through the tank.
- the auxiliary heater heats only the water in the upper 0.6m of the vessel.

**System/Installation:**

The freeze protection system eliminates the performance losses associated with heat exchange between the collectors and the heat storage. The mounting of the freeze protection system is, however, rather laborious.

**Component Performance:** (Collector, Heat Exchangers, Storage, Pipework, Valves, Fittings, Pumps, Auxiliary)

The heat loss coefficient of the collectors was found to be about 40% higher than calculated (6W/m$^2$ $^o$C versus 4.2W/m$^2$ $^o$C)

A number of burners were taken out of the burner bed in order to reduce the capacity of the boiler from the nominal 24kW to14kW, a value only slightly beyond the maximum capacity required theorecically.
The indicated efficiency of the boiler is 75 at full load, referred to the upper heating value of natural gas (80% referred to the lower heating value).

| PERFORMANCE EVALUATION/CONCLUSIONS | REF. | NL | EI | 77 | 78 | 21 |
|---|---|---|---|---|---|---|

**Comparison of Measured with Predicted Performance**
(State reasons for any discrepancy and summarise the main factors affecting performance)

A comparison between the design calculations and the actual results shows the net heat demand for heating + hot water (30447kWh) to have been 45% higher than the 21000kWh clcualted for a normal year.
Further the heat loss coefficient of the collectors was found to be about 40% higher than calculated. Nevertheless the amount of solar heat used (12492kWh) is almost equal to the 12900kWh that were expected. The main cause of this somewhat surprising result is the large share of solar heat in the (larger) heat demand in spring and early autumn. This compensates for the smaller output of the solar system in mid-winter.

**Occupants/Response**

The comfort requirements for the house were fully met during the responding period.

**Conclusions**

- see comparison of measured and predicted performance.
- the accuracy of the measurements is judged to have been satisfactory. The daily heat balances usually agree within 10% - sudden heat changes in the ambient temp. causing the largest deviations. Presumably the large thermal time constant of the house has a large part in these discrepancies.

**Future Work**

The discrepancies mentioned require further investigation which may yield somewhat different results and additional explanations. It was, however, judged worthwhile to not the interaction between the heat demand of the house and the output of the solar system. The influence of the heat capacity of the collector on the efficiency measurements also deserves attention. The intention is to continue the measurements and observations during two more heating seasons.

## P1 MONTHLY PERFORMANCE

REF: NL EI 77 78 21

| Month | Year | No Days Data | Average Temp °C: Internal | Average Temp °C: External | SOLAR ENERGY kWh: Global Irradiation On Collector Area | Solar Energy Collected | Solar Energy Used: Space Heating Useful Losses Incl. | Space Heating Useful Losses Excl. | Domestic Hot Water | Input To Heat | Heat Pump Total Output kWh: Space Heating | Domestic Hot Water | AUXILIARY ENERGY kWh: Heat Pump | Space Heating | Domestic Hot Water | Pumps & Fans | Free Heat Gains kWh: Internal | Direct Solar | SUMMARY kWh: Total Load | %: Percent-age Solar | Syst. Eff. |
|---|---|---|---|---|---|---|---|---|---|---|---|---|---|---|---|---|---|---|---|---|---|
| Column No | | | 1 | 2 | 3 | 4 | 5 | 6 | 7 | 8 | 9 | 10 | 11 | 12 | 13 | 14 | 15 | 16 | 17 | 18 | 19 |
| J | 78 | 31 | | | 1518 | | 221 | 221 | (140) | - | | | - | 5025 | | 38 | 1044 | 357 | 5386 | 7 | 24 |
| F | 78 | 28 | | | 2766 | | 504 | 504 | (140) | - | | | - | 4060 | | 38 | 1044 | 604 | 4704 | 14 | 23 |
| M | 78 | 31 | | | 2795 | | 1400 | 1400 | (140) | - | | | - | 2363 | | 38 | 1044 | 954 | 3903 | 39 | 41 |
| A | 78 | 30 | | | 5479 | | 1981 | 1836 | (140) | - | | | - | 1075 | | 25 | 1044 | 1250 | 3196 | 65 | 39 |
| M | 78<br>77 | 15<br>16 | | | 7073 | | 1530 | 973 | (140) | - | | | - | 0 | | 16 | 0 | 0 | 1670 | 100 | 24 |
| J | 77 | 30 | | | 4971 | | 955 | 596 | (140) | - | | | - | 6 | | 3 | 0 | 0 | 1101 | 99 | 22 |
| J | 77 | 31 | | | 5170 | | 600 | 267 | (140) | - | | | - | 52 | | 4 | 0 | 0 | 792 | 93 | 14 |
| A | 77 | 31 | | | 4439 | | 517 | 168 | (140) | - | | | - | 0 | | 3 | 0 | 0 | 657 | 100 | 15 |
| S | 77 | 30 | | | 4111 | | 962 | 739 | (140) | - | | | - | 0 | | 3 | 0 | 0 | 1102 | 100 | 27 |
| O | 77 | 31 | | | 3615 | | 1271 | 1268 | 176 | - | | | - | 0 | | 14 | 1044 | 800 | 1447 | 100 | 40 |
| N | 77 | 7 | | | (1800) | | 460 | 460 | (140) | - | | | - | 1680 | | 36 | 1044 | 390 | 2280 | 26 | 33 |
| D | 77 | 31 | | | 1508 | | 375 | 375 | (140) | - | | | - | 3752 | | 37 | 1044 | 327 | 4267 | 12 | 34 |
| TOTAL | | | — | — | 46245 | | 10776 | 8807 | (1716) | - | | | - | 18013 | | 255 | 7308 | 4682 | 30505 | — | — |
| AVERAGE | | | | | | | | | | | | | | | | | | | | 41 | 27 |

TOTAL: 10523 kWh

| PROJECT DESCRIPTION | ZOETERMEER HOUSE 1 | REF. | NL | ZO | 78 | 78 | 22 |
|---|---|---|---|---|---|---|---|

**Main Participants:**
Principal:Vereningde Bedrijven Bredero N.V. Utrecht; Architecture: Bouwcentrum, Rotterdam, Werkgroep Kokon Architecten B.V. Rotterdam;
Design Installations:Technisch Physische Dienst TNO-TH, Delft, Installatie Techniek Bredero B.V., Utrecht, Bouwcentrum, Rotterdam; Construction:Bredero's Bouwbedrijf Nederland B.V. Utrecht, Installatie Techniek Bredero B.V. Utrecht;
Financial Support:ECN (Energieonderzoek Centrum Nederland), which has co-ordination of the Dutch National Solar Energy Program; Contact Person:Ir. C. den Ouden, Technisch Physische Dienst TNO-TH, Postbox 155, Delft.

**Project Description:**
This project contains four subsidy-built houses in the municipality of Zoetermeer. To acquire as much experience as possible during the planned trial period following the delivery of the houses, a different installation was installed in each house. Three liquid collector systems and one air collector system are used in combination with three different conventional heating systems connected thereto.

**Project Objectives:**
The aim of these four experimental solar houses is to come to the solution of a number of problems, which has not yet been solved when the project started. These problems lie in the field of heat transfer, the solar installation itself, material knowledge, architectural approach and town and country planning. The particular aspect of these four houses is that these problems are investigated systematically and in coherence to one another by applying four heat distrubution systems and two different collectors. Moreover, it had to be found out whether these houses could be built within the confines of the subsidised building sector. Altogether these has been strived for a complete integration of the building and the solar heating installation of the dwellings in a technically and economically justified way. After the delivery of the houses in August 1977 a detailed monitoring program was started in January 1978. This program will be continued for at least two years.

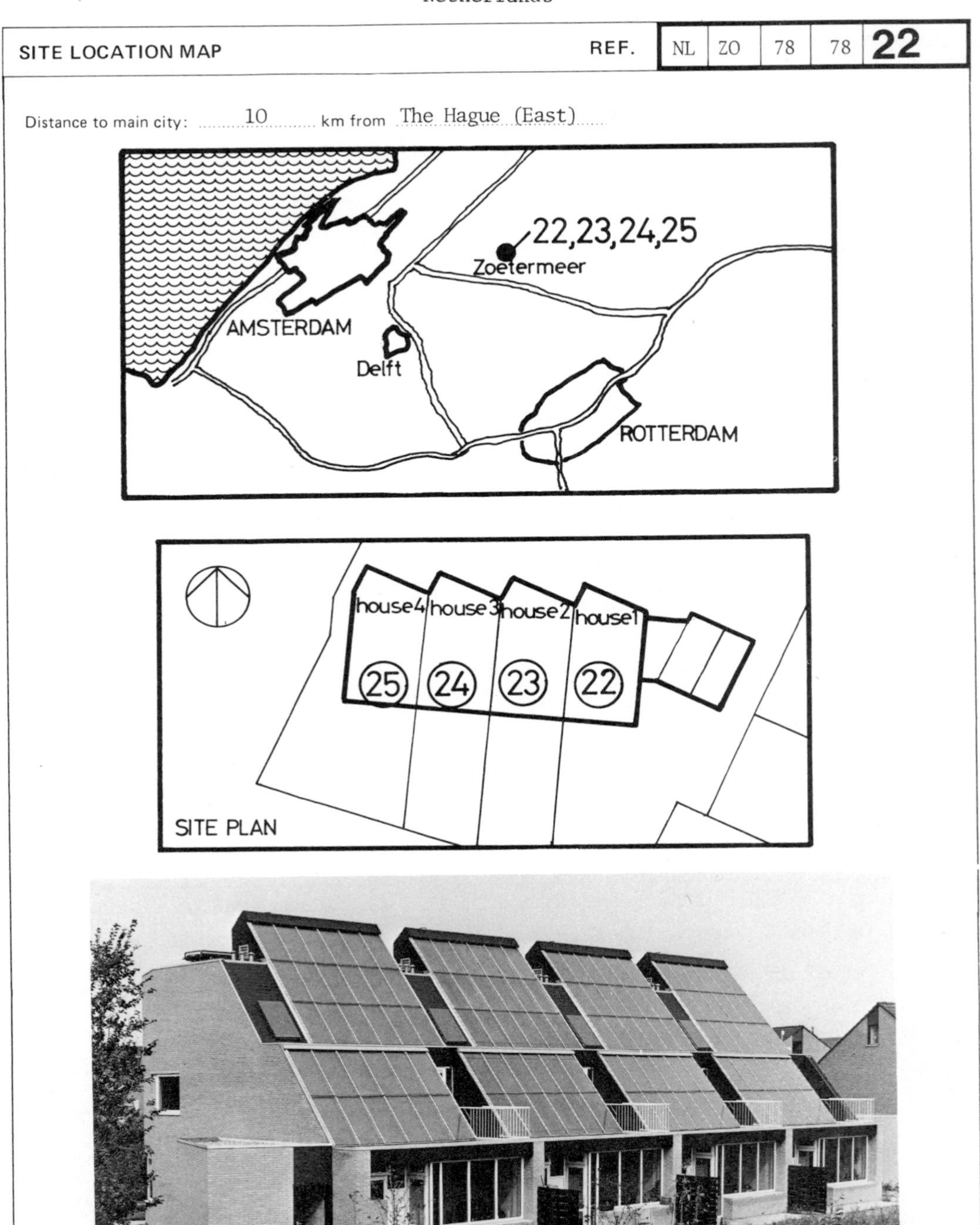
SITE LOCATION MAP
REF. NL ZO 78 78 22
Distance to main city: 10 km from The Hague (East)
22,23,24,25
Zoetermeer
AMSTERDAM
Delft
ROTTERDAM
house4
house3
house2
house1
25
24
23
22
SITE PLAN

| SUMMARY SHEET | REF. | NL | ZO | 78 | 78 | 22 |
|---|---|---|---|---|---|---|

**Design Data**

**A1 CLIMATE**

1) Source of Data: Royal Dutch Meteorological Institute in De Bilt
2) Latitude: 52 ° 8 ′ Longitude: 5 ° 7 ′ Altitude: -2 metres
3) Global Irradiation (horiz plane): 976 kWh/m²·year % Diffuse: 64
4) Degree Days: 3644 Base Temp: 18 °C
5) Sunshine Hours: July: 225 January: 73 Annual: 1400

**B1 BUILDING**

1) Building Type: 200 kg/m³ No Occupants: 2
2) Floor area: 130 m² Heated Volume: 375 m³
3) Design Temperature: External: -10 °C Internal: 20 °C
4) Ventilation Rate: 0.5 + 0.1*V wind a.c.h. Vol. Heat Loss: 0.85 W/m³·K
5) Space Heat Load: 16,400 kWh Hot Water Load: 2000 kWh

**C1 SYSTEM**

1) Absorber Type: flat plate, spectral selective, stainless steel.
2) Collector Area 35 m² (Aperture) Coolant: water glycol
3) Orientation: 180° Tilt: 50° Glazing: single
4) Storage Volume: 2.0 m³ Heat Emitters: radiators + underfloor
5) Auxiliary System: gas fired boiler (SH) electric (DHW) Heat Pump: no

**PERFORMANCE MONITORING**

1) Is there a Computer Model? Yes 2) Start Date for Monitoring Programme: 1 January 1978
3) Period for which Results Available: 1 January to 31 December 1978
4) No of Measuring Points per house: 38
5) Data Acquisition System: Microprocessor controlled system including: 130 analogue inputs, 25 digital inputs modem, magnetic tape recorder, microprocessor controlled filter.

| | Space Heating | Hot Water | Total |
|---|---|---|---|
| 6) Predicted | 26 % of 16504 kWh | 51 % of 2017 kWh | 29 % of 18541 kWh |
| 7) Measured | 26 % of 18115 kWh | 35 % of 1011 kWh | 27 % of 19126 kWh |

8) Solar Energy Used: Including useful losses 144.8 kWh/m²·year
Excluding losses: 93.3 kWh/m²·year

9) System Efficiency: $\frac{\text{Solar Energy Used}}{\text{Global Irradiation on collector}} \times 100 =$ 14 %

## A2 LOCAL CLIMATE

REF. NL ZO 78 78 **22**

1 – Average Cloud cover: 5.5 Octas

2 – Average daily max temperatures (July) : 20.8 °C

Average daily min temperatures (Jan) : -1.3 °C

3 – Source of Weather Data:
The data used in our simulation studies have been supplied by the Royal Dutch Meteorological Institute, which is situated in De Bilt, rather central in the country. Out of 10 years (1961-1970), 1962 appeared to give an average value for the solar contribution, which was the main reason to use this year in our simulation studies.

4 – Micro Climate/Site Description:
The Dutch climate is characterised by rather a large diversity. The differences between the continental and maritime climate are noticeable in a relatively small area. In the western part of the country the climate has a strong maritime tendency with a moderate character (rather low temperatures in summer and relatively high temperatures in winter and also rather strong western winds). Somewhat up-country the air soon gets a more continental character (less wind, higher temperature differences between summer and winter and even over a day's period).

| MONTH | YEAR | Irradiation on horizontal plane | | Degree Days | Precipitation | Average Ambient Temperature | Average Relative Humidity | Average Wind Speed | Prevailing Wind Direction |
|---|---|---|---|---|---|---|---|---|---|
| | | Global | Diffuse | Base: 18 °C | | | | | |
| JAN | 1962 | 20 | 14 | 451 | 73 | 3.6 | 88 | 4.8 | S.W |
| FEB | 1962 | 39 | 27 | 426 | 54 | 2.9 | 81 | 4.7 | N.E |
| MARCH | 1962 | 76 | 51 | 497 | 41 | 2.0 | 80 | 3.3 | N.N.E |
| APRIL | 1962 | 98 | 63 | 300 | 37 | 8.3 | 79 | 4.6 | S.W |
| MAY | 1962 | 119 | 79 | 249 | 50 | 10.3 | 73 | 3.6 | S.W |
| JUNE | 1962 | 169 | 96 | 142 | 58 | 14.2 | 80 | 3.0 | S.W |
| JULY | 1962 | 132 | 91 | 113 | 71 | 15.2 | 80 | 2.7 | W.N.W |
| AUGUST | 1962 | 127 | 76 | 94 | 91 | 15.8 | 80 | 3.3 | S.W |
| SEPT | 1962 | 97 | 59 | 155 | 68 | 13.5 | 80 | 2.7 | S.W |
| OCT | 1962 | 59 | 36 | 226 | 63 | 11.2 | 87 | 2.0 | S.S.W |
| NOV | 1962 | 19 | 17 | 414 | 54 | 4.3 | 90 | 2.3 | E.N.E |
| DEC | 1962 | 21 | 15 | 579 | 62 | -0.7 | 86 | 3.1 | E.N.E |
| Totals | | 976 | 623 | 3644 | 723 | | | | |
| Averages | | 81 | 52 | | | 8.4 | 86 | 3.3 | S.W |
| Units | – | kWh/m²·month | | °C days | mm | °C | % | m/s | – |

| B2 BUILDING DESCRIPTION | REF. | NL | ZO | 78 | 78 | **22** |
|---|---|---|---|---|---|---|

**Design Concepts**

The building of this type of house required a new approach. For the energy collection a number of special technical components had to be fitted in. For example, a place had to be found for the $2m^3$ storage tank, whereas the south side roof area should be sufficiently large to accommodate $40m^2$ of solar heat collectors. Moreover, the installation differs for each house. The manner in which a solution is found for the roof area with the solar collectors deserves special attention.

**Construction**

1 — Glazing: 14.5 % Area: 9.4(S)+16 $m^2$ 'U' value: 1.8 $W/m^2 \cdot K$

2 — 'U' values: Walls: 0.58 $W/m^2 \cdot K$ Floor: 0.70 $W/m^2 \cdot K$ Roof/loft: 2.7 $W/m^2 \cdot K$

3 — Thermal mass: 200 $kg/m^3$ Annual Energy Demand: 18,400 kWh

Middle class subsidy-built private house; concrete framework and masonry claddings outwards; special design to create place for the solar collectors (see views); the windowframes are plastic cavity construction.

**Energy Conservation Measures**

Energy saving was realised by giving extra attention to the insulation of the house. The cavity walls were provided with 5cm rockwool insulation blankets (k-value:0.58 $W/m^2.K$). The ground floor was insulated with 6cm rockwool insulation placed immediately under the cement finished floor (k-value:0.50 $W/m^2.K$). The windows in the north side facade are smaller than usual, whereas all the windows of house are provided with special double glazing (thermo + k-value:1.8 $W/m^2.K$). Finally, full consideration is given to the architectonic design.

Free Heat Gains: Internal: 2760 kWh/year Direct Solar: 2220 kWh/year

**Section Through House** Showing main solar components

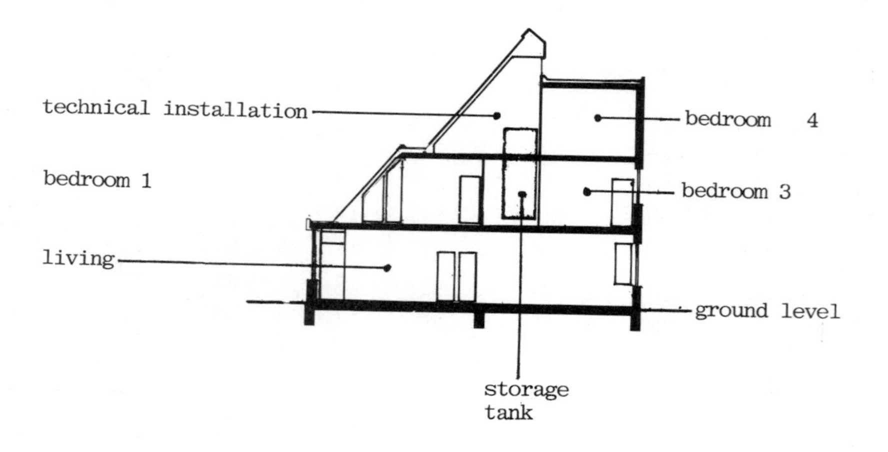

| C2 SYSTEM DESCRIPTION | REF. | NL | ZO | 78 | 78 | **22** |
|---|---|---|---|---|---|---|

**Collector/Storage**

Primary circuit, water content:1.2 litres/m$^2$ collector area;
insulation:5cm glass-wood (collector), 10cm glass-wool (storage),
1.5cm ploymethane (pipework); frost protection:ethylene glycol (30%);
overheating protection:overload heat exchanger; corrosion protection:inhibitor(1%)

**Space Heating/Domestic Hot Water**

Auxiliary(SH):gas fired boiler, (DHW):electric boiler;
heat emitters:radiator+floor heating, temp range:rad. 20-55°C, F.h. 20-40°C;
DHW insulation:3.5 polymethane, set temp: 85°C(Jan-June),55°C(June-Dec)

**Control Strategy/Operating Modes**

The primary circuit including 35m$^2$ flat plate collectors of corrugated steel, with a spectral selective coating (tin oxide on black enamel) gives off the heat to the water in the heat storage tank(volume 2m$^3$). The pump in this circuit is operated by means of a differential temperature controller (one sensor on the backside of an absorberplate, the other in the output of the helix). If T 5°C the pump starts and T 2°C it stops, a non-return valve avoids backflow. The thus heated water of which the temperature is limited up to 80°C, is passed by means of a circulation pump directly through the radiators of a conventional central heating installation. A floor heating system is connected in parallel thereto (only for the ground floor). The control of the radiator heating installation is performed by means of a weather-responsive control. When the temperature of the water in the tank drops below a level required for a sufficient heat supply to the radiators, which are of a slightly greater surface area than usual in connection with the lower supply temperature, two three-way valves are automatically operated to alter the radiator circuit so that the circulation bypasses the heat storage tank to pass through an auxiliary heating apparatus in the form of a conventional gas fired boiler. As soon as solar radiation causes the temperature in the tank to reach a sufficient value, the initial mode of operation of the installation is restored, hence making optimal use of the free solar heat available.

**Diagram of System** Showing sensor locations, power ratings, storage volumes & flows.

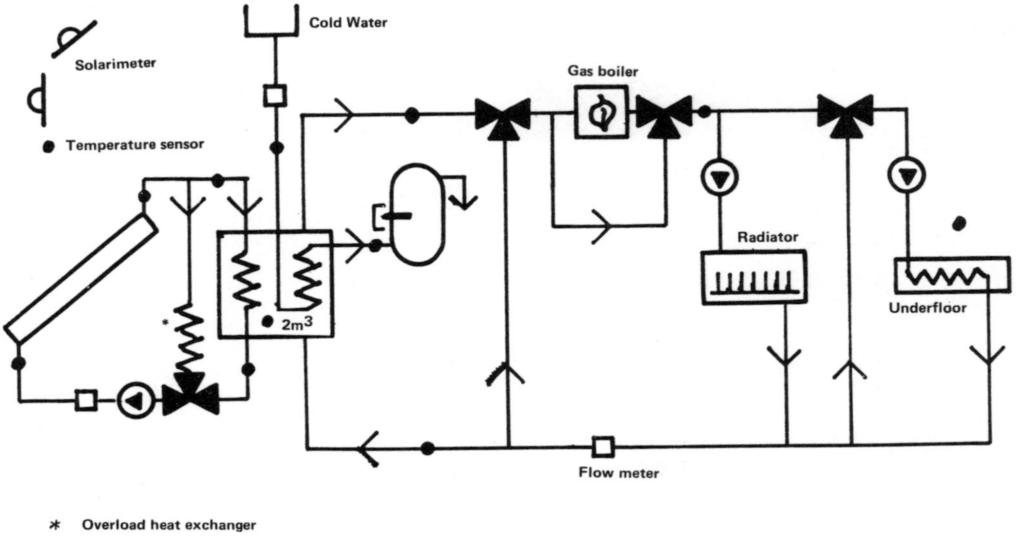

Cold Water
Solarimeter
Temperature sensor
Gas boiler
Radiator
Underfloor
2m$^3$
Flow meter

* Overload heat exchanger

| TECHNICAL APPRAISAL/PRACTICAL EXPERIENCE | REF. | NL | ZO | 78 | 78 | 22 |
|---|---|---|---|---|---|---|

**System/Design:**
Before discussing the design of the system one note has to be made. This installation has been designed in the first place to save auxiliary energy for the space heating and in the second place, because there is a solar heated storage tank in the house, the domestic hot water is preheated by solar energy. In the first one and a half years that the installation operates there has been noticed one point of design that works out negative on the system performance.

**System/Installation:**
The positive effect of the low temperature level operated floor heating on the performance does not work out, because of the way of operating this part of the installation. The temperature on which the floor heating operates in ordered by controlling the by-pass of the return water from this part of the syste. So the water temperature before the three-way valve is equal to the temperature of the waterflow through the radiators.

**Component Performance:** (Collector, Heat Exchangers, Storage, Pipework, Valves, Fittings, Pumps, Auxiliary)
After a month a non-return valve in the collector circuit was installed. It seemed that there was a negative flow during the night. First the three-way valves were thermic-driven types. These valves gave some problems (they did not open totally and worked very slowly) and were replaced by motor-driven types. Due to this problem during the months May to December there was a leakage of collected heat to the overload heat-exchanger, which had a negarive influence on the performance during this year. The performance of the total collector remained ca. 20% under the expected performance on base of measurements in our lavoratories. This can be due to the heat-losses of the inter-connections between the collectors and the behaviour of a total collector roof can differ a little bit from the behaviour of a stand-alone collector.

**System Operation/Controls, Electronics:**
The following set-points have been set during the year 1978: differential temperature between backside of avsorber-plate and borrom of the storage tank: Pump on/ T 5°C
Pump off/ T 2°C
Collected heat to the overload heat-exchanger if temperature in the top of the storage tank 80°C.

Relationship outdoor temperature-temperature of waterflow to the radiator system:

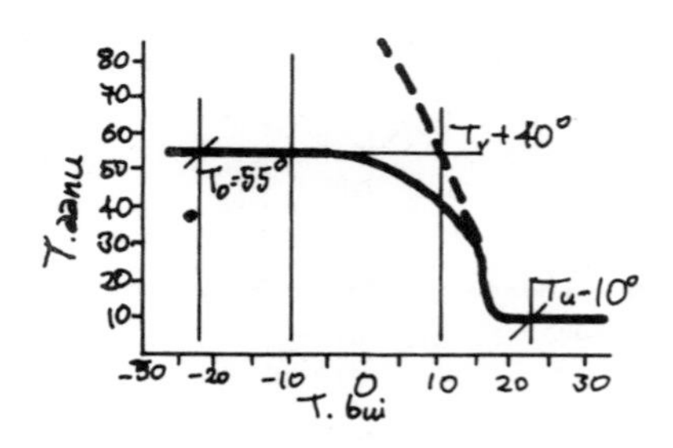


Relationship outdoor temperature-temperature of waterflow to the floor-heating system:

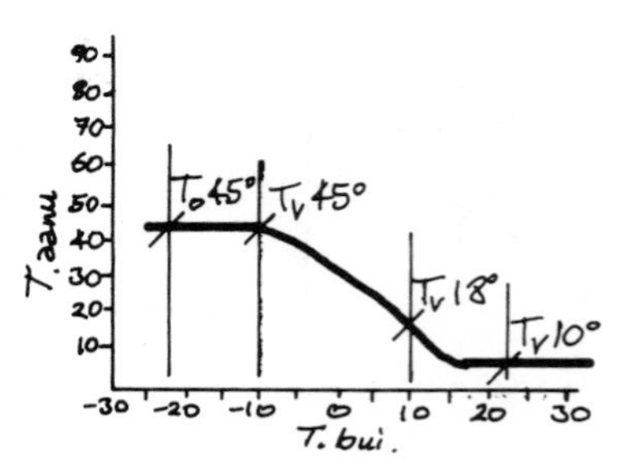

| PERFORMANCE EVALUATION/CONCLUSIONS | REF. | NL | ZO | 78 | 78 | 22 |
|---|---|---|---|---|---|---|

**Comparison of Measured with Predicted Performance**
(State reasons for any discrepancy and summarise the main factors affecting performance)

Some differences have been observed between the measured performance in the year 1978 and the calculated performance in our reference year 1962. Amont other things these differences can be caused by the following:

a) The years 1962 and 1978 are not identical as far as meteorological circumstances are concerned (in future all calculations will be carried out using the measured data of the year 1978).
b) The conduct of the primary system has been calculated on the basis of efficiency data deducted from alboratory tests at one collector. However, the conduct of a collector roof consisting of 30 of these collectors proved not to be an extrapolation of the conduct of only one collector. In fact in this case the measured collector efficiency is about 20% lower than the calculated efficiency.
c) Neither the conduct of all electronic controllers in the installation not the behaviour of the inhabitants can be inserted exactly into the computer model.
d) The daily amodunt of domestic hot water was assumed to be equal, whereas in reality this amount varies considerably (e.g. holidays). In practice the total is lower.

**Modifications to System**

The mounting of the temperature-sensor onto the absorber plate has been improved. The three-way valves of the installation have been replaced by another type of valve.

**Occupants/Response**

Due to stagnation losses of the storage tank the temperature inside the house may occasionally be high during the summer season.

**Conclusions**

During this period, provided that the 20% decrease in efficiency is taken into account, the amounts of energy collected by the collector installations correspond well with the amounts calculated with the computer model The system efficiency has proved to be lower than calculated. This is caused by heat losses of several installation part.

**Future Work**

The measurements will be continued until December 1979. It will be tried to bring the calculated and measured results into line with one another, even for one day.

## P1 MONTHLY PERFORMANCE

REF: NL ZO 78 78 22

| Month | Year | No Days Data | Average Temp °C Internal | Average Temp °C External | SOLAR ENERGY kWh: Global Irradiation On Collector Area | Solar Energy Collected | Solar Energy Used: Space Heating Useful Losses Incl. | Space Heating Useful Losses Excl. | Domestic Hot Water | Input To Heat | Heat Pump Total Output kWh: Space Heating | Domestic Hot Water | AUXILIARY ENERGY kWh: Heat Pump | Space Heating | Domestic Hot Water | Pumps & Fans | Free Heat Gains kWh: Internal | Direct Solar | SUMMARY kWh: Total Load | %: Percentage Solar | Syst. Eff. |
|---|---|---|---|---|---|---|---|---|---|---|---|---|---|---|---|---|---|---|---|---|---|
| Column No | | | 1 | 2 | 3 | 4 | 5 | 6 | 7 | 8 | 9 | 10 | 11 | 12 | 13 | 14 | 15 | 16 | 17 | 18 | 19 |
| J | 78 | 30 | 18.6 | 3.0 | 1224 | 220 | 200 | (140) | 20 | | | | | (2854) | 48 | 64 | 336 | 178 | 3122 | (7) | 18 |
| F | 78 | 24 | 19.4 | 1.1 | 1669 | 313 | 284 | (200) | 29 | | | | | (2829) | 73 | 59 | 325 | 294 | 3215 | (10) | 19 |
| M | 78 | 29 | 20.3 | 6.1 | 3333 | 872 | 835 | (560) | 37 | | | | | (1552) | 80 | 67 | 341 | 345 | 2504 | (35) | 26 |
| A | 78 | 28 | 20.8 | 7.5 | 4299 | 1206 | 1168 | 865 | 38 | | | | | 648 | 60 | 64 | 328 | 408 | 1914 | 62 | 28 |
| M | 78 | 31 | 22.0 | 13.0 | 4694 | 1364 | 1085x | 585 | 41 | | | | | 73 | 50 | 68 | 220 | 252 | 1249 | 90 | 24 |
| J | 78 | 30 | 22.8 | 16.5 | 5028 | 910 | 54x | 29 | 46 | | | | | 3 | 26 | 35 | 7 | 11 | 129 | 78 | 2 |
| J | 78 | 31 | 22.1 | 16.7 | 4780 | 695 | 81x | 52 | 5 | | | | | 0 | 59 | 20 | 14 | 14 | 145 | 59 | 2 |
| A | 78 | 31 | 23.0 | 16.9 | 4449 | 833 | 48x | 28 | 11 | | | | | 0 | 70 | 20 | 9 | 9 | 129 | 46 | 1 |
| S | 78 | 30 | 21.0 | 14.7 | 3260 | 732 | 144x | 89 | 35 | | | | | 18 | 37 | 64 | 33 | 30 | 234 | 76 | 5 |
| O | 78 | 31 | 20.4 | 11.3 | 2084 | 574 | 475x | 251 | 39 | | | | | 786 | 41 | 66 | 382 | 270 | 1341 | 38 | 25 |
| N | 78 | 30 | 20.0 | 7.0 | 1629 | 514 | 232x | 90 | 25 | | | | | 1883 | 54 | 62 | 387 | 257 | 2194 | 12 | 16 |
| D | 78 | 31 | 18.3 | 1.7 | 926 | 207 | 112x | 27 | 25 | | | | | 2751 | 62 | 63 | 375 | 149 | 2950 | 5 | 15 |
| TOTAL | | | — | — | 37375 | 8440 | 4718x | (2916) | 351 | | | | | 13397 | 660 | 654 | 2757 | 2217 | 19126 | — | — |
| AVERAGE | | | 20.7 | 9.6 | | | | | | | | | | | | | | | | (27) | 14 |

TOTAL:3267 kWh C.O.P =

x :there has been a leak to the overload heat exchanger

():the measurement of the mass flow in the heat-emitting circuit failed

| PROJECT DESCRIPTION ZOETERMEER HOUSE 2 | REF. | NL | ZO | 78 | 78 | **23** |
|---|---|---|---|---|---|---|

**Main Participants:**

The same at Project 22 House 1.

**Project Description:**

The same as Project 22 House 1.

**Project Objectives:**

The same as Project 22 House 1

## SITE LOCATION MAP

The same as Project 22 House 1.

## A2 LOCAL CLIMATE

The same as Project 22 House 1.

## B2 BUILDING DESCRIPTION

**Design Concepts**

The same as Project 22 House 1 apart from annual energy demand which is 17,520 kWh.

**Construction**

The same as Project 22 House 1.

**Energy Conservation Measures**

The same as Project 22 House 1.

**Section Through House** Showing main solar components

The same as Project 22 House 1.

## C2 SYSTEM DESCRIPTION

**Diagram of System** Showing sensor locations, power ratings, storage volumes & flows.

Cold water
Solarimeter
Gas boiler
sensor
$35\ m^2$
$2\ m^3$
Radiators
Flow meter

* Overload heat-exchanger

| C2 SYSTEM DESCRIPTION (CONT) | REF. | NL | Zo | 78 | 78 | 23 |
|---|---|---|---|---|---|---|

**Collector/Storage**
The same as Project 22 House 1.

**Space Heating/Domestic Hot Water**
The same as Project 22 House 1 apart from the heat emitters which are radiators only with a temperature range of 20 - 55°C.

**Other Points/Heat Pump:**

**Control Strategy/Operating Modes**
The primary circuit including 35m flat plate collectors of corrugated steel, with a spectral selective coating (tin oxide on black enamel) gives off the heat to the water in the heat storage tank (volume $2m^3$). The pump in this circuit is operated by means of a differential temperature controller (one sensor on the backside of an absorberplate, the other in the output of the helix. If $\Delta T > 5°C$ the pump starts and if $T < 2°C$ it stops, a non-return valve avoids backflow. The thus heated water of which the temperature is limited up to 80°C, is passed by means of a circulation pump directly through the radiators of a conventional central heating installation. The control of the radiator heating installation is performed by means of a conventional room thermostat. When the temperature of the water in the tank drops below a level required for a sufficient heat supply to the radiators, which are of a larger surface area than usual in connection with the lower supply temperature, two three-way valves are automatically operated to alter the radiator circuit so that the circulation bypasses the heat storage tank to pass through an auxiliary heating apparatus in the form of a conventional gas fired boiler. As soon as solar radiation causes the temperature in the tank to reach a sufficient value, the initial mode of operation of the installation is restored, hence making optimal use of the free solar heat available.

## TECHNICAL APPRAISAL/PRACTICAL EXPERIENCE

**System/Design:**
Before discussing the design of the system one note has to be made. This installation has been designed in the first place to save auxiliary energy for the space heating and in the second place, because there is a solar heated storage tank in the house, the domestic hot water is preheated by solar energy.

**System/Installation:**
During the measurements we concluded that thė stand still losses of the electric boiler for domestic hot water covered a considerable quantity of the total energy need of this installation part. In order to reduce this amount of energy a manually operated by-pass was applied to the boiler of this house. Moreover, the boiler should be switched off electrically when this by-pass is at work. Dur to the manual control the influence of this on the results of the measurements is not yet very large.

**Component Performance:** (Collector, Heat Exchangers, Storage, Pipework, Valves, Fittings, Pumps, Auxiliary)
After a month a non-return valve in the collector-circuit was installed. It seemed that there was a engative flow during the night. First the three-way valves were thermic-driven types. These valves gave some problems (they did not open totally and worked very slowly) and were replaced by motor-driven types. Due to this problem there may be a leakage of collected heat to the overload heat-exchanger, which, again might have had a negarive influence on the performance furing this year. The stationary efficiency of the total collector remained circa 20% under the expected performance on the basis of measurements in our laboratoried. This can be due to the heat losses of the interconnections between the collectors and the behaviour of a total collector can differ a little bit from the behaviour of a stand-alone collector. In February the contact between the sensor and the absorber plate was disturbed, which resulted in the fact that the collector pump was not switched on. The contact was restored and improved in March.

SUMMARY SHEET ZOETERMEER, House 2 REF. | N1 | Zo | 78 | 78 | **23**

**Design Data**

**A1 CLIMATE**

1) Source of Data: Royal Dutch Meteorological Institut in De Bilt
2) Latitude: 52 ° 8 ′ Longitude: 5 ° 7 ′ Altitude: -2 metres
3) Global Irradiation (horiz plane): 976 kWh/m²·year % Diffuse: 64
4) Degree Days: 3,644 Base Temp: 18 °C
5) Sunshine Hours: July: 225 January: 73 Annual: 1,400

**B1 BUILDING**

1) Building Type: 200 kg/m³ No Occupants: 4
2) Floor area: 130 m² Heated Volume: 375 m³
3) Design Temperature: External: -10 °C Internal: 20 °C
4) Ventilation Rate: 0.5 + 0.1*V wind a.c.h. Vol. Heat Loss: 0.85 W/m³·K
5) Space Heat Load: 13,200 kWh Hot Water Load: 4,320 kWh

**C1 SYSTEM**

1) Absorber Type: Flat plate, spectral selective, stainless steel
2) Collector Area 35. m² (Aperture) Coolant: water + glycol
3) Orientation: 180° Tilt: 50° Glazing: single
4) Storage Volume: 2.0 m³ Heat Emitters: Radiators
5) Auxiliary System: Gas fired boiler (SH) Electric (DHW) Heat Pump: --

**PERFORMANCE MONITORING**

1) Is there a Computer Model? Yes
2) Start Date for Monitoring Programme: 1st January 1978
3) Period for which Results Available: 1st January 1978 to 31st December 1978
4) No. of Measuring Points per house: 38
5) Data Acquisition System: Microprocessor controlled system including 130 analogue inputs, 25 digital inputs modem, magnetic tape recorder, microprocessor controlled filter.

| | Space Heating | Hot Water | Total |
|---|---|---|---|
| 6) Predicted | 33 % of 12196 kWh | 47 % of 4323 kWh | 37 % of 16519 kWh |
| 7) Measured | 25 % of 13583 kWh | 40 % of 3090 kWh | 28 % of 16673 kWh |

8) Solar Energy Used: Including useful losses 133.5 kWh/m²·year; Excluding losses: 96.1 kWh/m²·year

9) System Efficiency: $\frac{\text{Solar Energy Used}}{\text{Global Irradiation on collector}} \times 100 =$ 13 %

| TECHNICAL APPRAISAL/PRACTICAL EXPERIENCE (CONT) REF. | NL | Zo | 78 | 78 | 23 |
|---|---|---|---|---|---|

**System Operation/Controls, Electronics:**

The following set-points have been set during the year 1978: Differential temperature between backside of absorber-plate and bottom of the storage tank: Pump on /$\Delta T > 5^{o}C$
Pump off/$\Delta T < 2^{o}C$
(During the period March-December, these set-points have been set, but in practice both T's were much higher. It is possible that the contact between the sensor and the absorber-plate is still not correct.) Collected heat to the overload heat-exchanger, if temperature in the top of the storage tank 80$^{o}$C.

## PERFORMANCE EVALUATION/CONCLUSIONS

**Comparison of Measured with Predicted Performance**
(State reasons for any discrepancy and summarise the main factors affecting performance)

Some differences have been observed between the measured performance in the yeat 1978 and the calculated performance in our reference year 1962. Amont other things these differences can ba caused by the following:

a) The years 1962 and 1978 are not identical as far as meteorological circumstances are concerned (in future all calculations will be carried out using the measured data of the year 1978).

b) The conduct of the primary system has been calculated on the basis of efficiency data deducted from laboratory tests at one collector. However, the conduct of a collector roof consisting of 30 of these collectors proved not to be an extrapolation of the conduct of only one collector. In fact in this case the measured collector efficiency is about 20% lower than the calculated efficiency.

c) Neither the conduct of all electronic controllers in the installation not the behaviour of the inhabitants can be inserted exactly into the computer model.

d) The daily amount of domestic hot water was assumed to be equal, whereas in reality this amount varied considerably(eg holidays). So in practice the total is lower.

**Modifications to System**

The mounting of the temperature-sensor onto the absorber place has been improved; the three-way-valves of the installation have been replaced by another type of valve.

**Occupants/Response**

Due to stagnation losses of the storage tank the temperature inside the house may occasionally be high during the summer season. In summer, sometines, the auxiliary heating for DHW is bypassed (operated by hand) because the water is heated to 45 - 60$^{o}$C by solar energy in the storage tank.

**Conclusions**

During this period, provided that the 20% decrease in efficiency is taken into account, the amounts of energy collected by the collector installations correspond well with the amounts calculated with the computer mode. The system efficiency has proved to be lower than calculated. This is caused by heat losses of several installation parts.

**Future Work**

The measurements will be continued until December 1979. It will be tried to bring the calculated and measured results into line with one another, even for one day.

# P1 MONTHLY PERFORMANCE

REF: NL Z0 78 78 23

| Month | Year | No Days Data | Average Temp °C: Internal | Average Temp °C: External | SOLAR ENERGY kWh: Global Irradiation On Collector Area | Solar Energy Collected | Solar Energy Used: Space Heating Useful Losses Incl. | Space Heating Useful Losses Excl. | Domestic Hot Water | Input To Heat | Heat Pump Total Output kWh: Space Heating | Domestic Hot Water | AUXILIARY ENERGY kWh: Heat Pump | Space Heating | Domestic Hot Water | Pumps & Fans | Free Heat Gains kWh: Internal | Direct Solar | SUMMARY kWh: Total Load | %: Percentage Solar | Syst. Eff. |
|---|---|---|---|---|---|---|---|---|---|---|---|---|---|---|---|---|---|---|---|---|---|
| Column No | | | 1 | 2 | 3 | 4 | 5 | 6 | 7 | 8 | 9 | 10 | 11 | 12 | 13 | 14 | 15 | 16 | 17 | 18 | 19 |
| J | 78 | 30 | 3.0 | 17.9 | 1224 | 215 | 174 | (93) | 41 | | | | | (2008) | 200 | (44) | 312 | 178 | 2423 | (9) | 18 |
| F | 78 | 24 | 1.1 | 18.5 | 1669 | 25 | -7 | (0) | 32 | | | | | (2074) | 238 | (39) | 332 | 294 | 2337 | (1) | 1 |
| M | 78 | 29 | 6.1 | 19.5 | 3333 | 777 | 647 | (435) | 130 | | | | | (1018) | 270 | (45) | 320 | 345 | 2065 | (38) | 23 |
| A | 78 | 28 | 7.5 | 20.2 | 4399 | 984 | 845 | (508) | 139 | | | | | (538) | 164 | (43) | 285 | 408 | 1686 | (58) | 23 |
| M | 78 | 31 | 13.0 | 21.7 | 4694 | 1315 | (644) | (285) | 173 | | | | | (168) | 127 | (44) | 119 | 156 | 1112 | (73) | (17) |
| J | 78 | 30 | 16.5 | 22.8 | 5028 | 1155 | (192) | (73) | 141 | | | | | (0) | 23 | (20) | 39 | 45 | 356 | (94) | (7) |
| J | 78 | 31 | 16.7 | 22.8 | 4780 | 940 | 61 | 34 | 93 | | | | | 21 | 65 | 7 | 11 | 10 | 240 | 64 | 3 |
| A | 78 | 31 | 16.9 | 23.1 | 4449 | 794 | 0 | 0 | 107 | | | | | 0 | 28 | 4 | 0 | 0 | 135 | 79 | 2 |
| S | 78 | 30 | 14.7 | 21.3 | 3260 | 547 | 52 | 26 | 105 | | | | | 68 | 95 | 9 | 23 | 21 | 320 | 49 | 5 |
| O | 78 | 31 | 11.3 | 20.8 | 2084 | 506 | 380 | 341 | 126 | | | | | 377 | 160 | 42 | 470 | 270 | 1043 | 49 | 24 |
| N | 78 | 30 | 7.0 | 19.5 | 1629 | 414 | 332 | 246 | 82 | | | | | 1155 | 206 | 40 | 490 | 257 | 1775 | 23 | 25 |
| D | 78 | 31 | 1.7 | 18.7 | 926 | 184 | 118 | 87 | 66 | | | | | 2718 | 279 | 41 | 559 | 149 | 3181 | 6 | 20 |
| TOTAL | | | — | — | 37375 | 7856 | (3438) | (2128) | 1235 | | | | | (10145) | 1855 | (378) | 2960 | 2133 | 16673 | — | — |
| AVERAGE | | | 9.6 | | | | | | | | | | | | | | | | | (28) | (13) |

TOTAL: 3363 kWh | C.O.P =

( ): the measurement of the mass-flow in the heat-emitting circuit failed.

| PROJECT DESCRIPTION | ZOETERMEER HOUSE 3 | REF. | NL | ZO | 78 | 78 | **24** |
|---|---|---|---|---|---|---|---|

**Main Participants:**
The same as Project 22 House 1.

**Project Description:**
The same as Project 22 House 1.

**Project Objectives:**
The same as Project 22 House 1

## SITE LOCATION MAP

The same as Project 22 House 1.

## A2 LOCAL CLIMATE

The same as Project 22 House 1.

## B2 BUILDING DESCRIPTION

**Design Concepts**
The same as Project 22 House 1.

**Construction**

The same as Project 22 House 1 apart from annual energy demand which is 17,300 kWh.

**Energy Conservation Measures**
The same as Project 22 House 1.

## C2 SYSTEM DESCRIPTION

**Collector/Storage**

1 – Primary circuit. Water content: 45 litres/m² collector area

2 – Insulation. Collector: 10cm glass wool Storage: 10cm glass wool Pipework: 2.5cm glass wool (inside)

3 – Frost Protection: –

4 – Overheating Protection: –

5 – Corrosion Protection: –

**Space Heating/Domestic Hot Water**

1 – Auxiliary (S.H.): Gas fired air heater (DHW): electric boiler

2 – Heat Emitters: Air Temp. Range: 20 - 70°C

3 – DHW Insulation: 3.5 polymethane Set Temperature 85°C (Jan-June), 55°C (June-Dec)

**Control Strategy/Operating Modes**

This system employs air-cooled collectors through which the air from the house is passed directly when there is sufficient solar radiation and a neat-demand in the house. When the house has been heated to the required temperature, the air valves are operated auromatically so that the collected solar heat is given off in the heat exchanger and is subsequently transferred to the heat storage tank by the water, which, in this mode of operation, is circulated by means of a pump. As soon as solar radiation causes the temperature in the tank to reach a sufficient value (80°C), the fan in the collector-citcuit is stopped. In the absence of solar radiation and a heat demand in the house the energy stored in the tank is used. In this mode the operation of the heat exchanger is reversed. When this energy becomes insufficient the air-heater is switched on (the air is still preheated in the heat exchanger).

C2 SYSTEM DESCRIPTION (CONT) REF. | NL | ZO | 78 | 78 | **24**

**Diagram of System** Showing sensor locations, power ratings, storage volumes & flows.

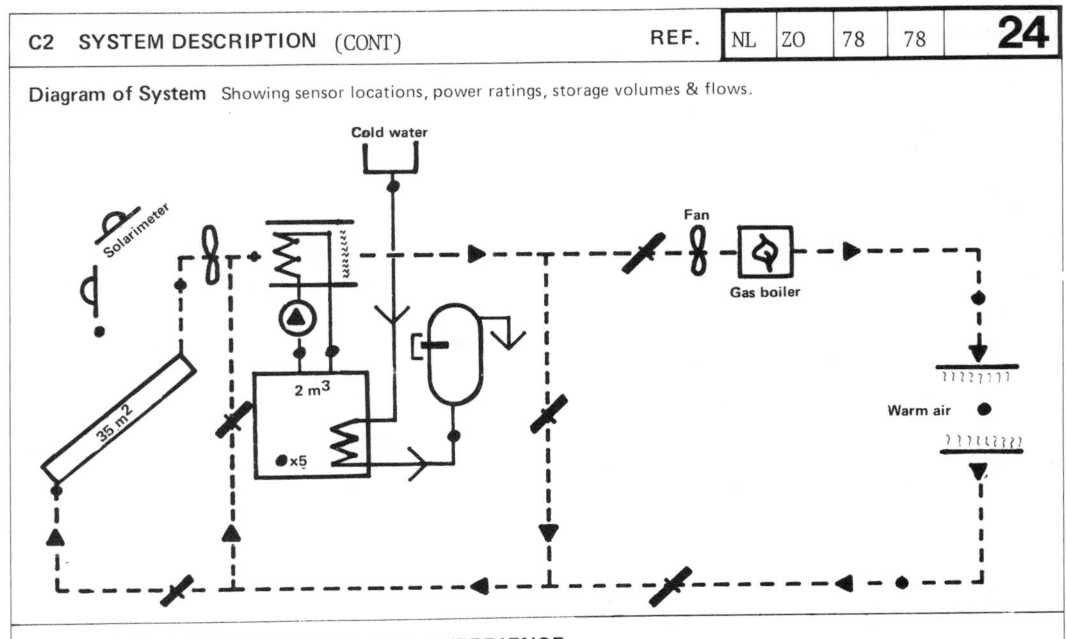


## TECHNICAL APPRAISAL/PRACTICAL EXPERIENCE

**System/Design:**

Before discussing the design of the system one note has to be made. This installation has been designed in the first place to save auxiliary energy for the space heating and in the second place, because there is a solar heated storage tank in the house, the domestic hot water is preheated by solar energy.

**System/Installation:**

Because of the lack of losses in the heat exchangers and in the storage system this system reaches its optimum proved that both solar radiation and heat demand are sufficiently avaiable. However, these losses will play an important part as soon as the heat cannot be lead to the house immediately. Also quite considerable are the piping losses which are due to the large sizes of the air-channels.

**Component Performance:** (Collector, Heat Exchangers, Storage, Pipework, Valves, Fittings, Pumps, Auxiliary)

Shortly after the completion of this house a white deposit appeared on the collector panes. At first we thought this deposit had been caused by gasses evaporating from the insulation material (glass wool). However, further analysis proved that the deposit had been caused by the applied tightening-kit. The gasses evaporating from this kit condensate on the cooler glass panes. This evaporation seems to be of a temporary naured, since the deposit does not seem to be increasing. Because of this deposit the transmission-coefficient of the glass has decreased from 0.84 to 0.76 which, again, results in a decrease of the collector efficiency.

**System Operation/Controls, Electronics:**

The following set-points have been set during the year 1978: differential temperature between air in the top of the collectors and water in the bottom of the storage tank: Fan on : $\Delta T > 5^{o}C$ , Fan off: $\Delta T < 2^{o}C$.
The primary fan is stopped if the temperature in the top of the storage tank $> 80^{o}C$.

| SUMMARY SHEET | ZOETERMEER, House 3 | REF. | Nl | Zo | 78 | 78 | 24 |
|---|---|---|---|---|---|---|---|

**Design Data**

**A1 CLIMATE**

1) Source of Data: Royal Dutch Meteorological Institut in De Bilt
2) Latitude: 52° 8′ Longitude: 5° 7′ Altitude: -2 metres
3) Global Irradiation (horiz plane): 976 kWh/m²·year % Diffuse: 64
4) Degree Days: 3,644 Base Temp: 18 °C
5) Sunshine Hours: July: 225 January: 73 Annual: 1,400

**B1 BUILDING**

1) Building Type: 200 kg/m³ No Occupants: 3
2) Floor area: 130 m² Heated Volume: 375 m³
3) Design Temperature: External: -10 °C Internal: 20 °C
4) Ventilation Rate: 0.5 + 0.1*V wind a.c.h. Vol. Heat Loss: 0.85 W/m³·K
5) Space Heat Load: 14,500 kWh Hot Water Load: 2,800 kWh

**C1 SYSTEM**

1) Absorber Type: Flat plate, spectral selective, stainless steel
2) Collector Area 35. m² (Aperture) Coolant: Air
3) Orientation: 180° Tilt: 50° Glazing: single
4) Storage Volume: 2.0 m³ Heat Emitters: Warm air
5) Auxiliary System: Gas fired boiler (SH) Electric (DHW) Heat Pump: --

**PERFORMANCE MONITORING**

1) Is there a Computer Model? Yes
2) Start Date for Monitoring Programme: 1st January 1978
3) Period for which Results Available: 1st January 1978 to 31st December 1978
4) No. of Measuring Points per house: 38
5) Data Acquisition System: Microprocessor controlled system including 130 analogue inputs, 25 digital inputs modem, magnetic tape recorder, microprocessor controlled filter.

| | Space Heating | Hot Water | Total |
|---|---|---|---|
| 6) Predicted | % of kWh | % of kWh | % of kWh |
| 7) Measured | 28 % of 12117 kWh | 23 % of 2710 kWh | 27 % of 14827 kWh |

8) Solar Energy Used: Including useful losses 114.8 kWh/m²·year; Excluding losses: 57.8 kWh/m²·year

9) System Efficiency: $\frac{\text{Solar Energy Used}}{\text{Global Irradiation on collector}} \times 100 =$ 11 %

| PERFORMANCE EVALUATION/CONCLUSIONS | REF. | NL | ZO | 78 | 78 | 24 |
|---|---|---|---|---|---|---|

**Comparison of Measured with Predicted Performance**
(State reasons for any discrepancy and summarise the main factors affecting performance)

Some differences have been observed between the measured performance in the year 1978 and the calculated performance in our reference year 1962. Among other things these differences can be caused by the following:

a) The years 1962 and 1978 are not identical as far as meteorological circumstances are concerned (in future all calculations will be carried out using the measured data of the year 1978).

b) The conduct of the primary system has been calculated on the basis of efficiency data deducted from laboratory tests at one collector. However, the conduct of a collector roof consisting of 30 of these collectors proved not to be an extrapolation of the conduct of only one collector. In fact in this case the measured collector efficiency is about 20% lower than the calculated efficiency.

c) Neither the conduct of all electronic controllers in the installation nor the behaviour of the inhabitants can be inserted exactly into the computer model.

d) The daily amount of domestic hot water was assumed to be equal, whereas in reality this amount varies considerably(eg holidays). So in practice the total is lower.

**Modifications to System**

The mounting of the temperature-sensor into the absorber plate has been improved.

**Occupants/Response**

Due to stagnation losses of the storage tank and the air-pipes the temperature inside the house may occasionally be high during the summer season.

**Conclusions**

The system efficiency has proved to be lower than calculated. This is caused by heat losses of several installation parts. In 1978 the air-flows of the installation were supposed to be constant. On ground of the results, however, it was decided to measure them constantly since there are doubts about the stability of these flows.

**Future Work**

The measurements will be continued until December 1979. It will be tried to bring the calculated and measured results into line with one another, even for one day.

## P1 MONTHLY PERFORMANCE

REF: NL Z0 78 78 24

| Month | Year | No Days Data | Average Temp °C Internal | Average Temp °C External | SOLAR ENERGY kWh: Global Irradiation On Collector Area | Solar Energy Collected | Solar Energy Used: Space Heating Useful Losses Incl. | Space Heating Useful Losses Excl. | Domestic Hot Water | Input To Heat | Heat Pump Total Output kWh: Space Heating | Domestic Hot Water | AUXILIARY ENERGY kWh: Heat Pump | Space Heating | Domestic Hot Water | Pumps & Fans | Free Heat Gains kWh: Internal | Direct Solar | SUMMARY kWh: Total Load | %: Percentage Solar | Syst. Eff. |
|---|---|---|---|---|---|---|---|---|---|---|---|---|---|---|---|---|---|---|---|---|---|
| Column No | | | 1 | 2 | 3 | 4 | 5 | 6 | 7 | 8 | 9 | 10 | 11 | 12 | 13 | 14 | 15 | 16 | 17 | 18 | 19 |
| J | 78 | 30 | 18.6 | 3.0 | 1224 | 287 | (266) | 55 | (21) | | | | | 2112 | 244 | 62 | 324 | 178 | 2643 | (11) | (23) |
| F | 78 | 24 | 19.4 | 1.1 | 1669 | 414 | 389 | 191 | 25 | | | | | 1797 | 207 | 59 | 306 | 294 | 2418 | 17 | 25 |
| M | 78 | 29 | 20.3 | 6.1 | 3333 | 799 | 763 | 214 | 36 | | | | | 706 | 255 | 58 | 320 | 345 | 1760 | 46 | 24 |
| A | 78 | 28 | 20.8 | 7.5 | 4299 | 1002 | 926 | 225 | 76 | | | | | 347 | 213 | 31 | 298 | 408 | 1562 | 64 | 23 |
| M | 78 | 31 | 22.0 | 13.0 | 4694 | 1095 | 167 | 112 | 92 | | | | | 134 | 186 | 31 | 17 | 23 | 579 | 45 | 6 |
| J | 78 | 30 | 22.8 | 16.5 | 5028 | 986 | 0 | 0 | 83 | | | | | 0 | 77 | 4 | 0 | 0 | 160 | 52 | 2 |
| J | 78 | 31 | 22.1 | 16.7 | 4780 | 574 | 13 | 10 | 56 | | | | | 17 | 154 | 6 | 2 | 2 | 240 | 29 | 1 |
| A | 78 | 31 | 23.0 | 16.9 | 4449 | 459 | 0 | 0 | 54 | | | | | 0 | 149 | 4 | 0 | 0 | 203 | 27 | 1 |
| S | 78 | 30 | 21.0 | 14.7 | 3260 | 342 | 37 | 32 | 82 | | | | | 12 | 130 | 5 | 8 | 7 | 261 | 46 | 4 |
| O | 78 | 31 | 20.4 | 11.3 | 2084 | 350 | 285 | 132 | 65 | | | | | 165 | 128 | 11 | 387 | 270 | 643 | 55 | 17 |
| N | 78 | 30 | 20.0 | 7.0 | 1629 | 431 | 405 | 298 | 26 | | | | | 776 | 156 | 55 | 436 | 257 | 1363 | 32 | 17 |
| D | 78 | 31 | 18.3 | 1.7 | 926 | 152 | 140 | 127 | 12 | | | | | 2660 | 183 | 54 | 466 | 149 | 2995 | 5 | 16 |
| TOTAL | | | — | — | 37375 | 6891 | 3391 | 1396 | 628 | | | | | 8726 | 2082 | 380 | 2564 | 1933 | 14827 | — | — |
| AVERAGE | | | 20.7 | 9.6 | | | | | | | | | | | | | | | | 27 | 11 |

TOTAL: kWh C.O.P =

| PROJECT DESCRIPTION ZOETERMEER HOUSE 4 | REF. | NL | ZO | 78 | 78 | 25 |
|---|---|---|---|---|---|---|

**Main Participants:**
The same as Project 22 House 1.

**Project Description:**
The same as Project 22 House 1.

**Project Objectives:**
The same as Project 22 House 1.

## SITE LOCATION MAP

The same as Project 22 House 1.

## A2 LOCAL CLIMATE

The same as Project 22 House 1.

## B2 BUILDING DESCRIPTION

**Design Concepts**
The same as Project 22 House 1.

**Construction**

The same as Project 22 House 1 apart from the annual energy demand which is 18,395 kWh.

**Energy Conservation Measures**
The same as Project 22 House apart from free heat gains which are 2,750 kWh/year (internal), and 3,650 kWh/year (direct solar).

## C2 SYSTEM DESCRIPTION

**Collector/Storage**
The same as Project 22 House 1.

**Space Heating/Domestic Hot Water**

1 – Auxiliary (S.H.): Gas fired air heater (DHW): Electric boiler

2 – Heat Emitters: Air heating Temp. Range: Air 20-70°C

3 – DHW Insulation: 3.5 polymethane Set Temperature 85°C (Jan-June) 55°C (June-Dec)

**Control Strategy/Operating Modes**
The primary circuit including 35m² glat plate collectors of corrugated steel, with a spectral selective coating (tin oxide on black enamel) gives off the heat to the water in the heat storage tank (volume 2m³). The pump in this circuit is operated by means of a differential temperature controller (one sensor on the backside of an absorber plate, the other in the output of the helix). If $\Delta T > 5°C$ the pump starts and if $\Delta T < 2°C$ it stops, a non-return valve avoids backflow. The thus heater water of which the temperature is limited up to 80°C, is passed by means of a circulation pump through a heat exchanger in a conventional air heating installation. This installation normally functions as a directly gas-fired heater-unit to theat the house to the required temperature. When sufficient energy is stored in the tank, the auxiliary heating installation is dosconnected and this energy is used to heat the house.

| C2 SYSTEM DESCRIPTION (CONT) | REF. | NL | ZO | 78 | 78 | 25 |
|---|---|---|---|---|---|---|

**Diagram of System** Showing sensor locations, power ratings, storage volumes & flows.

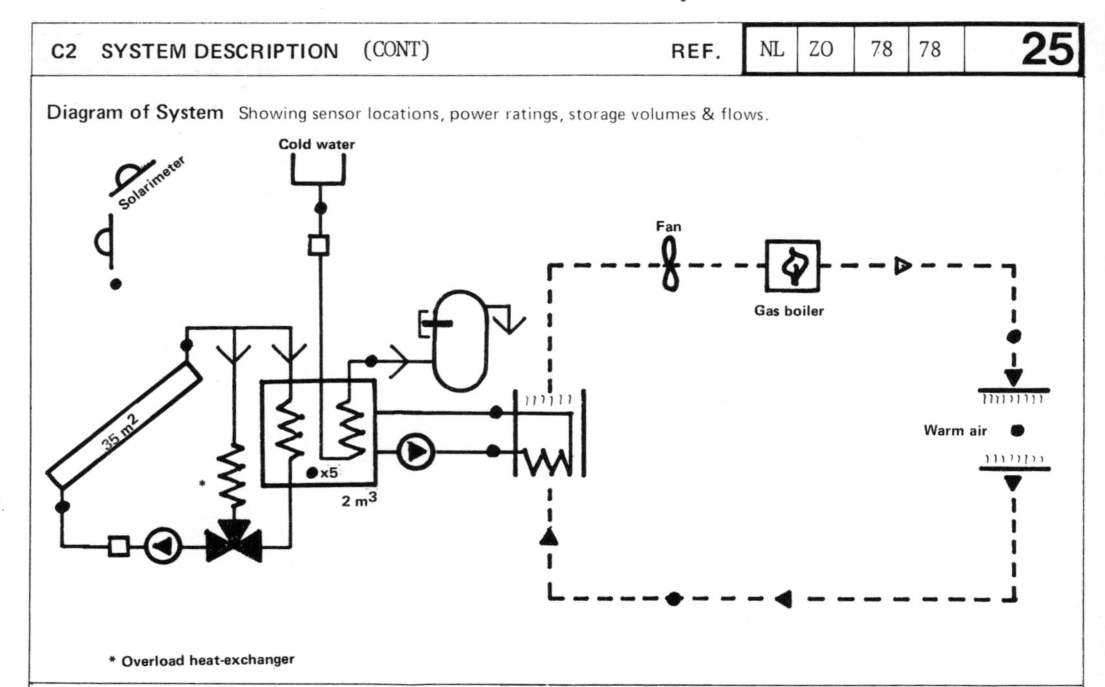


## TECHNICAL APPRAISAL/PRACTICAL EXPERIENCE

**System/Design:**
Before discussing the design of the system one note has to be made. This installation has been designed in the first place to save auxiliary energy for the space heating and in the second place, because there is a solar heated storage tank in the house, the domestic hot water is preheated by solar energy.

**Component Performance:** (Collector, Heat Exchangers, Storage, Pipework, Valves, Fittings, Pumps, Auxiliary)
After a month a non-return valve in the collector-circuit was installed. It seemed that there was a negative flow during the night. First the three-way valve was a thermic-driven type. This valve gave some problems (it did not open totally and worked very slowly) and were replaced by motor-driven types. Due to this problem during this year there might have been a leakage of collected heat to the overload heat-exchanger, which, again, might have had a negative influence on the performance during this year. The performance of the total collector remained ca. 20% under the expected performance on base of measurements in our laboratories. This can be due to the heat losses of the interconnections between the collectors and the behaviour of a total collector roof can differ a little bit from the behaviour of a stand-alone collector.

**System Operation/Controls, Electronics:**
The following set-points have been set during the year 1978: differential temperature between backside of absorber plate and borrom of the storage tank: Pump on $/\Delta T > 5^{o}C$
Pump off $/\Delta T < 2^{o}C$
Collected heat to the overload heat exchanger if temperature in the top of the storage tank 80°C.
The air heating system is operated by means of a thermostat with two set points. If the air temperature in the living room is higher than the higher set point the fan does not work, until the temperature passes this set point. The fan and pump are started and the storage tank gives off the heat to the heat-exchanger. When this heat becomes insufficient, the air-heater is switched on.

| SUMMARY SHEET | ZOETERMEER, House 4 | REF. | N1 | Zo | 78 | 78 | 25 |
|---|---|---|---|---|---|---|---|

**Design Data**

**A1 CLIMATE**

1) Source of Data: Royal Dutch Meteorological Institut in De Bilt

2) Latitude: 52° 8′ Longitude: 5° 7′ Altitude: -2 metres

3) Global Irradiation (horiz plane): 976 kWh/m$^2$·year % Diffuse: 64

4) Degree Days: 3,644 Base Temp: 18 °C

5) Sunshine Hours: July: 225 January: 73 Annual: 1,400

**B1 BUILDING**

1) Building Type: 200 kg/m$^3$ No Occupants: 2 (Oct-Dec: 1)

2) Floor area: 130 m$^2$ Heated Volume: 375 m$^3$

3) Design Temperature: External: -10 °C Internal: 20 °C

4) Ventilation Rate: 0.5 + 0.1*V wind a.c.h. Vol. Heat Loss: 0.85 W/m$^3$·K

5) Space Heat Load: 16,400 kWh Hot Water Load: 2,000 kWh

**C1 SYSTEM**

1) Absorber Type: Flat plate, spectral selective, stainless steel

2) Collector Area 35. m$^2$ (Aperture) Coolant: water + glycol

3) Orientation: 180° Tilt: 50° Glazing: single

4) Storage Volume: 2.0 m$^3$ Heat Emitters: Warm air

5) Auxiliary System: Gas fired boiler (SH)<br>Electric (DHW) Heat Pump: --

**PERFORMANCE MONITORING**

1) Is there a Computer Model? Yes 2) Start Date for Monitoring Programme: 1st January 1978

3) Period for which Results Available: 1st January 1978 to 31st December 1978

4) No. of Measuring Points per house: 38

5) Data Acquisition System: Microprocessor controlled system including 130 analogue inputs, 25 digital inputs modem, magnetic tape recorder, microprocessor controlled filter.

| | Space Heating | Hot Water | Total |
|---|---|---|---|
| 6) Predicted | 30 % of 16366 kWh | 51 % of 2017 kWh | 32 % of 18383 kWh |
| 7) Measured | 31 % of 17711 kWh | 28 % of 1931 kWh | 30 % of 19642 kWh |

8) Solar Energy Used: Including useful losses 194.4 kWh/m$^2$·year; Excluding losses: 133.8 kWh/m$^2$·year

9) System Efficiency: $\frac{\text{Solar Energy Used}}{\text{Global Irradiation on collector}} \times 100 =$ 16 %

| PERFORMANCE EVALUATION/CONCLUSIONS | REF. | NL | ZO | 78 | 78 | 25 |
|---|---|---|---|---|---|---|

**Comparison of Measured with Predicted Performance**
(State reasons for any discrepancy and summarise the main factors affecting performance)

Some differences have been observed between the measured performance in the year 1978 and the calculated performance in our reference year 1962. Among other things these differences can be caused by the following:

a) The years 1962 and 1978 are not identical as far as meteorological circumstances are concerned (in future all calculations will be carried out using the measured data of the year 1978).

b) The conduct of the primary system has been calculated on the basis of efficiency data deducted from laboratory tests at one collector. However, the conduct of a collector roof consisting of 30 of these collectors proved not to be an extrapolation of the conduct of only one collector. In fact in this case the measured collector efficiency is about 20% lower than the calculated efficiency.

c) Neither the conduct of all electronic controllers in the installation not the behaviour of the inhabitants can be inserted exactly into the computer model.

d) The daily amount of domestic hot water was assumed to be equal, whereas in reality this amount varies considerably (eg holidays of the inhabitants).

**Modifications to System**

The mounting of the temperature-sensor onto the absorber plate has been improved. A three-way valve of the installation has been replaced by naother type of valve.

**Occupants/Response**

Due to stagnation losses of the storage tank the temperature inside the house may occasionally be high during the summer season.

**Conclusions**

During this period, provided that the 20% decrease in efficiency is taken into account, the amounts of energy collected by the collector installation correspond well with the amounts calculated with the computer model. The system efficiency has proved to be lower than calculated. This is caused by heat losses of several installation parts. It seems that the lower tmeperature level of the air-heating gives a better total performance than in the houses with radiators and/or floorheating.
In 1978 the air-flows of the installation were supposed to be constant. On ground of the results, however, it was decided to measure them constantly since there are doubts about the stability of these flows.

**Future Work**

The measurements will be continued until December 1979. It will be tried to bring the calculated and measured results into line with one another, even for one day.

## P1 MONTHLY PERFORMANCE

REF: NL Z0 78 78 25

| Month | Year | No Days Data | Average Temp °C: Internal | Average Temp °C: External | SOLAR ENERGY kWh: Global Irradiation On Collector Area | Solar Energy Collected | Solar Energy Used: Space Heating Useful Losses Incl. | Solar Energy Used: Space Heating Useful Losses Excl. | Solar Energy Used: Domestic Hot Water | Solar Energy Used: Input To Heat | Heat Pump Total Output kWh: Space Heating | Heat Pump Total Output kWh: Domestic Hot Water | AUXILIARY ENERGY kWh: Heat Pump | AUXILIARY ENERGY kWh: Space Heating | AUXILIARY ENERGY kWh: Domestic Hot Water | AUXILIARY ENERGY kWh: Pumps & Fans | Free Heat Gains kWh: Internal | Free Heat Gains kWh: Direct Solar | SUMMARY kWh: Total Load | SUMMARY %: Percentage Solar | SUMMARY %: Syst. Eff. |
|---|---|---|---|---|---|---|---|---|---|---|---|---|---|---|---|---|---|---|---|---|---|
| Column No | | | 1 | 2 | 3 | 4 | 5 | 6 | 7 | 8 | 9 | 10 | 11 | 12 | 13 | 14 | 15 | 16 | 17 | 18 | 19 |
| J | 78 | 30 | 18.6 | 3.0 | 1224 | 298 | 282 | 101 | 16 | | | | | 2694 | 160 | 44 | 257 | 179 | 3152 | 9 | 24 |
| F | 78 | 24 | 19.4 | 1.1 | 2669 | 485 | 467 | 223 | 18 | | | | | 2105 | 142 | 43 | 227 | 295 | 2732 | 18 | 29 |
| M | 78 | 29 | 20.3 | 6.1 | 3333 | 920 | 881 | 334 | 39 | | | | | 1158 | 162 | 29 | 218 | 347 | 2240 | 41 | 28 |
| A | 78 | 28 | 20.8 | 7.5 | 4299 | 1099 | 1036 | 390 | 63 | | | | | 445 | 129 | 20 | 196 | 410 | 1673 | 66 | 26 |
| M | 78 | 31 | 22.0 | 13.0 | 4694 | 1226 | 728 | 541 | 88 | | | | | 199 | 99 | 18 | 50 | 111 | 1114 | 73 | 17 |
| J | 78 | 30 | 22.8 | 16.5 | 5028 | 933 | 323 | 244 | 91 | | | | | 71 | 63 | 13 | 20 | 48 | 548 | 76 | 8 |
| J | 78 | 31 | 22.1 | 16.7 | 4780 | 570 | 245 | 212 | 47 | | | | | 112 | 127 | 14 | 26 | 33 | 531 | 55 | 6 |
| A | 78 | 31 | 23.0 | 16.9 | 4449 | 428 | 71 | 61 | 41 | | | | | 34 | 90 | 11 | 7 | 11 | 236 | 47 | 3 |
| S | 78 | 30 | 21.0 | 14.7 | 3260 | 481 | 259 | 226 | 53 | | | | | 183 | 100 | 14 | 39 | 54 | 595 | 52 | 10 |
| O | 78 | 31 | 20.4 | 11.3 | 2084 | 544 | 492 | 339 | 52 | | | | | 554 | 91 | 20 | 240 | | 1189 | 46 | 26 |
| N | 78 | 30 | 20.0 | 7.0 | 1629 | 457 | 433 | 410 | 24 | | | | | 1135 | 91 | 27 | 255 | 259 | 1683 | 27 | 28 |
| D | 78 | 31 | 18.3 | 1.7 | 926 | 217 | 202 | 218 | 15 | | | | | 3602 | 130 | 46 | 324 | 150 | 3949 | 5 | 23 |
| TOTAL | | | — | — | 37375 | 7658 | 5419 | 3299 | 547 | | | | | 12292 | 1384 | 299 | 1859 | 2169 | 19642 | — | — |
| AVERAGE | | | 20.7 | 9.6 | | | | | | | | | | | | | | | | 30 | 16 |

TOTAL: 3846 kWh | C.O.P =

| PROJECT DESCRIPTION | BEBBINGTON | REF. | UK | BE | 79 | 80 | 26 |
|---|---|---|---|---|---|---|---|

**Main Participants:**

Merseyside Improved Housing Association
Barry Natton, Chief Executive
Hugh Evans, Director

Scheme and Experiment designed by:
Peter Greenwood, Director of Building Science, University of York
Howard Ward, Head of Civil Engineering, Hong Kong Polytechnic
In Association With:
Paterson Macaulay and Owens
Chartered Architects
Monitoring In Association With:
Pilkington Brothers Limited
Mr A D Cunningham, R & D Laboratories, Lathom, Nr Ormskirk, Lancashire
and
Department of Energy
Dr G Long, ETSU, Building 10.28, AERE, Harwell, Oxfordshire, OX11 ORA

**Project Description:**

The scheme consists of 14 houses. There are two blocks of five terraced houses, one with solar walls and the other of the same plan form but built to the 1976 building regulations. Two further semi-detached houses face south-east and two south-west, all having solar walls.

**Project Objectives:**

The main aim in designing the experiment, that is the layout and the detail design and construction of the houses, was for the purpose of monitoring the scheme to compare the thermal performance of the solar houses with the traditionally designed houses of the same plan form on the same site. A comprehensive monitoring scheme is to be carried out for two years after the houses are occupied and will cover all aspects which effect the performance of the dwellings. The areas to be monitored include: meteorological data; internal environment temperatures; thermal performance of the structural elements in both the solar and the traditional control houses; performance of the solar wall/heating system and energy consumption/cost in both the solar and the traditional control houses. Other studies will include daylighting, artificial lighting and, of vital importance, the influence the houses have on life style and the acceptance of the houses by the occupants. The monitoring of the houses over the two heating seasons will provide evidence of the physical performance of the houses on a long term basis under normal operating conditions. Short term in depth experiments on selected houses will help to isolate and quantify specific parameters influencing the performance of the dwelling. The data obtained will be compared with the results of the prediction studies in order to develop more accurate predicition techniques.

SITE LOCATION MAP | REF. | UK | BE | 79 | 80 | **26**

– Distance to main city: 7 km from Liverpool

– Please indicate height of nearest obstructions

The development is situated in the village of Higher Bebington, Wirral, NGR 31,85 84,80

Fruit trees 6 m high approx.
Adjacent buildings 12 m high approx.

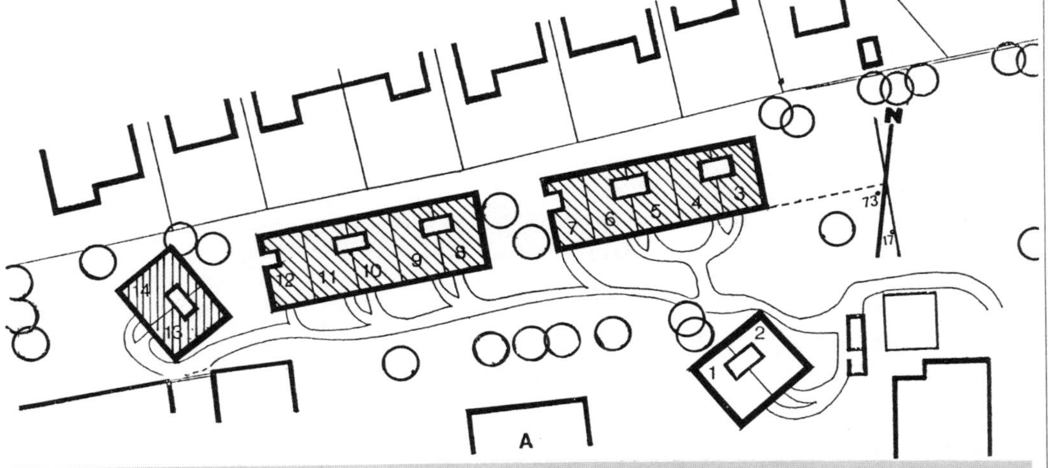

Photograph of Project

| SUMMARY SHEET BEBINGTON | REF. | GB | Be | 79 | 80 | 26 |
|---|---|---|---|---|---|---|

**Design Data**

**A1 CLIMATE**

1) Source of Data: Bidston Observatory
2) Latitude: 53 ° 2 ′ Longitude: 3 ° 5 ′ Altitude: 60 metres
3) Global Irradiation (horiz plane): (835) kWh/m²·year % Diffuse:
4) Degree Days: 2,531 Base Temp: 18.3 °C
5) Sunshine Hours: July: 181.1 January: 49.6 Annual: 1,451

**B1 BUILDING**

1) Building Type: Domestic terrace and Semi-detached single storey No Occupants: 1 or 2senior citizens
2) Floor area: 50 m² Heated Volume: 117.5 m³
3) Design Temperature: External: -1 °C Internal: 21 °C
4) Ventilation Rate: 0.5 to 4. a.c.h. Heat Loss: 3.3 kW Solar ) 4.9 kW Tradit.) @ 2 ach
5) Space Heat Load: kWh Hot Water Load: kWh

**C1 SYSTEM**

1) Absorber Type: Solar Wall, natural black brick
2) Collector Area 19 m² (Aperture) Coolant: Air
3) Orientation: 163° Tilt: 90° Glazing: Double 4:12:4 mm
4) Storage Volume: m³ Heat Emitters: Solar wall/ Warm air
5) Auxiliary System: Electric Heat Pump: --

**PERFORMANCE MONITORING**

1) Is there a Computer Model? Yes 2) Start Date for Monitoring Programme: 1979
3) No. of Measuring Points per house: varying between 5 and 65
4) Data Acquisition System: 330 channel datalert 80 data logger (BOC) hard wired to PDP11 computer (DEC).

| | Space Heating | Hot Water | Total |
|---|---|---|---|
| 6) Predicted | % of kWh | % of kWh | % of kWh |
| 7) Measured | % of kWh | % of kWh | % of kWh |

8) Solar Energy Used: Including useful losses kWh/m²·year
Excluding losses: kWh/m²·year

9) System Efficiency: $\frac{\text{Solar Energy Used}}{\text{Global Irradiation on collector}} \times 100 =$ %

## A2 LOCAL CLIMATE

REF. UK BE 79 80 **26**

Mean monthly temp (July 1976-78):15.9°C; Mean max temp(July 1976-78):19.3°C
Mean monthly temp (Jan 1976-78):4.53°C; Mean min temp(Jan 1976-78):2.66°C

Source of Weather Data: Institute of Oceanographic Sciences, Bidston Observatory, Birkenhead.

Micro Climate/Site Description: Elevated island site surrounded on three sides by dwelling houses and open to the east.

| | Jan | Feb | Mar | Apr | May | Jun | Jul | Aug | Sep | Oct | Nov | Dec | Tots. |
|---|---|---|---|---|---|---|---|---|---|---|---|---|---|
| Precipitation mm | 58.4 | 44.4 | 44.0 | 43.9 | 52.5 | 51.5 | 67.7 | 78.2 | 70.4 | 77.7 | 70.2 | 66.4 | 725.3 |
| Average ambient temperatures °C | 4.3 | 4.5 | 5.8 | 8.1 | 11.2 | 14.0 | 15.4 | 15.4 | 13.5 | 10.2 | 6.9 | 5.1 | 9.5 |

## B2 BUILDING DESCRIPTION

### Design Concepts

This scheme is a single storey sheltered housing project for senior citizens. The total environment concept is a numerate process of design with climate; good design-spaces and detail construction; orientation and the use of direct insolation; insulation to make large savings and alternative technology to give the highest gains at the lowest cost.

### Construction

1. Glazing: 18.77%; Area: 7.72 $m^2$ ; 'U' value:3.0 $W/m^2.K$
2. 'U' values: Walls: 0.29 $W/m^2.K$; Floor: 0.32 $W/m^2.K$; Roof/loft: 0.16 $W/m^2.K$
3. Thermal mass: 2200 $kg/m^3$.

### Energy Conservation Measures

Solar wall ducted warm air heating system. Double glazing 3:6:3 mm; 100mm dritherm in rear and end walls; draught stripping windows and doors; 80mm fibreglass roof board(externally); 50mm fibreglass slab under floor.

## C2 SYSTEM DESCRIPTION

### Collector/Storage

Seven hour time lag in solar black body structure.
Primary circuit:fluid:warm air.

### Space Heating/Domestic Hot Water

1. Auxiliary(SH):4.5kW storage fan heater; (DHW):two stage 3kW elec. immersion heater.
2. Heat emitters:1.5kW convector panel heater(Brm),0.5kW convector panel heater(hall) 0.75kW ingra red wall heater (bathroom).

Economy 7 tarriff on whole house:00.30-07.30 GMT, 01.30-08.30 BST.

### Control Strategy/Operating Modes

Solar warm air heating: Manual control by occupants ; ventilators to control ducted warm air and external blinds to control reception of radiation by the solar wall.

Electric auxiliary heating system: Controlled by time clocks and thermostats operated by the occupants.

| C2 SYSTEM DESCRIPTION | REF. | UK | BE | 79 | 30 | 26 |
|---|---|---|---|---|---|---|

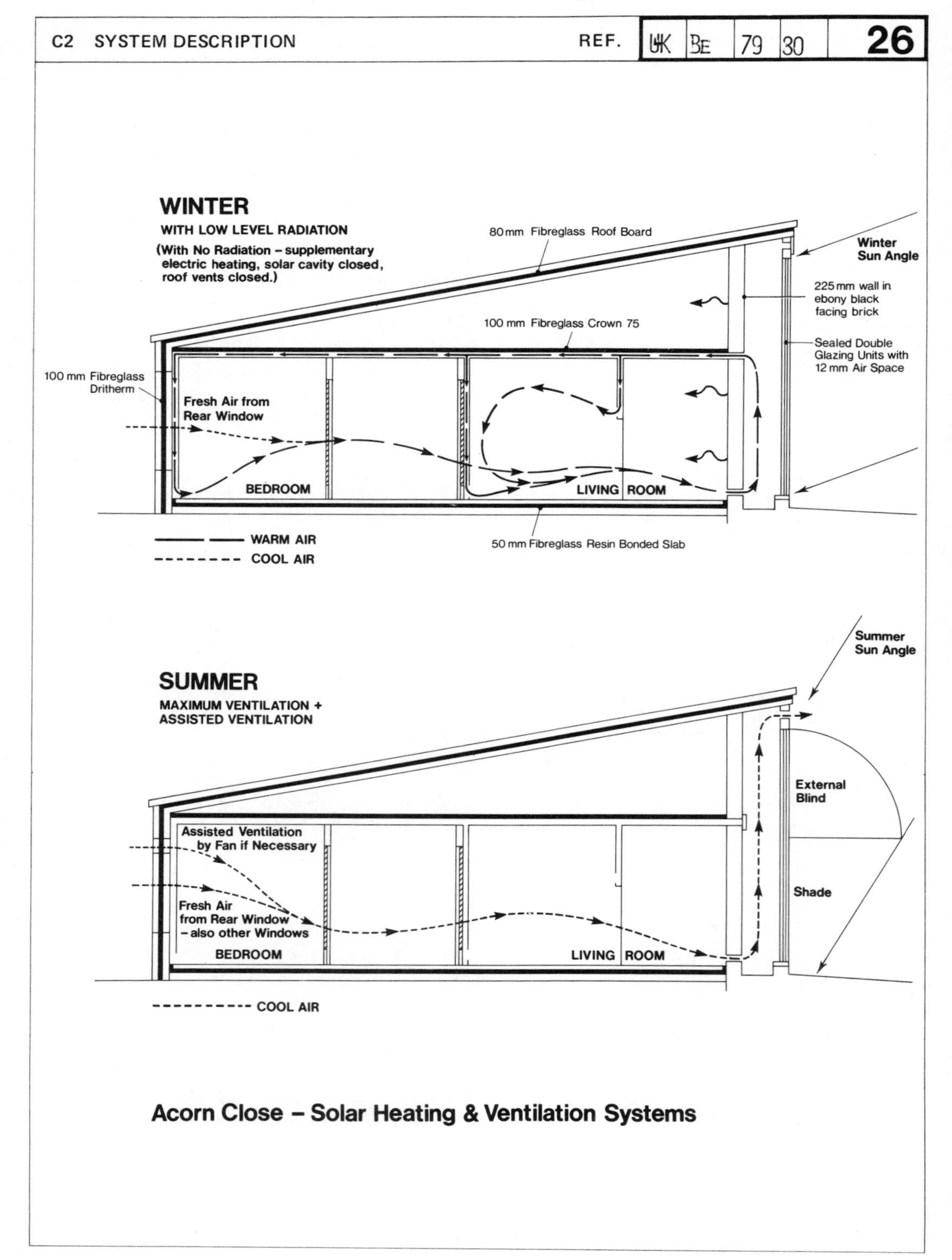


**Acorn Close – Solar Heating & Ventilation Systems**

| TECHNICAL APPRAISAL/PRACTICAL EXPERIENCE | REF. | UK | BE | 79 | 80 | 26 |
|---|---|---|---|---|---|---|

**System/Design:**

The solar heated houses have one outer wall covered by a double glazed outer skin. Solar radiation is absorbed by the high density black brick wall and re-radiated heat into the house on the storage heater principle. Also, by means of natural convection through a ducted system, the warm air heated by the high density wall is circulated throughout the house. The highly insulated structure, including double glazing, has the objective of minimising heat loss from the building.

**System/Installation:**

The installation of the system is an integral part of the construction of the houses. Only traditional building materials were employed.

**stem Operation/Controls, Electronics:**

Hand operated ventilators to control input of ducted air; hand operated external dutch glinds to control radiation received by the solar wall; hand operated 7.5" single stage aerofoil fan to draw outside air through the house for cooling purposes.

## PERFORMANCE EVALUATION/CONCLUSIONS

**Occupants/Response**

In general occupants have reacted favourably to their new houses.

**Future Work**

A monitoring programme at present under way will evaluate the thermal performance of the houses and any associated benefits over a two year period.

| PROJECT DESCRIPTION | CARDIFF | REF. | UK | CA | 79 | 79 | 27 |
|---|---|---|---|---|---|---|---|

**Main Participants:** Solar Energy Unit, University College, Cardiff.

Professor B.J. Brinkworth, Mechanical Engineering Department,
University College,
Newport Road,
Cardiff, CF2 1TA.

Mr. T.T.Lewis, Project Supervisor.

Mr. P. Taylor, Monitoring and Instrumentation.

ENERGY TECHNOLOGY SUPPORT UNIT OF DEPARTMENT OF ENERGY

**Project Description:** The project consists of a solar water heating system for four student flats (20 students). 45 Copper panels give a total effective collection area of 25 $m^2$. The array can be separated into three banks of 12 panels and one of 9 panels for the purpose of comparison of three types of panel coating (black chrome, copper oxide and matt black paint). A commercially available control unit is fitted. A direct arrangement is used with the domestic water passing through the solar collector panels, and returning to a 1$m^3$ storage tank made of galvanised steel lined with epoxy coating. The system was designed with patent glazing as part of the roof; the panels themselves are accessed from inside the roof. An air space of about 40 mm is maintained between the panels and the glazing, rear losses are minimised by a 25 mm rockwool mat, backed by a 32 mm polyisocyanurate foam slab faced with aluminium foil.

**Project Objectives:**

The objective of the programme are as follows:-

1. To monitor sufficient system variables to provide data for checking
   (a) assumptions made in the construction of theoretical models.
   (b) predictions of performance
   (c) optimisation and sensitivity analysis.
2. To provide information for economic assessments.
3. To provide information on long-term changes in system performance, and on reliability and durability of components in service.

| SITE LOCATION MAP | REF. | UK | CA | 79 | 79 | 27 |
|---|---|---|---|---|---|---|

Distance to main city: 2 km from centre of Cardiff. (Direction:NNW).

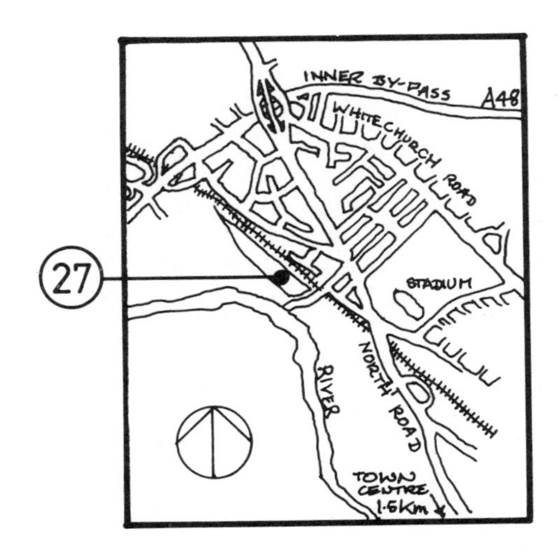

**SUMMARY SHEET** CARDIFF **REF.** UK | CA | 79 | 79 | **27**

**Design Data**

**A1 CLIMATE**

1) Source of Data: Meteorological Office, Cardiff, Wales Airport.

2) Latitude: 51° 24′N Longitude: 03° 21′ W Altitude: 220 metres

3) Global Irradiation (horiz plane): 977.5 $kWh/m^2 \cdot year$ % Diffuse: 58.8

4) Degree Days: - Base Temp: - °C

5) Sunshine Hours: July: 202.7 January: 47.8 Annual: 1,571

**B1 BUILDING**

1) Building Type: 4 Storey flats No Occupants: 20 (4 flats, 5 students in each)

2) Floor area: $m^2$ Heated Volume: $m^3$

3) Design Temperature: External: °C Internal: 18 °C

4) Ventilation Rate: a.c.h. Vol. Heat Loss: $W/m^3 \cdot K$

5) Space Heat Load: kWh Hot Water Load: 18,000 kWh

**C1 SYSTEM**

1) Absorber Type: Flat copper sheet (3 types: black chrome, copper oxide & black paint)

2) Collector Area 25 $m^2$ (Aperture) Coolant: Water

3) Orientation: 130.5° East Tilt: 28° Glazing: single 6mm wired glass.

4) Storage Volume: 0.8 $m^3$ Heat Emitters: -

5) Auxiliary System: Immersion heaters Heat Pump: -

**PERFORMANCE MONITORING**

1) Is there a Computer Model? Yes

2) Start Date for Monitoring Programme: February 1979

3) Period for which Results Available: February 1979 to May 1979

4) No. of Measuring Points per house: 11

5) Data Acquisition System: Micro data M200 L Data Logger with M200 U interface unit.

| | Space Heating | Hot Water | Total |
|---|---|---|---|
| 6) Predicted | ... % of ... kWh | 30-35 % of 4,000 kWh * | ... % of ... kWh |
| 7) Measured | ... % of ... kWh | 29 % of 6,701 kWh * | ... % of ... kWh |

*measurement period : Feb 79 - June 79 inclusive

8) Solar Energy Used: Including useful losses ... $kWh/m^2 \cdot year$

Excluding losses: 77.7 (5 months) $kWh/m^2 \cdot year$

9) System Efficiency: $\frac{\text{Solar Energy Used}}{\text{Global Irradiation on collector}} \times 100 =$ 17.3 %

54

## A2 LOCAL CLIMATE

REF. | UK | CA | 79 | 79 | **27**

Average daily max temps(July: 19.6°C; average daily min temps (Jan): 2.6°C.

Source of Weather Data: Meteorological Office, Cardiff-Wales Airport.

Micro Climate/Site Description: Cardiff (population 300,000)is situated in South Wales on the south-facing coast of the River Severn Estuary. Hilly country provides the narrow coastal plain with shelter from North winds.

| | | Jan | Feb | Mar | Apr | May | Jun | Jul | Aug | Sep | Oct | Nov | Dec | Tots. | Avgs. |
|---|---|---|---|---|---|---|---|---|---|---|---|---|---|---|---|
| Irradiation on horiz. plane kWh/m2.month | Global | 18.1 | 34.5 | 73.2 | 95 | 145.5 | 145 | 149.8 | 118.8 | 91.7 | 60.3 | 28.3 | 17.2 | 997.5 | 81.5 |
| | Diffuse | 13.4 | 22.4 | 43.9 | 59.9 | 82.9 | 84.1 | 85.4 | 72.5 | 50.4 | 30.2 | 17.3 | 12.2 | 574.6 | 47.9 |
| Average ambient temp.°C | | 3.5 | 4.0 | 6.0 | 8.5 | 11.5 | 14.5 | 16 | 15.5 | 13.5 | 11 | 7 | 4.5 | - | - |
| Average wind speed | | 6 | 6 | 5 | 5 | 4 | 4 | 4 | 4 | 4 | 5 | 5 | 6 | - | - |

## B2 BUILDING DESCRIPTION

**Design Concepts**

High density student flats, constructed 1976. There are 5 persons to each flat. The solar domestic hot water system serves one wing of a 3-wing block (i.e. 4 flats). Each flat has one hot water cylinder, with one shower and three hot water taps.

**Construction**

Annual energy demand:18,000 kWh (DHW)

**Energy Conservation Measures**

10mm insulation on pipes leading to the hot water cylinders. Showers are proveded instead of baths.

## C2 SYSTEM DESCRIPTION

**Collector/Storage**

- Primary circuit. Water content: 1.9 litres/m² collector area
- Insulation. Collector: 60 mm Storage: 120 mm Pipework: 10 mm
- Frost Protection: Automatic drain down of panels when fin temperature reaches 3°C
- Overheating Protection: Thermostat (60°) on preheat tank. Panels drain automatically at 90°
- Corrosion Protection: All non-ferrous pipework. The galvanised steel preheat tank has an epoxy coating internally.

**Space Heating/Domestic Hot Water**

- Auxiliary (S.H.): - (DHW): 4 Immersion heaters
- DHW Insulation: calorifiers:20 mm rockwool, Al-faced.Calorifier outlet pipes:none Set Temperature: 55°C

**Other Points/Heat Pump:**

The calorifiers are situated in airing cupboards – the auxilliary heating is left switched on continuously.

**Control Strategy/Operating Modes**

A commercially available differential controller causes water to circulate when the collector fin temperature exceeds the store temperature by 3°C . When this is reduced to approximately 1.5° circulation ceases

C2 SYSTEM DESCRIPTION (CONT) REF. UK CA 79 79 27

**Diagram of System** Showing sensor locations, power ratings, storage volumes & flows.

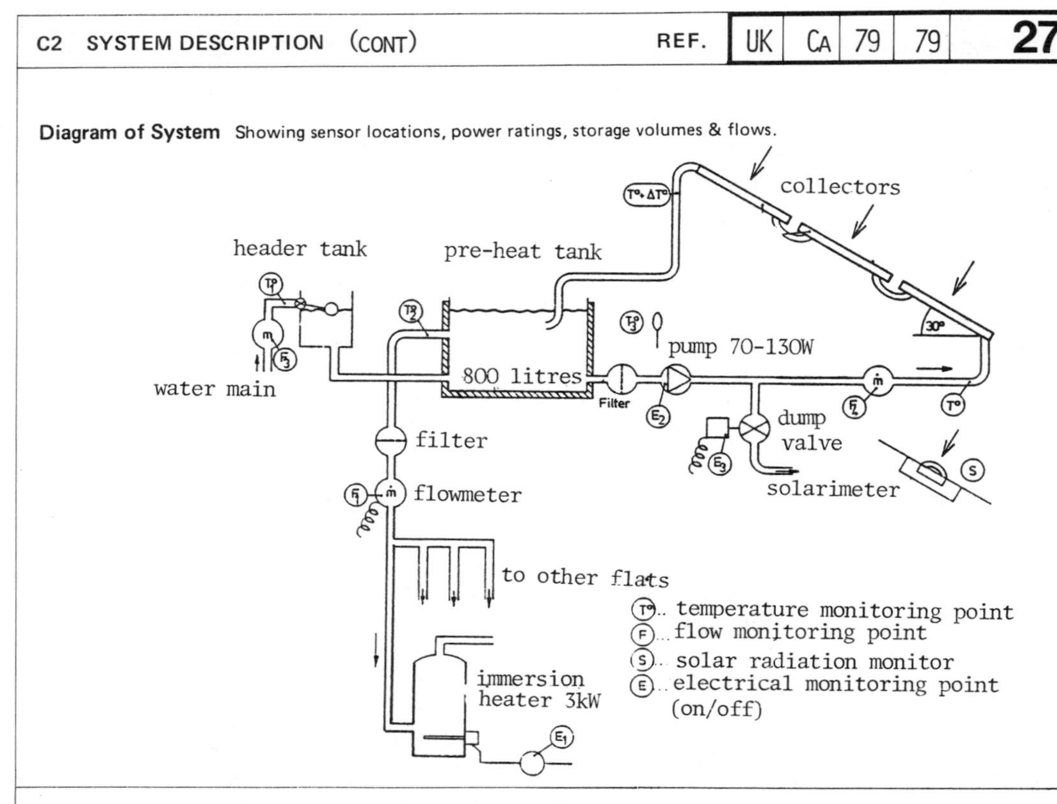


TECHNICAL APPRAISAL/PRACTICAL EXPERIENCE

**System/Design:**
The roof glazing was installed during construction of the flats so that the integrity of the roofing would not be lost during installation of the collectors. These were fitted later from inside the roof space. The pipework is arranged for the independent control of the water flows to panels having different surface coatings.

**System/Installation:**
At the time in installation the use of ball-valves in hot water tanks was **not** recommended. A separate header tank was installed for maintaining the level in the preheat tank (see diag.). The preheat tank comproses one half of the water storage for the 4 flats beneath. Thus additional floor strengthening was not required.

**Component Performance:** (Collector, Heat Exchangers, Storage, Pipework, Valves, Fittings, Pumps, Auxiliary)
The circulator pump bearings required replacement after approximately 18 months' service. This was due to incomplete purging of air from the pump, which is mounted horizontally rather than vertically.
Slight degradation of the properties of the selective panel surfaces has been noted.

**System Operation/Controls, Electronics:**
The control box failed in February 1979 and was replaced by the company concerned. The thermostat used for draining the panels at low temperatures has given good service: it has remained true to its set point since it was installed in 1977. During Summer 1978 the 60° temperature limiting thermostat on the preheat tank malfunctioned. Its true set point drifted down to approximately 50°C.

PERFORMANCE EVALUATION/CONCLUSIONS REF. UK CA 79 79 **27**

### Comparison of Measured with Predicted Performance

(State reasons for any discrepancy and summarise the main factors affecting performance)

The major factor causing variations in performance of the system is the fluctuating student population of the flats. Full demand only occurs during academic terms, and the periods covered by these do not coincide with full months; this must be borne in mind when inspecting the monthly performance figures.

Comparison of the auxiliary heating figures with those from two identical blocks of flats on the same site showed good agreement with the predicted performance during January '78 - July '78. However, it was found that consumption varied widely from flat to flat, so the results using this method are dependent on the usage habits of the students in the flats concerned.

During August 1978 a test run was performed under conditions of controlled load (water consumption). The results were summarised graphically (see accompanying sheet). Under normal demand conditions (20 persons x 45 litres/day) the system efficiency (i.e. the solar energy used divided by the incident solar energy) is 0.29 (± 0.05) compared to 0.32 predicted.

### Modifications to System

1. Separate feed pipes to the individual flats were installed October 1978, so that the water consumption of each flat could be measured more easily
2. The setting on the drain down thermostat was reduced from 5°C to 3°C, thus reducing the number of unnecessary draining operations

### Occupants/Response

There is a change of occupants each year, and the presence of the solar water heating system is not thought to significantly affect their water usage habits. However more information is needed on this subject. Some students have shown great interest in the system.

### Conclusions

1. The losses from the calorifiers are quite large (120 watts continuously). This represents the major loss from the system, although it does provide useful heat in the airing cupboards. Temperature setting should be reduced and calorifier output pipes lagged. Lagging on pipes between preheat tank and calorifiers needs improving.
2. There is some degree of stratification in the preheat tank. The possibility of maximising this and developing a better control strategy should be considered.
3. Draining of the panels in Winter should be minimised (both because of water wastage and the time taken for the system to purge itself of air afterwards).

### Future Work

Present monitoring programme continues until June 1981.

**P1 MONTHLY PERFORMANCE** REF: UK CA 79 79 27

| Month | Year | No Days Data | Average Temp °C | | SOLAR ENERGY kWh | | | | | | Heat Pump Total Output kWh | | AUXILIARY ENERGY kWh | | | | Free Heat Gains kWh | | SUMMARY | | |
|---|---|---|---|---|---|---|---|---|---|---|---|---|---|---|---|---|---|---|---|---|---|
| | | | | | | | Solar Energy Used | | | | | | | | | | | | kWh | % | |
| | | | | | | | Space Heating Useful Losses | | | | | | | | | | | | | | |
| | | | Internal | External | Global Irradiation On Collector Area | Solar Energy Collected | Incl. | Excl. | Domestic Hot Water | Input To Heat | Space Heating | Domestic Hot Water | Heat Pump | Space Heating | Domestic Hot Water | Pumps & Fans | Internal | Direct Solar | Total Load | Percent-age Solar | Syst. Eff. |
| Column No | | | 1 | 2 | 3 | 4 | 5 | 6 | 7 | 8 | 9 | 10 | 11 | 12 | 13 | 14 | 15 | 16 | 17 | 18 | 19 |
| J | | | | | | | | | | | | | | | | | | | | | |
| F | 79 | 16 | | | 742.5 | | | | 139 | | | | | | 949 | | | | 1088 | 12.8 | 18.7 |
| M | 79 | 31 | | | 1686 | | | | 407 | | | | | | 1123 | | | | 1530 | 26.6 | 24.0 |
| A | 79 | 30 | | | 2933 | | | | 259 | | | | | | 729= | | | | 988 | 26.2 | 8.8 |
| M | 79 | 26 | | | 2534 | | | | 430 | | | | | | 1179 | | | | 1609 | 26.7 | 17.0 |
| J | 79 | 29 | | 15 | 3367 | | | | 708= | | | | | | 779 | | | | 1486 | 47.6 | 21.0 |
| J | 79 | 22 +9 | | 18 | 2706 +1031 | | | | 0 +251 | ==(flats unoccupied) ===(simulated usage) | | | | | not in use | | | | | | 24.4 |
| A | | | | | | | | | | | | | | | | | | | | | |
| S | | | | | | | | | | | | | | | | | | | | | |
| O | | | | | | | | | | | | | | | | | | | | | |
| N | | | | | | | | | | | | | | | | | | | | | |
| D | | | | | | | | | | | | | | | | | | | | | |
| TOTAL | | 132 | — | — | 11263 | | | | 1943 | | | | | | 4758 | | | | 6701 | — | — |
| AVERAGE | | | | | | | | | | | | | | | | | | | | 29.0 | 17.3 |

TOTAL: 1943 kWh

C.O.P =

= Vacation dates: March 16 - 24 April, during this period, the occupancy figures are lower resulting in reduced hot water demand.

==Summer vacation starts 1st July

===From July 23 a measured mass of water was removed daily from pre-heat tank. Thus, July has been EXCLUDED from totals and averages.

| PROJECT DESCRIPTION | LONDON | REF. | UK | Lo | 79 | 80 | 28 |
|---|---|---|---|---|---|---|---|

**Main Participants:**

| | |
|---|---|
| Owner | London Borough of Southwark |
| Design of Solar Water Heating System: | South London Consortium for Local Authority Research and Development. |
| Construction: Rehabilitation | Donald James - Chartered Surveyors<br>A. Coldman & Sons - Contractor |
| Financial Support | U.K. Department of Energy<br>(100% of monitoring costs) |
| Monitoring | South London Consortium,<br>125, Camberwell Road,<br>London, SE5 OHB.<br>Telephone: (01) 701 0326<br>Contact: A.J. Kirk Dip Arch RIBA - Chief Dev. Arch.<br>Dr. L. Makkar - Principal Research Officer. |

**Project Description:**

The project involves solar water heating systems installed in 14, 3-storey terraced houses while they were being rehabilitated for the London Borough of Southwark. The solar systems were completed in June 1979. Thirteen houses have 5m$^2$ of collector panel, and one has 10m$^2$. Thirteen houses have storage cylinders of 440 litres (96 galls) and one house has a different system with one storage cylinder of 159 litres (35 galls) for the solar heated water and a separate cylinder of the same size heated with an immersion heater. In all cases the cylinders are on the semi basement floor - next to the bathroom.

**Project Objectives:**

The Project objective is the detailed monitoring of the heat gain from the solar systems in relation to the usage patterns of the occupants of the houses. The monitoring programme measures the heat input to the cylinder from the solar panel, the heat supplied by the supplementary immersion heater, and the heat used by the occupants. Heat losses are then allowed for to give the useful heat supplied by the panels. In three houses more detailed usage monitoring is being carried out, to give some indications on patterns of hot water usage - timing, amounts and temperatures. Weather data is also being collected. One house, the one with the double panel, is remaining vacant for the time being to house the monitoring equipment and for demonsration purposes.
The monitoring period is 2½ years, starting July, 1979.

| SITE LOCATION MAP | REF. | UK | Lo | 79 | 8C | 28 |
|---|---|---|---|---|---|---|

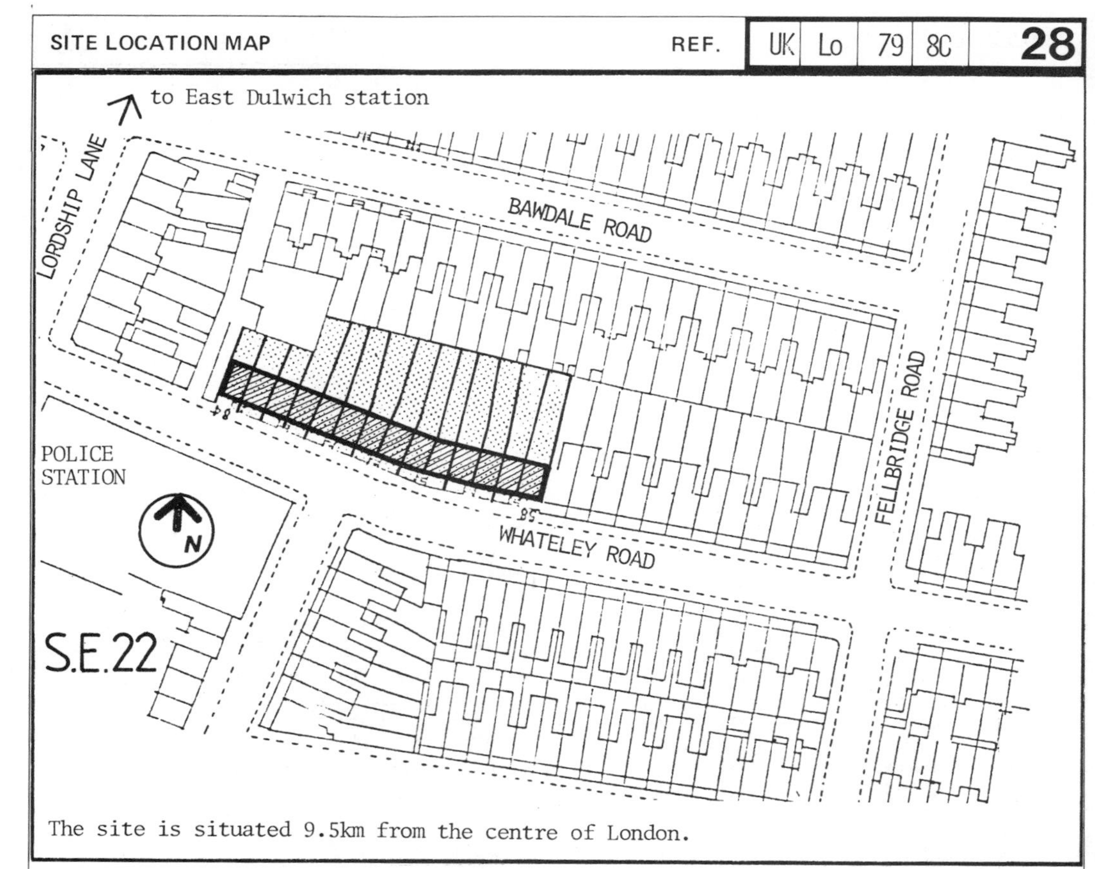


The site is situated 9.5km from the centre of London.

Photograph of Project

| SUMMARY SHEET | LONDON | REF. | UK | Lo | 79 | 80 | 28 |
|---|---|---|---|---|---|---|---|

**Design Data**

## A1 CLIMATE

1) Source of Data: Kew
2) Latitude: 54° Longitude: ° Altitude: metres
3) Global Irradiation (horiz plane): 911 kWh/m²·year % Diffuse: 58
4) Degree Days: 2600 Base Temp: 18 °C
5) Sunshine Hours: July: 186 January: 53 Annual: 1489

## B1 BUILDING

1) Building Type: single family houses No Occupants: 4-5
2) Floor area: 106.50 m² Heated Volume: n/a m³
3) Design Temperature: External: n/a °C Internal: n/a °C
4) Ventilation Rate: n/a a.c.h. Vol. Heat Loss: n/a W/m³·K
5) Space Heat Load: n/a kWh Hot Water Load: 4,320 kWh (240 litres DHW/day (4 people))

## C1 SYSTEM

1) Absorber Type: copper, tube & fin, matt black paint finish.
2) Collector Area 5 m² (Aperture) Coolant: non-toxic m-propylene glycol 50% with distilled water
3) Orientation: Tilt: 30° Glazing: single
4) Storage Volume: 0.509 m³ Heat Emitters: n/a
5) Auxiliary System: electric immersion heater Heat Pump: none

## PERFORMANCE MONITORING

1) Is there a Computer Model? Yes
2) Start Date for Monitoring Programme: July 1979
3) Period for which Results Available: July 1978 to June 1981
4) No. of Measuring Points per house: Varies from 7 to 27
5) Data Acquisition System: Micro-processor controlled data-logger. Some points logged at fixed intervals. Some high-speed conditional logging. Immediate integration of water flow and temperatures to give heat for each occasion of hot water usage.

| | Space Heating | Hot Water | Total |
|---|---|---|---|
| 6) Predicted | % of kWh | 45% % of 4,320 kWh | % of kWh |
| 7) Measured | % of kWh | % of kWh | % of kWh |

8) Solar Energy Used: Including useful losses kWh/m²·year
Excluding losses: kWh/m²·year

9) System Efficiency: Solar Energy Used / Global Irradiation on collector x 100 = %

## A2 LOCAL CLIMATE

REF. UK Lo 79 80 **28**

Micro Climate/Site Description: KEW. 1969 Data.

| | | Jan | Feb | Mar | Apr | May | Jun | Jul | Aug | Sep | Oct | Nov | Dec | Totals |
|---|---|---|---|---|---|---|---|---|---|---|---|---|---|---|
| Irradiation on horiz. plane | Global | 19 | 34 | 47 | 113 | 121 | 159 | 150 | 108 | 75 | 54 | 26 | 11 | 911 |
| kWh/m2.month | Diffuse | 14 | 21 | 30 | 60 | 77 | 82 | 77 | 70 | 46 | 31 | 16 | 9 | 533 |
| Average ambient temp. °C | | 6.4 | 2.3 | 4.5 | 8.5 | 12.1 | 14.3 | 17.9 | 16.9 | 14.7 | 13.1 | 6.3 | 3.7 | 10.1(ave) |

## B2 BUILDING DESCRIPTION

### Design Concepts

The main criterion was to keep the collector panels as inconspicuous as possible, and as the slate roofs were being totally recovered, it was possible to fit the panels onto the rafters and cover them with patent glazing resulting in an almost continuous roof line. The underneath of the collector is insulated with mineral wool quilt.

### Section Through House Showing main solar components

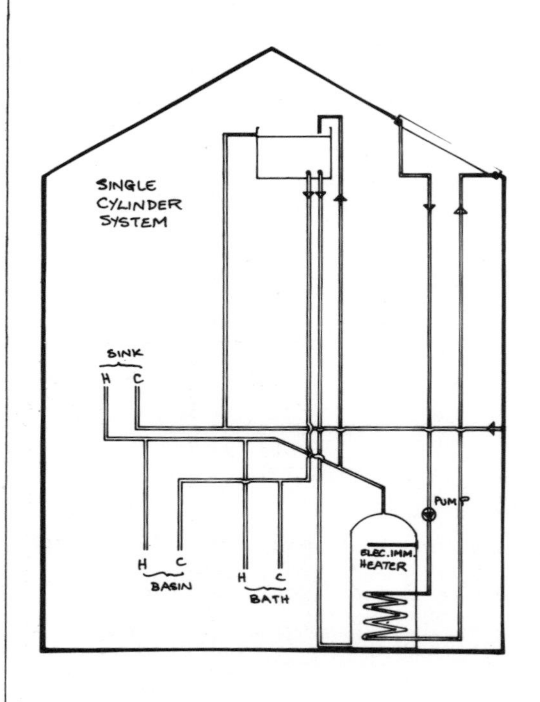


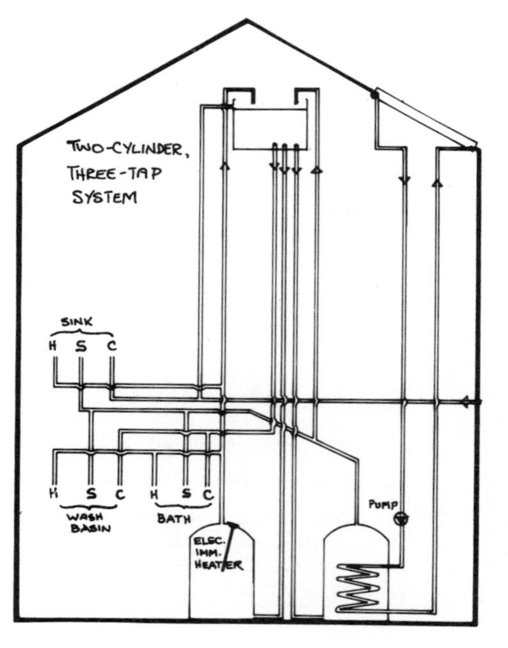

**C2 SYSTEM DESCRIPTION** REF. UK Lo 79 80 **28**

**Collector/Storage**

1 — Primary circuit. Water content: ........ litres/$m^2$ collector area

2 — Insulation. Collector: 100mm Storage: 50mm Pipework: 25mm

3 — Frost Protection: 50% m-propylene glycol

4 — Overheating Protection: safety valve, expansion vessel, large storage cylinder

5 — Corrosion Protection: ........

**Space Heating/Domestic Hot Water**

1 — Auxiliary (S.H.): n/a (DHW): elec. immersion heater

2 — Heat Emitters: ........ Temp. Range: ........

3 — DHW Insulation: 25mm Set Temperature: 60°C

**Control Strategy/Operating Modes**

The pump in the primary circuit is controlled by a differential temperature controller which has a sensor at the panel and at the cylinder. The differential is 2°C.
In all the houses with a single cylinder the primary coil is located in the bottom half of the cylinder and the immersion heater is fitted horizontally 1/3 down from the top. Switching of the immersion heater is at the discretion of the user. In the house with two cylinders, the bath, washbasin and sink all have three taps; hot, cold & 'solar'. The 'solar' tap is opened first and then either hot (from the second cylinder fitted with the immersion heater) or cold water added as required. The control of the immersion heater is again left to the user.

**Diagram of System** Showing sensor locations, power ratings, storage volumes & flows.

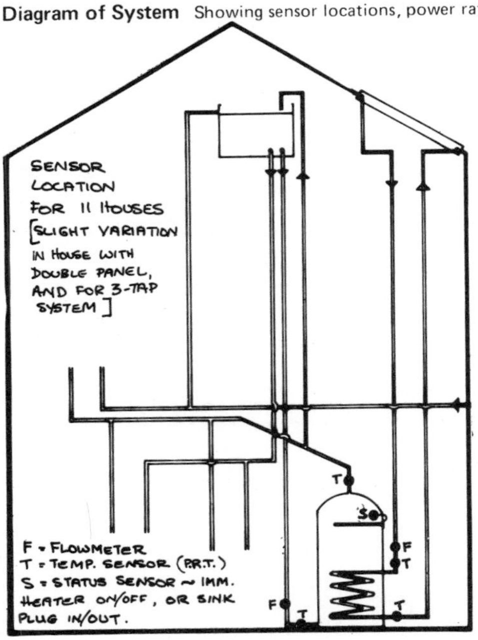

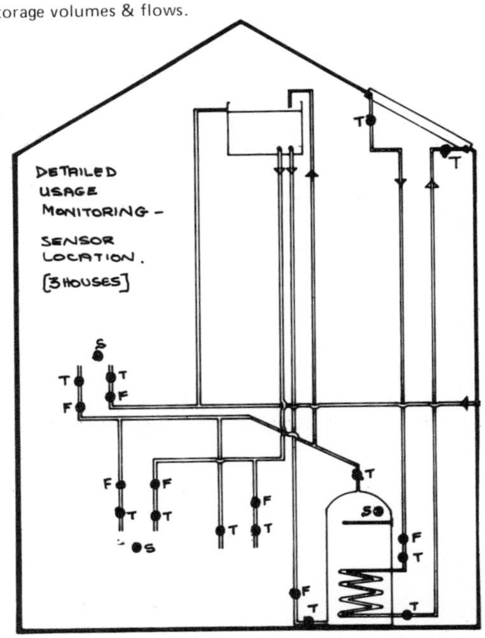

| TECHNICAL APPRAISAL/PRACTICAL EXPERIENCE | REF. | UK | Lo | 79 | 80 | 28 |
|---|---|---|---|---|---|---|

**System/Installation:**

1. The joints between sections of panels and panels to pipework need very careful attention.

2. When using patent glazing, some air circulation between glass and panel should be allowed to prevent condensation, but this should be the minimum so as not to cool the panel surface unneccesarily.

3. Initial problems with air locks in several of the systems - due either to incomplete joints or fixing of flowmeters after system had been filled.

**Component Performance:** (Collector, Heat Exchangers, Storage, Pipework, Valves, Fittings, Pumps, Auxiliary)

4. The wiring of the differential controller seemed to prove difficult and pipes were not always lagged in places where they were hidden!

| PROJECT DESCRIPTION | MACCLESFIELD, UK | REF. | UK | Ma | 77 | 78 | 29 |
|---|---|---|---|---|---|---|---|

Main Participants:

Granada Television Ltd (Mr B Trueman)

Electricity Council Research Centre (Capenhurst)
(A Mould J Siviour F. Stephen)

School of Architecture, University of Manchester
(Mr D R Wilson)

Mr and Mrs Grant, owners of the property.

Project Description:

The project was designed and built during 1975/76 and was used by Granada Television for a series of programmes on currently available energy conservation techniques. It was based on the conversion and rehabilitation of an existing 2 - storey Coach house, to provide living accommodation for a family of four people. The building was well insulated, and the internal planning was arranged to take full advantage of microclimate characteristics.

Project Objectives:

1) To study the possibility of upgrading existing buildings, using un-skilled labour, commercially available products, and to minimise energy consumption.

2) To illustrate a range of energy conserving measures applied to an existing dwelling, including an active solar system, heat pump, mechanical ventilation and heat recovery.

Note

Limitations in the monitoring system mean that comprehensive data is not available, with the result that some figures have been estimated.

| SITE LOCATION MAP | MACCLESFIELD | REF. | UK | Ma | 77 | 78 | 29 |
|---|---|---|---|---|---|---|---|

0
100 m
FARM
Solar House
PRESTBURY ROAD
PRIVATE ROAD
to Macclesfield

Photograph of Project

Note: No casual visitors.

**SUMMARY SHEET** MACCLESFIELD **REF.** UK | Ma | 77 | 78 | **29**

**Design Data**

**A1 CLIMATE**

1) Source of Data: Capenhurst, Cheshire
2) Latitude: 53 ° ′ Longitude: ° ′ Altitude: metres
3) Global Irradiation (horiz plane): 874 kWh/m²·year % Diffuse:
4) Degree Days: 2211 Base Temp: 15.5 °C
5) Sunshine Hours: July: 152 January: 37 Annual: 1241

**B1 BUILDING**

1) Building Type: 2 storey det. house — No Occupants: 4
2) Floor area: 114 m² — Heated Volume: 360 m³
3) Design Temperature: External: 0 °C — Internal: 19.5 °C
4) Ventilation Rate: 0.7 a.c.h. — Vol. Heat Loss: 0.69 W/m³·K
5) Space Heat Load: 22,000 kWh — Hot Water Load: 4.500 kWh

**C1 SYSTEM**

1) Absorber Type: Open Trickle, Black Anodised Aluminium
2) Collector Area 42 m² (Aperture) — Coolant: Water
3) Orientation: 223 Tilt: 34 — Glazing: Single 4mm
4) Storage Volume: 6.1 m³ — Heat Emitters: Radiators
5) Auxiliary System: Solid Fuel Boiler 3.5Kw Heat Recovery — Heat Pump: 1.5 Kw

**PERFORMANCE MONITORING**

1) Is there a Computer Model? No
2) Start Date for Monitoring Programme: July 1977
3) No. of Measuring Points per house: 24
4) Data Acquisition System: Chart Recorder.

| | Space Heating | Hot Water | Total |
|---|---|---|---|
| 6) Predicted | % of kWh | % of kWh | 13 % of 26500 kWh |
| 7) Measured | % of kWh | % of kWh | 24 % of 12450 kWh |

8) Solar Energy Used: Including useful losses kWh/m²·year
Excluding losses: 70.6 kWh/m²·year

9) System Efficiency: Solar Energy Used / Global Irradiation on collector x 100 = 9 %

56

| A2 LOCAL CLIMATE | REF. | UK | Ma | 77 | 78 | 29 |
|---|---|---|---|---|---|---|

1 — Average Cloud cover: .............................................. Octas

2 — Average daily max temperatures (July) : 19.9. °C

Average daily min temperatures (Jan) : 0.8 °C

3 — Source of Weather Data: Average over 30 years, Macclesfield (M)
Data from Manchester airport (MA)
Global insulation measured in Capenhurst (March 74 to Feb 75) (c)

4 — Micro Climate/Site Description:

- No particular microclimatic differences between Macclesfield and Capenhurst (hence agreement of meteorological data is very good)
- Manchester data less representative.

| | | C | | | MA | M | MA | MA | |
|---|---|---|---|---|---|---|---|---|---|
| MONTH | YEAR | Irradiation on horizontal plane: Global | Irradiation on horizontal plane: Diffuse | Degree Days Base 15.5 °C | Precipitation | Average Ambient Temperature | Average Relative Humidity | Average Wind Speed. | Prevailing Wind Direction |
| JAN | | 15.6 | | 359 | 77 | 2.8 | 89 | 5.5 | |
| FEB | | 27.4 | | 323 | 53 | 3.1 | 89 | 5.5 | |
| MARCH | | 63.4 | | 304 | 45 | 5.5 | 82 | 5. | |
| APRIL | | 98.7 | | 222 | 49 | 7.9 | 78 | 5. | |
| MAY | | 128. | | 139 | 57 | 11.1 | 74 | 4.5 | |
| JUNE | | 145. | | – | 61 | 14.2 | 77 | 4.5 | |
| JULY | | 136.3 | | – | 79 | 15.7 | 79 | 4.5 | |
| AUGUST | | 109.4 | | – | 81 | 15.5 | 81 | 4.5 | |
| SEPT | | 72.6 | | 79 | 67 | 13.2 | 82 | 5. | |
| OCT | | 42.5 | | 155 | 78 | 9.5 | 84 | 5. | |
| NOV | | 21.3 | | 280 | 80 | 6.1 | 86 | 5. | |
| DEC | | 14. | | 350 | 72 | 4.0 | 86 | 5.5 | |
| Totals | | 874.2 | | 2,211 | 799 | – | – | – | |
| Averages | | | | | | 9.5 | 85 | 5 | |
| Units | – | kWh/m²· month | | °C days | mm | °C | % | m/s | – |

B2 BUILDING DESCRIPTION MACCLESFIELD, UK REF. UK Ma 77 78 **29**

**Design Concepts**

The building has external insulation, low fabric losses, high thermal mass , mechanical ventilation and heat recovery.

**Construction**

1 — Glazing: 10 % Area: ........ $m^2$ 'U' value: 2.5 $W/m^2 \cdot K$

2 — 'U' values: Walls: 1.36 $W/m^2 \cdot K$ Floor: 0.36 $W/m^2 \cdot K$ Roof/loft: 0.22 $W/m^2 \cdot K$

3 — Thermal mass: ........ $kg/m^3$ Annual Energy Demand: 26,500 kWh

**Energy Conservation Measures**

A conservatory has been built on the South West face of the house, with pebble-bed heat storage.

The first floor of the house is also insulated $U = 0.5\ W/m^2K$.

Free Heat Gains: Internal: 3660 kWh/year Direct Solar: 4640 kWh/year

**Section Through House** Showing main solar components

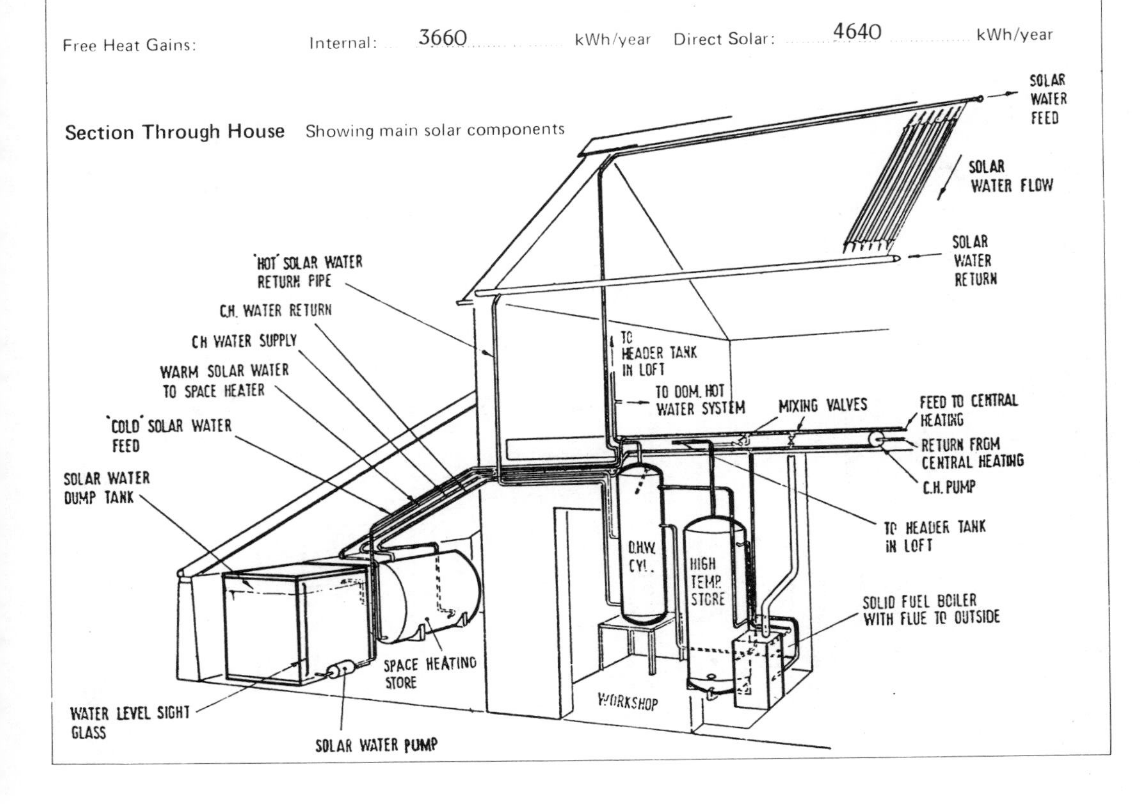

| C2 SYSTEM DESCRIPTION | MACCLESFIELD | REF. | Uk | Ma | 77 | 78 | 29 |
|---|---|---|---|---|---|---|---|

**Collector/Storage**

1 – Primary circuit. Water content: 47.6 litres/m² collector area

2 – Insulation. Collector: ........ Storage: ........ Pipework: ........

3 – Frost Protection: ........

4 – Overheating Protection: Metal Purlins, Natural Ventilation of air space behind collectors

5 – Corrosion Protection: ........

**Space Heating/Domestic Hot Water**

1 – Auxiliary (S.H.): Solid Fuel & HP (DHW): Immersion Heater & Solid Fuel

2 – Heat Emitters: Radiators Temp. Range: 45°C - 25°C

3 – DHW Insulation: 100mm Set Temperature: 60°C

**Other Points/Heat Pump:** Heat Pump up-grades heat from Dump Tank to Space Heat Tank (right operation)

**Control Strategy/Operating Modes**

Return water from Collector goes first to DHW, then to Space Heating, and finally to Dump Tank, By-Pass valves avoid DHW or SH tanks if water temperature is too low. Heat Pump operates at night until Dump Tank reaches 5°C. Solid Fuel Boiler charges High Temperature Store. HT store water outlet is mixed to feed radiators. Return water to store is limited to 25°C. Primary Pump is switched on at 15 min. intervals to check for incident energy, and is left on, or switched off depending on temperature rise across collector.

**Diagram of System** Showing sensor locations, power ratings, storage volumes & flows.

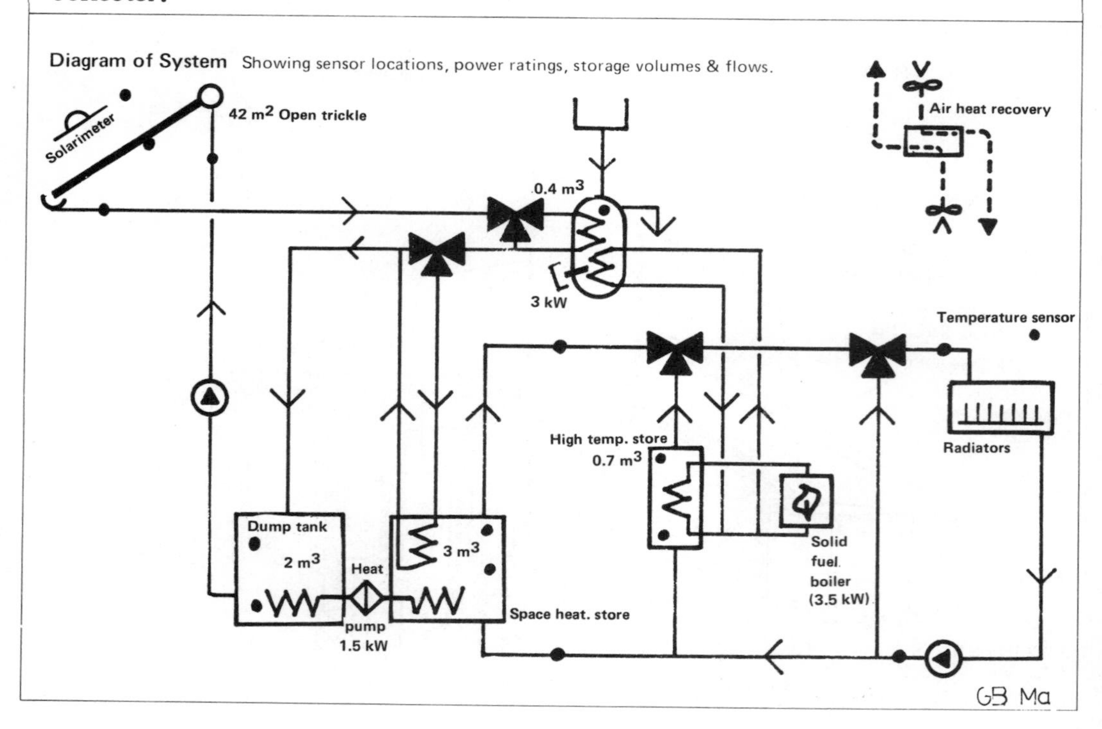

| TECHNICAL APPRAISAL/PRACTICAL EXPERIENCE | REF. | UK | Ma | 77 | 78 | 29 |
|---|---|---|---|---|---|---|

**System/Design:**

- To ensure an even distribution of water supply to collectors required high velocity jets impinging on collectors (ridge pipe)

**System/Installation:**

- Solar roof was assembled on the floor and lifted (by parts) to the roof. This was a difficult process, and the resulting collector was difficult to align in a flat plane.

**Component Performance:** (Collector, Heat Exchangers, Storage, Pipework, Valves, Fittings, Pumps, Auxiliary)

Collector: condensation on underside of glass under certain weather conditions. High velocity jets (ridge pipe) have worn the anodised surface at points of impact and may erode panels. Glazing is very good, but Si-rubber seals are too tenacious and glass panels are difficult to remove intact.

Pipework: Difficult joint between collectors and polyprop. Return pipe resulted in leaks. These leaks (plus evaporation) require topping up of primary circuit and expenditure on inhibitors. Filter cartridges are another expense.

Heat Pump: The sump immersion heater in the oil (designed to avoid condensation when the HP is not operating) is wasteful of energy (1KW/day)

Thermostatic valves: The valve settings were numbered and it proved difficult to obtain a long-term set temperature, the valves having to be adjusted from time to time.

**System Operation/Controls, Electronics:**

Lack of a throttling device at the outlet of high temperature store resulted in large temperature swings on inlet to radiators (25oC to 40oC) and quicker - than expected depletion rates of store.
One flow meter (flap type) failed to provide accurate results.

| PERFORMANCE EVALUATION/CONCLUSIONS | REF. | UK | Ma | 77 | 78 | 29 |
|---|---|---|---|---|---|---|

**Comparison of Measured with Predicted Performance**
(State reasons for any discrepancy and summarise the main factors affecting performance)

The system performed generally as predicted. Some differences with design figures may come from:

- collector flow rate lower than expected
- lack of heat emitter in living room required increased temperature inlet to radiators 40°C.
- freezing of some external pipes (poorly insulated and exposed to wind and moisture) prevented collection in January and part of February)
- regular tests for incident energy on collector resulted in losses (overcast days)

**Modifications to System**

- high temperature store fitted with a throttling device in order to regulate discharge rate. Immersion heater added as alternative/supplement to solid fuel boiler
- A thermostatic switch has been added to stop space heating circulating pump when no heat required. (in addition to existing radiator thermostatic valves)
- One fan convector unit (included in design) has now been installed. Air is convected at ceiling height in order to avoid feeling of discomfort when' using low temperature air (down to 27°C).

**Occupants/Response**

- natural and mechanical ventilation is very un-obtrusive and adequate
- low temperature heat emitters very safe for children
- heat recovery system is very satisfactory

**Conclusions**

- heat pump greatly improved collection efficiency
- low efficiency of solid fuel boiler (40%) gives a cost per KWh comparable with off peak electricity
- low energy requirements of the house have resulted in low solar energy contribution
- Large thermal mass and good insulation improve comfort levels by reducing internal temperature swings, typically 1°C over 24 hrs.
- the mixing valve and controller which limits return water to high temp store to 25°C gave very satisfactory results

**Future Work**

- Monitoring of main variables continued in order to obtain building energy balance and check effect of modifications.

## P1 MONTHLY PERFORMANCE

REF: GB MA 77 78 29

| Month | Year | No Days Data | Average Temp °C Internal | Average Temp °C External | SOLAR ENERGY kWh: Global Irradiation On Collector Area | Solar Energy Collected | Solar Energy Used: Space Heating Useful Losses Incl. | Space Heating Useful Losses Excl. | Domestic Hot Water | Input To Heat | Heat Pump Total Output kWh: Space Heating | Domestic Hot Water | AUXILIARY ENERGY kWh: Heat Pump | Space Heating | Domestic Hot Water | Pumps & Fans | Free Heat Gains kWh: Internal | Direct Solar | SUMMARY* kWh: Total Load | %: Percent-age Solar | Syst. Eff. |
|---|---|---|---|---|---|---|---|---|---|---|---|---|---|---|---|---|---|---|---|---|---|
| Column No | | | 1 | 2 | 3 | 4 | 5 | 6 | 7 | 8 | 9 | 10 | 11 | 12 | 13 | 14 | 15 | 16 | 17 | 18 | 19 |
| J | 78 | | 18.4 | 2.6 | 651 | | | 0 | 0 | 0 | | 0 | 0 | 1348 | 430 | (30) | (430) | (150) | (1808) | 0 | 0 |
| F | 78 | | 17.5 | 2.6 | 1260 | | | 0 | 0 | 15 | | 26 | 11 | 1124 | 390 | (40) | (370) | (250) | (5580) | (1) | (1) |
| M | 78 | | 18.8 | 5.8 | 2945 | | | 119 | (0) | 266 | | 470 | 204 | 472 | | (220) | (410) | (400) | (1281) | (30) | (13) |
| A | 78 | | 18.6 | 6.5 | 4140 | | | 116 | (27) | 229 | | 405 | 176 | 198 | | (210) | (380) | (570) | (956) | (39) | (9) |
| M | 78 | | 20.4 | 11.4 | 5022 | | | 38 | (360) | 124 | | 219 | 95 | 0 | 40 | (200) | (340) | (650) | (860) | (61) | (10) |
| J | 78 | | 20.5 | 14.0 | 5460 | | | 0 | 309 | 0 | | 0 | 0 | 0 | 41 | (190) | (210) | (690) | (540) | (57) | (6) |
| J | 78 | | 20.3 | 14.2 | 5208 | | | 0 | 302 | 0 | | 0 | 0 | 0 | 48 | (190) | (300) | (650) | (540) | (56) | (6) |
| A | 78 | | 20.0 | 16.0 | 4402 | | | 0 | 310 | | | 0 | 0 | 0 | 40 | (190) | (310) | (570) | (540) | (57) | (7) |
| S+ | 77 | 8 | 21.0 | 14.7 | 813+ | | | | (90+) | (27+) | | (48+) | (21+) | 0 | (20+) | (50+) | (90+) | (130+) | (208+) | (56+) | (14+) |
| O | 77 | | 19.8 | 11.5 | 1953 | | | 45 | (48) | (45) | | (80) | (35) | 102 + | (45) | (140) | (380) | (310) | (460) | (30) | (7) |
| N | 77 | | 19.3 | 5.1 | 990 | | | (0) | (0) | (271) | | (480) | (209) | 1136 + | (55) | (120) | (220) | (150) | (1790) | (15) | (27) |
| D | 77 | | 19.2 | 5.3 | 620 | | | (0) | (0) | (226) | | (400) | (174) | 1294 + | (55) | (140) | (220) | (124) | (1890) | (12) | (36) |
| TOTAL | | | — | — | 33464 | | | (318) | (1446) | (1203) | | (2130) | (925) | (6857) | | (1700) | (3660) | (4640) | (12450) | — | — |
| AVERAGE | | | 19.5 | 9.2 | | | | | | | | | | | | | | | | (24) | (9) |

TOTAL: 2967 kWh | C.O.P = 2.3

* Solar Input To Heat Pump Included As Solar Contribution To Total Load

* Total Output From Heat Pump Included As Contribution To Total Load

(3) calculated from measurements at Capenhurst;
(15) occupancy and 40% of domestic electricity consump.; (+) 8 days occupancy only (September)

| PROJECT DESCRIPTION MACHYNLLETH, UK | REF. | UK | Mach | 78 | 79 | 30 |
|---|---|---|---|---|---|---|

**Main Participants:**

Design, Construction and Monitoring by staff of the National Centre for Alternative Technology, Machynlleth, Wates. Tel Machynlleth 2400.

Robert W Todd - Engineer
Roderick James - Architect

**Project Description:**

This project concerned the conversion and rehabilitation of an existing Slate Cutting Shed during 1975/76, to provide office and exhibition space for NCAT. The building was fitted with extensive thermal insulation, $100m^2$ of 'Open-Trickle' Collectors and a $100m^3$ seasonal capacity water store, situated outside the building, and below ground level.

**Project Objectives:**

NCAT is concerned with demonstrating technologies suited to a long-term sustainable lifestyle, Resource conservation and the use of renewable energy sources are key aspects of their work. The site is not coupled to the electricity supply grid.

The Solar Heated Exhibition Building and Office was designed in 1975/76 as part of an overall plan to make the site (which includes living accommodation for 15 people) completely independent of fossil fuels. The aim was therefore for 100% solar contribution to space heating, with costs kept to a minimum.

The heating system commenced operation in 1977. Performance monitoring was considered essential to ensure that the system was operating as intended and to pinpoint necessary modifications.

Limitations in the monitoring system have resulted in an incomplete performance evaluation. Nevertheless, useful data has been presented on the energy balance of the store from July 1978 until March 1979.

| SITE LOCATION MAP MACHYNLLETH, UK. | REF. | UK | Mach 78 | 79 | 30 |
|---|---|---|---|---|---|

– Distance to main city: 30 km from ABERYSTWYTH

– Please indicate height of nearest obstructions

The site is overshadowed to the south and easy by hills, so that in Mid-winter, direct radiation is only available after noon. There are also some trees which cause overshadowing of the collectors to the South-west at low solar altitudes.

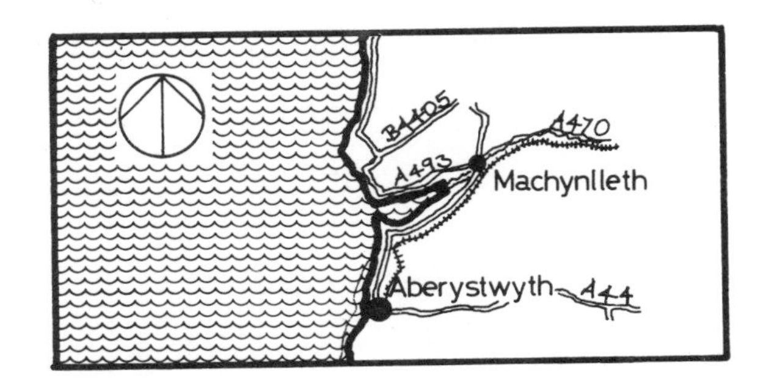

Photograph of Project

**SUMMARY SHEET** MACHYNLLETH, UK — **REF.** UK | Mach 78 | 79 | **30**

**Design Data**

**A1 CLIMATE**

1) Source of Data: Aberporth (40 miles) Aberystwyth (20 miles)
2) Latitude: 52 ° 30′ Longitude: 3 ° 49′ Altitude: 80 metres
3) Global Irradiation (horiz plane): 1030 kWh/m$^2$·year % Diffuse: 60
4) Degree Days: 2053 Base Temp: 15.5 °C
5) Sunshine Hours: July: 141 January: 43 Annual: 1300

**B1 BUILDING**

1) Building Type: Office & Exhibition No Occupants: Are 2-3
2) Floor area: 170 m$^2$ Heated Volume: 460 m$^3$
3) Design Temperature: External: -1 °C Internal: 15.5 °C
4) Ventilation Rate: 0.5 a.c.h. Vol. Heat Loss: 0.59 W/m$^3$·K
5) Space Heat Load: 13,300 kWh Hot Water Load: N.A. kWh

**C1 SYSTEM**

1) Absorber Type: Open Trickle, Matt Black Aluminium
2) Collector Area 70 + 30 m$^2$ (Aperture) Coolant: Water
3) Orientation: 180° Tilt: 34° + 55° Glazing: Double
4) Storage Volume: 100 + 2 m$^3$ Heat Emitters: Under floor
5) Auxiliary System: 7kw Wood Burning Stove Heat Pump: No

**PERFORMANCE MONITORING**

1) Is there a Computer Model? No
2) Start Date for Monitoring Programme: July 1978
3) Period for which Results Availalbe: 1978 to 1979
4) No. of Measuring Points per house: 17
5) Data Acquisition System: Foster Cambridge 6 Channel Chart Recorder

| | Space Heating | Hot Water | Total |
|---|---|---|---|
| 6) Predicted | 100 % of 13300 kWh | % of kWh | % of kWh |
| 7) Measured | % of kWh | % of kWh | % of kWh |

8) Solar Energy Used: Including useful losses ........ kWh/m$^2$·year
Excluding losses: ........ kWh/m$^2$·year

9) System Efficiency: $\frac{\text{Solar Energy Used}}{\text{Global Irradiation on collector}} \times 100 =$ %

A2 LOCAL CLIMATE MACHYNLLETH, UK REF. UK Mach 78 79 **30**

- Average daily max temps (July):18.1$^{o}$C; Average daily min temps (Jan):1.8$^{o}$C

– Source of Weather Data:

Aberporth (64 Km) Aberystwyth and Llety-Evan-Hen (32Km)

For design purposes radiation data was estimated from sunshine hours in the manner suggested by Duffie and Beckman and a model developed by the Polytechnic of Central London using real weather data was used for comparison.

– Micro Climate/Site Description:

The site is in a narrow valley and is fairly sheltered from wind. Nearby wind speed averages 5m/sec but on the site 2.5m/sec is more typical.

Shadowing from nearby hill causes prolonged freeze-ups in December and January which are not typical of the surrounding area. This suggests that the heating load in these months may be higher than usual for the area (the degree days given are for Aberporth). A further difference with Aberporth is the relatively large amount of cloud cover and for this reason sunshine hours for Llety-Evan-Hen have been used.

Data from Aberystwyth site measurements suggest annual precipitation 2,000mm

* Based on Aberporth cloudiness

** Estimate on site

| MONTH | Long Term Ave. YEAR | Irradiation on horizontal plane | | Degree Days | +++ | Average Ambient | Average Relative | Average Wind | Prevailing Wind |
|---|---|---|---|---|---|---|---|---|---|
| | | Global | Diffuse | Base 15.5C | Precipitation | Temperature | Humidity | Speed | Direction |
| JAN | | 19.5 | 14.6 | 323 | 94 | 4.5 | | | |
| FEB | | 39.2 | 23.5 | 301 | 74 | 4.3 | | | |
| MARCH | | 77.5 | 38.7 | 292 | 81 | 6.2 | | | |
| APRIL | | 129 | 64.5 | 228 | 59 | 8.0 | | | |
| MAY | | 148.8 | 81.8 | 156 | 60 | 10.9 | | | |
| JUNE | | 162 | 81 | - | 76 | 13.5 | | | |
| JULY | | 161 | 88.5 | - | 88 | 14.9 | | | |
| AUGUST | | 130 | 71.5 | - | 108 | 15.2 | | | |
| SEPT | | 81 | 44.5 | 72 | 81 | 13.5 | | | |
| OCT | | 43.4 | 26.1 | 138 | 119 | 10.6 | | | |
| NOV | | 24 | 16.8 | 239 | 106 | 7.6 | | | |
| DEC | | 15.5 | 11.3 | 304 | 116 | 5.8 | | | |
| Totals | | 1030 | 562.8 | 2053 | 1062 | | | | |
| Averages | | | | | | 9.6 | | 2.5++ | S.W. |
| Units | – | kWh/m$^2$· month | | °C days | mm | °C | % | m/s | – |

| B2 BUILDING DESCRIPTION | MACHYNLLETH, UK | REF. | UK | Mach | 78 | 79 | 30 |
|---|---|---|---|---|---|---|---|

**Design Concepts**

The geometry of the existing building influenced the design in some respects.

**Construction**

1 – Glazing: 30 % Area: 51.1 $m^2$ 'U' value: 2.5 $W/m^2 \cdot K$

2 – 'U' values: Walls: 0.2 $W/m^2 \cdot K$ Floor: 0.2 $W/m^2 \cdot K$ Roof/loft: 0.19 $W/m^2 \cdot K$

3 – Thermal mass: 60 $kg/m^3$ Annual Energy Demand: 13,300 kWh
Space Heating only

Collectors are insulated with 200mm urethane Building Walls/Roof/Floor with 150mm polystyrene Storage Tank with 300mm polystyrene on walls and 600mm polystyrene on top.

**Energy Conservation Measures**

Some windows ($20m^2$) are fitted with OKALUX shutters. A relatively low minimum internal temperature of 15.5°C was chosen for the winter period, although incidental gains increase this during the day. The fairly high Mean Radiant Temperature was expected to make the low air temperature acceptable.

Free Heat Gains: Internal: 1,100 kWh/year Direct Solar: 5000 kWh/year
Space Heating Season only.

**Section Through House** Showing main solar components

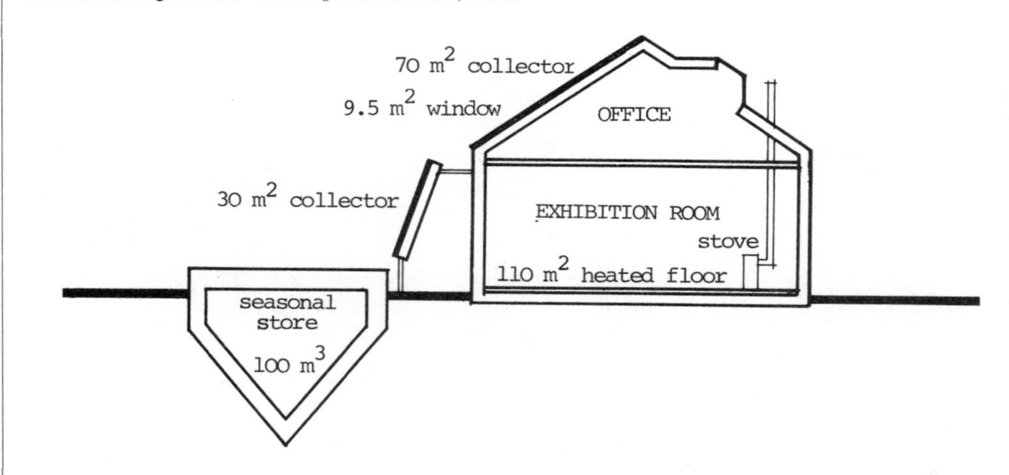

| C2 SYSTEM DESCRIPTION MACHYNLLETH, UK. | REF. | UK | Mach | 78 | 79 | 30 |
|---|---|---|---|---|---|---|

**Collector/Storage**

1 — Primary circuit. Water content: 1000 or 2 litres/$m^2$ collector area

2 — Insulation. Collector: 200mm Storage: 300mm Pipework:

3 — Frost Protection: Primary Pump disconnected

4 — Overheating Protection: None

5 — Corrosion Protection: Plastic and Aluminium Pipes

**Space Heating/Domestic Hot Water**

1 — Auxiliary (S.H.): Wood Stove 7kw (DHW): N/A

2 — Heat Emitters: Under floor Temp. Range: 25°C - 15.5°C

3 — DHW Insulation: N/A Set Temperature: N/A

**Other Points/ Control Strategy/Operating Modes** High time constant of underfloor heating requires water circulation at night. Auxiliary operates independently by radiation and free from stove and flue.

Small Tank Temp is limited between 25°C - 40°C and feeds the underfloor heating system directly.
Collector water is circulated as first priority into the Small Tank, and alternatively into the Large Tank. The large tank can also heat the small tank if necessary and possible. Underfloor Surface Temperature is varied between 21°C (T out 0°C) and 15.5°C (T out = 15.5°C).
When both stores are depleted, underfloor water is kept circulating down to 15.5°C, to reduce the cooling rate of the building.

**Diagram of System** Showing sensor locations, power ratings, storage volumes & flows.

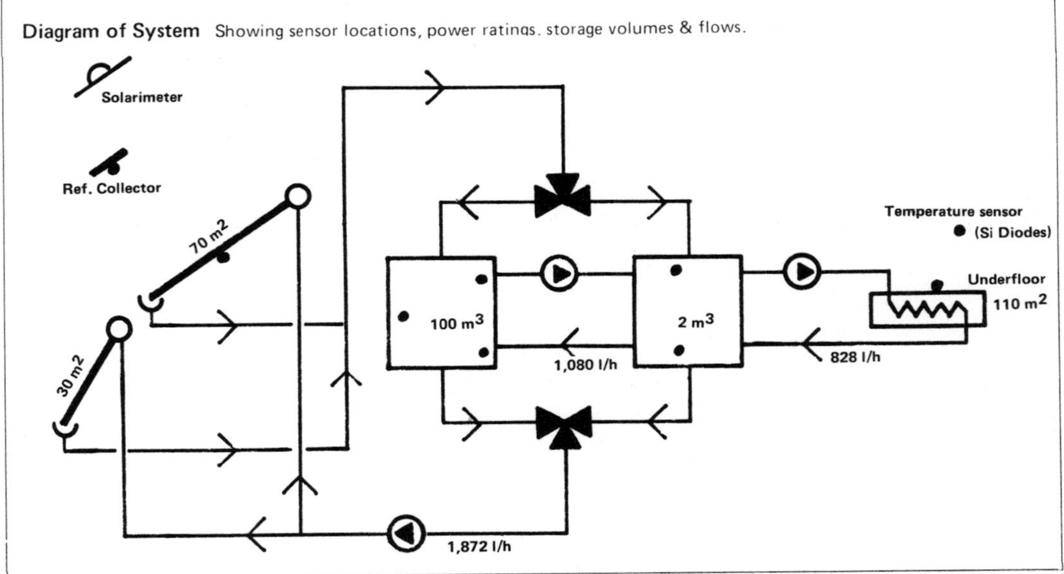

| TECHNICAL APPRAISAL/PRACTICAL EXPERIENCE | REF. | UK | Mach | 78 | 79 | 30 |
|---|---|---|---|---|---|---|

**System/Design:** Main problems encountered in the design of building:

1. Choice of the level of insulation and therefore the building energy demand.
2. Trade off between : large collector/small store v. small collector/large store. storage insulation v. collector and storage size.
3. Need to achieve a low overall cost for collector/storage.

**System/Installation:** Difficulties with trickle collector :

1. Adequate sealing to avoid vapour transmission and providing access to ridge pipe for cleaning.
2. Avoid leaks from plastic gutter joints - difficult due to high thermal expansion.

Difficulties with <u>storage</u> : excavating and smoothing hole for seasonal storage very labour intensive.

**Component Performance:** (Collector, Heat Exchangers, Storage, Pipework, Valves, Fittings, Pumps, Auxiliary)

Collector : poor efficiency, particularly at high temp. Escape of vapour into the building caused condensation problems. High static temperatures damaged collector insulation. Cracking of several panes of glass where they are supported by clips. Blocking of 15% of ridge pipe holes after 2 years operation. Rain penetrating patent glazing.

Underfloor distribution : several water leaks resulting in waste of heated water. These were due to the nylon pipes in the floor contracting after the heating system had operated for some time. On the full width of the building the contraction amounted to around 5cm which was sufficient to break several connections between the floor pipes and the distribution pipes.

Storage : Water pressure forced sheet of butyl rubber into a crack insulation bursting the butyl sheet with consequent loss of storage water.

**System Operation/Controls, Electronics:**

Loss of power on several occasions during the summer resulting in reduced collection and main store failing to reach desired temperature.

PERFORMANCE EVALUATION/CONCLUSIONS REF. | UK | Mach | 78 | 79 | 30

**Comparison of Measured with Predicted Performance**
(State reasons for any discrepancy and summarise the main factors affecting performance)

- The system failed to provide 100% solar heating as expected. This is interpreted as the result of low main storage temperatures resulting from low collection efficiency, loss of water, failure to operate on the basis of 100m$^2$ of collectors and summer power failures.
- All data presented here was obtained with only 70m$^2$ of collector operational.

**Modifications to System**

- Improvement of power supply.
- Under floor connections improved to avoid thermal stresses.+
- Sealing around edge of collector to avoid vapour problems.+
- It is intended to increase the collector flow rate in an attempt to keep collector outlet temp down and improve summer collection efficiency.
- Collector insulation replaced by polyisocyanurate.+
- Storage insulation cemented with polyisocyanurate and lined in corners with Fipec matting to avoid liner bursting. +

**Occupants/Response**

- Over the period Feb to Nov the level of comfort was generally acceptable except for one or two areas i.e. near a much used outside door, and at the base of the east end window where cold air draughts are a problem.

**Conclusions**

- During 1978 the building maintained an internal temperature above 15.5C from the end of February to the end of November without the use of any auxiliary heating. The underfloor heating system behaves well and successfully heats the building with the low water temperature specified in the design.
- During Dec. 1978 and Jan/Feb 1979 the system has contributed to background heating, in particular avoiding low night time temperatures, and the wood stove has been used during he day.

**Future Work**

- It is hoped that the modifications mentioned, and the use of the full 100m$^2$ collector rather than just 70m2, will significantly improve the performance in 1979/80

+ modifications already carried out.

- Future projects to use enclose waterways collectors.

| PROJECT DESCRIPTION | MILTON KEYNES | REF. | UK | MK | 75 | 76 | 31 |
|---|---|---|---|---|---|---|---|

**Main Participants:**

BUILT ENVIRONMENT RESEARCH GROUP of POLYTECHNIC OF CENTRAL LONDON
Professor R. Maw, 35 Marylebone Road, London, N.W.1.

MILTON KEYNES DEVELOPMENT CO.
Mr. J. Doggart, Wavendon Tower, Wavendon, Milton Keynes, Mk 17.

HOUSING DEVELOPMENT DIRECTORATE of the DEPARTMENT OF ENVIRONMENT.

**Project Description:**

The project consists of a single, three bedroom middle terraced house, of standard construction, fitted with 35m$^2$ of collectors designed to provide 59% of the annual energy demand.

**Project Objectives:**

- To study the possibility of using solar energy for providing part of the space heating and domestic hot water in standard UK housing conditions and to assess possible mass market applications.
- A standard local authority house was used and all modifications reuqired for installing the solar heating system were examined for the relative cost and estimated benefit.
- The rather poor insulation qualities of the house were not improved, in order to compare the performance of the house and system with the conventional houses on the estate, and establish the fuel savings that are possible by installing a solar heating system into a conventional house of a design similar to many houses being built in Milton Keynes at the time (1974). The original occupants were asked not to allow the presence of the system to influence their energy usage behaviour.
- As a result of the first two years operation, the objectives have been modified. Following extensive system modifications, the work has centred upon experimentation with the system operating parameters in order to improve the performance. The new occupants have been encouraged to adapt their energy use to the availability of solar energy. New insulation has also been added to the house.

| SITE LOCATION MAP | MILTON KEYNES | REF. | UK | MK | 75 | 76 | 31 |
|---|---|---|---|---|---|---|---|

– Distance to main city: 75 km from LONDON (Direction: N-NW)

– Please indicate height of nearest obstructions

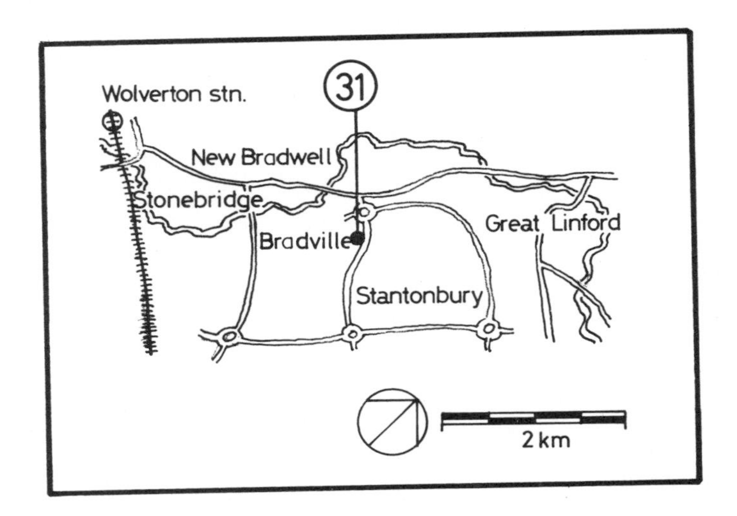

Photograph of Project

| SUMMARY SHEET | MILTON KEYNES | REF. | UK | MK | 75 | 76 | 31 |
|---|---|---|---|---|---|---|---|

**Design Data**

**A1 CLIMATE**

1) Source of Data: Kew & Cambridge Met Office.
2) Latitude: 52° 03′ Longitude: 0° 45′ Altitude: 85 metres
3) Global Irradiation (horiz plane): 911 kWh/m²·year % Diffuse: 58
4) Degree Days: 2,022 Base Temp: 18 °C
5) Sunshine Hours: July: 186 January: 53 Annual: 1,489

**B1 BUILDING**

1) Building Type: Mid terrace house No Occupants: 3
2) Floor area: 85 m² Heated Volume: 234 m³
3) Design Temperature: External: -1 °C Internal: 18 °C
4) Ventilation Rate: 1 a.c.h. Vol. Heat Loss: 1.5 W/m³·K
5) Space Heat Load: 6,750 kWh Hot Water Load: 6,500 kWh

**C1 SYSTEM**

1) Absorber Type: Al-Rollbond, anodised black (abs: 0.82) *
2) Collector Area 35 m² (Aperture) Coolant: water + glycol + inhib.
3) Orientation: 170 Tilt: 30 Glazing: single, 4mm reduced Fe.
4) Storage Volume: 2 x 2.1 = 4.2 m³ Heat Emitters: forced air convector.
5) Auxiliary System: gas boiler (SH) immersion heater (DHW) Heat Pump: --

* Indicates significant changes after March 1978.

**PERFORMANCE MONITORING**

1) Is there a Computer Model? Yes 2) Start data for Monitoring Programme: April 1975
3) Period for which results available: September 1975 to March 1979.
4) No. of Measuring Points per house: 23
5) Data Acquisition System: 12 channel "dot" chart recorder (Foster Cambridge).

| | Space Heating | Hot Water | Total |
|---|---|---|---|
| 6) Predicted | 50 % of 6,750 kWh | 68 % of 6,500 kWh | 59 % of 13250 kWh |
| 7) Measured ** | 22 % of 3603 kWh | 41 % of 3871 kWh | 32 % of 7474 kWh |
| ** | 40 % of 5923 kWh | 64 % of 2956 | 48 % of 8879 kWh |

8) Solar Energy Used: Including useful losses ** 68/122 kWh/m²·year
Excluding losses: ** 62/112 kWh/m²·year

9) System Efficiency: $\frac{\text{Solar Energy Used}}{\text{Global Irradiation on collector}} \times 100 =$ 6/12 % **

**:2 years data

| A2 LOCAL CLIMATE | MILTON KEYNES | REF. | UK | MK | 75 | 76 | 31 |
|---|---|---|---|---|---|---|---|

1 – Average Cloud cover: 5.6 Octas

2 – Average daily max temperatures (July) : 22 °C

Average daily min temperatures (Jan) : 0,5 °C

3 – Source of Weather Data:

C : data from Cambridge, km from site. Average data from 1931-1960.

K : data from Kew (75 km S-SE from site)

K*: The data used for the original design corresponds to Kew, year 69 (K*). This year was characterised by a mean temp. below average and bright sunshine 4% below average

4 – Micro Climate/Site Description:

Milton Keynes is a low density new town, located in a flat plain in mid-south-east England (approximately 75 km n-nw from London).

No particular micro-climatic characteristics are expected to modify significantly the data obtained in Kew or Cambridge.

| | | K* | | | C | K* | K | K | |
|---|---|---|---|---|---|---|---|---|---|
| MONTH | YEAR | Irradiation on horizontal plane Global | Diffuse | Degree Days Base: 18 °C | Precipitation | Average Ambient Temperature | Average Relative Humidity | Average Wind Speed | Prevailing Wind Direction |
| JAN | | 19 | 14 | 346 | 49 | 6.4 | 88% | 3 | |
| FEB | | 34 | 21 | 304 | 35 | 2.3 | 84% | 3 | |
| MARCH | | 47 | 30 | 282 | 33 | 4.5 | 79% | 3 | |
| APRIL | | 113 | 60 | 197 | 45 | 8.5 | 73% | 3 | |
| MAY | | 121 | 77 | 113 | 46 | 12.1 | 73% | 3 | |
| JUNE | | 159 | 82 | – | 37 | 14.3 | 70% | 3 | |
| JULY | | 150 | 77 | – | 61 | 17.9 | 72% | 2.5 | |
| AUGUST | | 108 | 70 | – | 48 | 16.9 | 75% | 2.5 | |
| SEPT | | 75 | 46 | 56 | 52 | 14.7 | 80% | 2 | |
| OCT | | 54 | 31 | 132 | 50 | 13.1 | 85% | 2.5 | |
| NOV | | 26 | 16 | 256 | 53 | 6.3 | 88% | 3 | |
| DEC | | 11 | 9 | 336 | 42 | 3.7 | 89% | 3.5 | |
| Totals | | 911 | 533 | 2022 | 551 | | | | |
| Averages | | | | | | 10.1 | 79.7% | 2.8 | SW |
| Units | – | kWh/m²·month | | °C days | mm | °C | % | m/s | – |

**B2 BUILDING DESCRIPTION** REF. | UK | MK | 75 | 76 | **31**

**Design Concepts**

- Standard design Corporation House
- Three bedroom, 5 person middle terrace
- Monopitch roof, 30 from horiz., facing 170 south

**Construction**

1 – Glazing: ............ % Area: 18.5* $m^2$ 'U' value: ............ $W/m^2 \cdot K$

2 – 'U' values: Walls: .75 $W/m^2 \cdot K$ Floor: 1.7 $W/m^2 \cdot K$ Roof/loft: .55 $W/m^2 \cdot K$

3 – Thermal mass: 93 $kg/m^3$ Annual Energy Demand: 13,250 kWh

* = 8.5(S) + 10(N)

- Timber frame - lightweight cladding

**Energy Conservation Measures**

- Insulation to current building regulations (1973 standards - eg single glazing, no shutters, draught strippings, 1" fibreglass in walls and loft. Floor composed of carpet and 4" concrete and 4" brick).

Free Heat Gains: Internal: 5,100 kWh/year (S/H: 3,250) Direct Solar: 5,900 kWh/year (S/H: 3,320)

**Section Through House** Showing main solar components

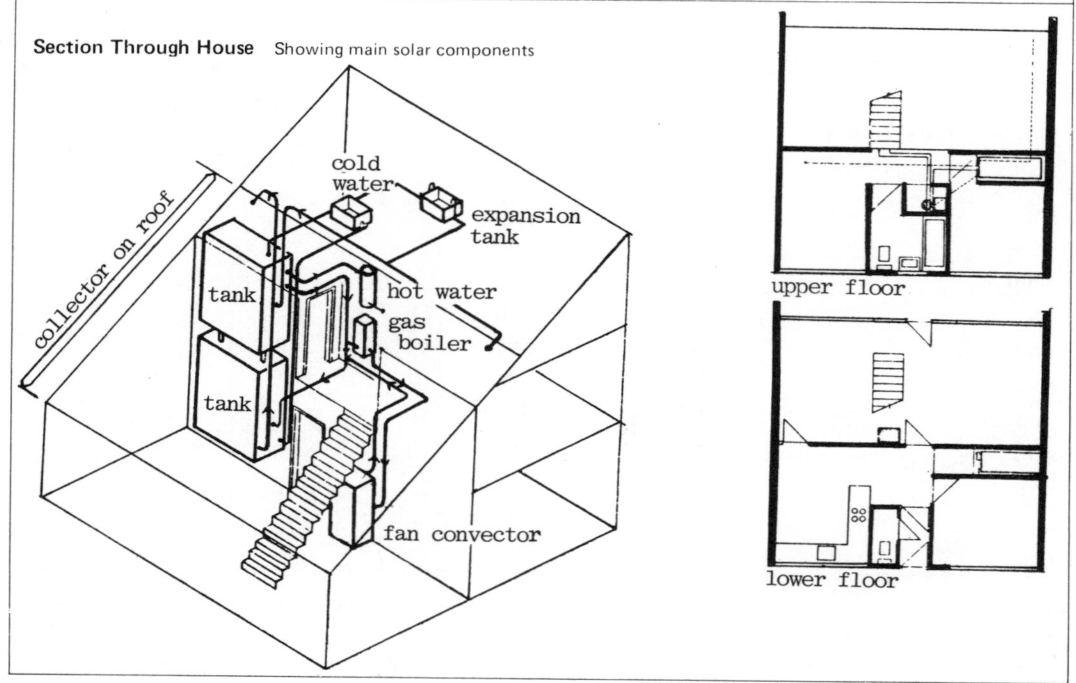

| C2 SYSTEM DESCRIPTION MILTON KEYNES | REF. | UK | MK | 75 | 76 | 31 |
|---|---|---|---|---|---|---|

**Collector/Storage**

1 – Primary circuit. Water content: 3.08 litres/$m^2$ collector area

2 – Insulation. Collector: 75mm fibreglass Storage: 100mm fibreglass Pipework:

3 – Frost Protection: 5% glycol. If ambient temp.falls below $-1^{o}C$ water circulates overnight

4 – Overheating Protection: Overnight circulation for storage temp. over $85^{o}C$

5 – Corrosion Protection: 55% inhibitors, Cu pipework (stainless steel for storage to collector pipe) Al foil sacrificial element, rubber interconnections for collectors.

**Space Heating/Domestic Hot Water**

1 – Auxiliary (S.H.): 6.5kW gas boiler (DHW): 3kW immersion heater

2 – Heat Emitters: .4kW fan + 2 finned heat exch Temp. Range: inlet: $40^{o}C$ outlet: $27^{o}C$

3 – DHW Insulation: Set Temperature: $65^{o}C$

**Other Points/Heat Pump:** – Direct system. Primary circuit includes $4.2m^2$ of storage.
– Modifications introduced after 03/78 specified later.

**Control Strategy/Operating Modes**

- DHW preheated in 4 heat exchangers inside storage tank. Temp topped electrically up to the set temperature.
- Gas boiler tops temp of inlet water to S.H. unit up to $40^{o}C$ (if necessary)
- Water circulates through collectors if temp rise is $5^{o}C$

**Diagram of System** Showing sensor locations, power ratings, storage volumes & flows.

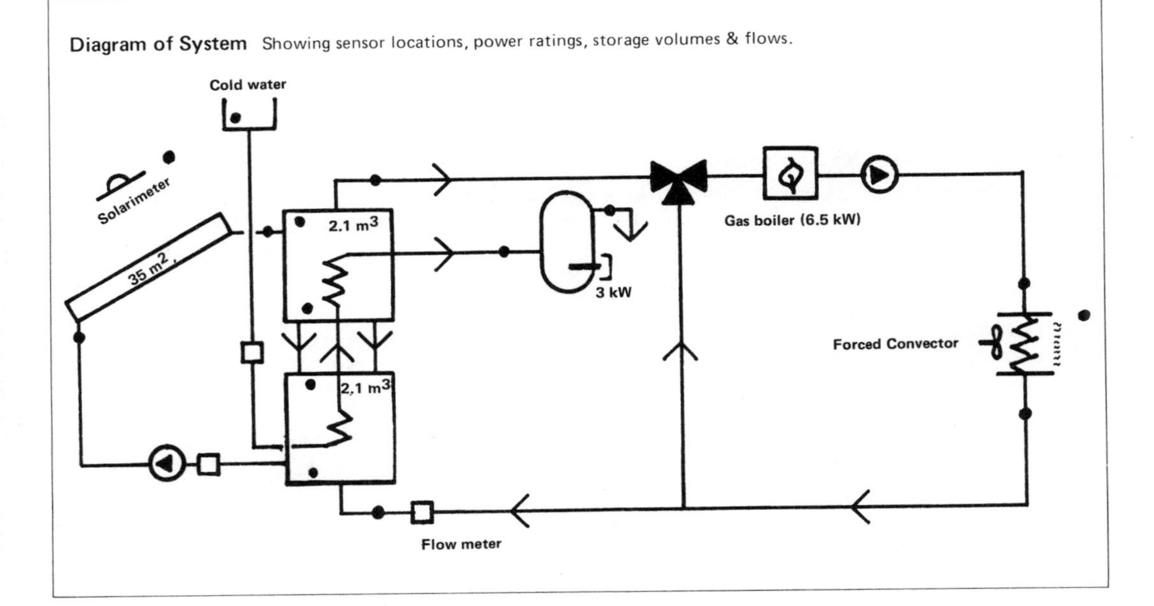

| TECHNICAL APPRAISAL/PRACTICAL EXPERIENCE | REF. | UK | MK | 75 | 76 | 31 |
|---|---|---|---|---|---|---|

**System/Design:**

1. Retro-fit of a solar system conditioned a number of issues (e.g. use of rectangular storage tanks, long pipe runs, etc.)

2. Design of DHW pre-heater exchangers should be simpler and more substantial than for the present design where pre-heat is obtained in four 4" pipes with brazed inter-connections.

**System/Installation:**

1. The positionment of the upper storage tank ($2.1m^3$) required a special ramp and a large team of men.

2. Upper floor had to be substantially strengthened.

**Component Performance:** (Collector, Heat Exchangers, Storage, Pipework, Valves, Fittings, Pumps, Auxiliary)

Collector: 1 collector leaking.
Insulation was increased so as to reduce air gas between the absorber and insulation.
Poor absorptivity of black anodised surface.

Heat Exchanger (for DHW pre-heat) 1 heat exchanger presented a leakage which required draining down the storage and replacing the pre-heaters.

Storage: Gaps in the insulation resulted in higher losses and over-heating in summer.

Valves: A pre-mix valve used to prevent too high DHW temperatures jammed open (summer 76).

**System Operation/Controls, Electronics:**

1. Two bi-metallic thermostats varied their set point due to overheating. This resulted in high heat losses due to unnecessary overnight circulation. (Oct, Nov, 75)

2. Heat losses due to reverse thermosyphon.

3. DHW Flow dimished as a result of build up of corrosion products at the inlet and outlet of the pre-heat cylinders.

PERFORMANCE EVALUATION/CONCLUSIONS REF. | UK | MK | 75 | 76 | **31**

### Comparison of Measured with Predicted Performance

(State reasons for any discrepancy and summarise the main factors affecting performance)

SYSTEM: The first two years of operation of the house have given a much lower solar contribution to total energy demands than expected. This has been interpreted as a result of over optimistic estimates, a poor collector performance (anodised surface had lower absorptivity than expected), poor use of low grade heat, and the malfunction of some control thermostats. As a result, several changes were introduced (completed March 1978)

### Modifications to System

1. Collector painted with matt black paint (est.absorptivity.98).
2. Solar operation has been completed separated from gas operation. The system operates now in a number of modes according to tank temperature. For T - $35^{o}$C solar heating only. For T - $25^{o}$C auxiliary heating only. For $35^{o}$C T - $25^{o}$C auxiliary and solar are alternated, the role of solar being merely to reduce rate of cooling down of building.*
3. Concentration of ethylene glycol increased from 5% to 17% to reduce heat losses due to overnight circulation for prost protection.
4. DHW temperature setting reduced from $65^{o}$C to $50^{o}$C.
5. Power rating of fan convector unit reduced to 0.2kW.
6. Non-return valve inserted in collector circuit. Lagging improved in DHW and loft. Thermostats changed and checked regularly.

### Occupants' Response

1. They asked for some way of quantitive measurement of the state of the system (eg. storage thermometer). This should help them to take maximum advantage of the solar system.
2. Noise due to convector fan, particularly when starting (solved after modifications)
3. Low air temperatures from fan convector unit (after modifications) and the absence of radiating surfaces, may produce feeling of discomfort.

### Conclusions

1. It is strongly recommended to separate solar from auxiliary heating modes.
2. DHW temperature setting should be reduced to $50^{o}$C or lower.
3. Electricity consumption of the system should be minimised (eg fen power rating)
4. The use of electronic temperature controllers is strongly recommended.
5. a non-return valve is advisable for preventing reverse thermosyphon.
6. Indirect systems seem more economical.
7. Double glazing of collection (or selective and single) may be advisable.
8. Cylindrical tanks are cheaper, and should be considered if the house is designed with the system.

### Future Work

- Monitoring program continued until August 1979.
- Final report (IEA format) to appear in June 1979.
- Mark 2 Solar House - design stage.

* This system enables full solar heating at tank temperatures well below $40^{o}$C in mild conditions, and enables the use of storage temperatures as low as $24^{o}$C.

## P1 MONTHLY PERFORMANCE

REF: UK MK 75 75 31

| Month | Year | No Days Data | Average Temp °C | | SOLAR ENERGY kWh | | | | | | Heat Pump Total Output kWh | | AUXILIARY ENERGY kWh | | | | Free Heat Gains kWh | | SUMMARY | | |
|---|---|---|---|---|---|---|---|---|---|---|---|---|---|---|---|---|---|---|---|---|---|
| | | | | | | | Solar Energy Used | | | | | | | | | | | | kWh | % | |
| | | | | | | | Space Heating Useful Losses | | | | | | | | | | | | | | |
| | | | Internal | External | Global Irradiation On Collector Area | Solar Energy Collected | Incl. | Excl. | Domestic Hot Water | Input To Heat | Space Heating | Domestic Hot Water | Heat Pump | Space Heating | Domestic Hot Water | Pumps & Fans | Internal | Direct Solar | Total Load | Percentage Solar | Syst. Eff. |
| Column No | | | 1 | 2 | 3 | 4 | 5 | 6 | 7 | 8 | 9 | 10 | 11 | 12 | 13 | 14 | 15 | 16 | 17 | 18 | 19 |
| J | 76 | 31 | 17.6 | 5.5 | 1352 | 281 | 27 | 24 | 68 | - | | | - | 586 | 324 | 101 | 300 | 240 | 1005 | 9 | 7.0 |
| F | 76 | 29 | 16.6 | 4.3 | 1580 | 352 | 31 | 31 | 46 | - | | | - | 583 | 296 | 92 | 470 | 300 | 956 | 8 | 4.9 |
| M | 76 | 24 | 17.4 | 4.7 | 3274 | 673 | 285 | 202 | 122 | - | | | - | 291 | 270 | 81 | 490 | 570 | 968 | 42 | 12.4 |
| A | 76 | 30 | 17.2 | 8.7 | 4444 | 927 | 241 | 172 | 148 | - | | | - | 61 | 189 | 35 | 570 | 560 | 639 | 61 | 8.7 |
| M | 76 | 31 | 19.6 | 13.6 | 5268 | 871 | 0 | 0 | 212 | - | | | - | 0 | 132 | 0 | 490 | 670 | 344 | 62 | 4.0 |
| J | 76 | 30 | 23.5 | 19.9 | 6351 | 1224 | 0 | 0 | 197 | - | | | - | 0 | 66 | 0 | 520 | 660 | 263 | 75 | 3.1 |
| J | 76 | 31 | 24.8 | 21.0 | 5972 | 989 | 0 | 0 | 231 | - | | | - | 0 | 20 | 0 | 530 | 630 | 251 | 92 | 3.9 |
| A | 76 | 23 | 23.0 | 18.5 | 5772 | 1024 | 0 | 0 | 211 | - | | | - | 0 | 49 | 0 | 310 | 650 | 260 | 81 | 3.7 |
| S | 75 | 30 | 22.0 | 14.2 | 4119 | 754 | 12 | 12 | 143 | - | | | - | 4 | 104 | 2 | 250 | 630 | 263 | 59 | 3.8 |
| O | 75 | 22 | 19.0 | 10.0 | 3188 | 523 | 117 | 64 | 122 | - | | | - | 125 | 191 | 30 | 410 | 500 | 555 | 43 | 7.5 |
| N | 75 | 24 | 18.0 | 5.8 | 1321 | 360 | 70 | 56 | 67 | - | | | - | 504 | 299 | 88 | 390 | 300 | 940 | 14 | 10.4 |
| D | 75 | 31 | 17.0 | 4.7 | 928 | 202 | (1) | 1 | 35 | - | | | - | 665 | 329 | 92 | 370 | 220 | 1030 | 4 | 3.9 |
| TOTAL | | | — | — | 43659 | 8180 | 784 | 562 | 1602 | - | | | - | 2819 | 2269 | 521 | (5100) | (5900) | 7474 | — | — |
| AVERAGE | | | 19.6 | 10.9 | SH:20206 | | SH:751 | | | | | | | | | | SH:(3250) | SH: (3320) | | 32 | 5.5 |

TOTAL: 2164 kWh

(3):measured on horizontal plane at Cardington and calculated for $34.5 m^2$ on inclined plane data corrected for missing days;

(5):+ unwanted heat losses; Nov. to Feb all tank losses useful; March April and Oct only a proportion of losses useful.

## P1 MONTHLY PERFORMANCE

REF: UK MK 78 79 31

| Month | Year | No Days Data | Average Temp °C Internal | Average Temp °C External | SOLAR ENERGY kWh: Global Irradiation On Collector Area | Solar Energy Collected | Solar Energy Used: Space Heating Useful Losses Incl. | Space Heating Useful Losses Excl. | Domestic Hot Water | Input To Heat | Heat Pump Total Output kWh: Space Heating | Domestic Hot Water | AUXILIARY ENERGY kWh: Heat Pump | Space Heating | Domestic Hot Water | Pumps & Fans | Free Heat Gains kWh: Internal | Direct Solar | SUMMARY kWh: Total Load | %: Percentage Solar | Syst. Eff. |
|---|---|---|---|---|---|---|---|---|---|---|---|---|---|---|---|---|---|---|---|---|---|
| Column No | | | 1 | 2 | 3 | 4 | 5 | 6 | 7 | 8 | 9 | 10 | 11 | 12 | 13 | 14 | 15 | 16 | 17 | 18 | 19 |
| J | 79 | 25 | 17.5 | -0.5 | 1263 | 341 | 149 | 136 | 98 | - | | | - | 1225 | 196 | 91 | | | 1668 | 15 | 19.6 |
| F | 79 | 28 | 18.1 | 0.4 | 1601 | 472 | 260 | 234 | 106 | - | | | - | 1050 | 187 | 93 | | | 1603 | 23 | 22.9 |
| M | 79 | 31 | 18.8 | 4.4 | 2767 | 861 | 576 | 495 | 191 | - | | | - | 347 | 140 | 87 | | | 1254 | 61 | 27.7 |
| A | 78 | 18 | 19.0 | 6.1 | 3554 | 1137 | 577 | 221 | 188 | - | | | - | 66 | 81 | 81 | | | 912 | 84 | 21.5 |
| M | 78 | 27 | 21.1 | 11.8 | 5178 | 1450 | 221 | - | 233 | - | | | - | 0 | 43 | 37 | | | 497 | 91 | 8.8 |
| J | 78 | 30 | 21.9 | 13.9 | 4789 | 816 | | - | 189 | - | | | - | - | 5 | - | | | 194 | 97 | 3:9 |
| J | 78 | 30 | 21.8 | 15.5 | 4644 | 875 | | - | 192 | - | | | - | - | 13 | - | | | 205 | 94 | 4.1 |
| A | 78 | 30 | 22.1 | 15.0 | 4406 | 749 | | - | 169 | - | | | - | - | 1 | - | | | 170 | 99 | 3.8 |
| S | 78 | 30 | 22.3 | 13.3 | 3992 | 764 | | - | 190 | - | | | - | - | 2 | - | | | 192 | 99 | 4.8 |
| O | 78 | 31 | 22.1 | 11.3 | 2512 | 674 | 143 | 102 | 168 | - | | | - | 3 | 47 | 10 | | | 361 | 86 | 12.4 |
| N | 78 | 30 | 19.3 | 7.4 | 1635 | 536 | 303 | 223 | 116 | - | | | - | 88 | 148 | 39 | | | 655 | 64 | 25.6 |
| D | 78 | 31 | 17.6 | 3.5 | 925 | 250 | 149 | 120 | 70 | - | | | - | 766 | 183 | 72 | | | 1168 | 18.8 | 23.7 |
| TOTAL | | | — | — | 37266 | 9025 | 2378 | 2030 | 1910 | - | | | - | 3545 | 1046 | 510 | | | 8879 | — | — |
| AVERAGE | | | 20.2 | 8.6 | | | | | | | | | | | | | | | | 48.3 | 11.5 |

SH:1170

SH:1025

**TOTAL:** 3940 **kWh**

April: monthly data estimated from 18 days recorded data; (14): excludes primary circuit.